CONSCIOUSNESS
CONTEMPORARY SCIENCE

Consciousness
in
Contemporary Science

EDITED BY

A. J. Marcel

M R C Applied Psychology Unit, Cambridge

AND

E. Bisiach

Istituto di Clinica Neurologica,
University of Milan

CLARENDON PRESS · OXFORD

Oxford University Press, Walton Street, Oxford OX2 6DP
Oxford New York Toronto
Delhi Bombay Calcutta Madras Karachi
Kuala Lumpur Singapore Hong Kong Tokyo
Nairobi Dar es Salaam Cape Town
Melbourne Auckland Madrid
and associated companies in
Berlin Ibadan

Oxford is a trade mark of Oxford University Press

Published in the United States
by Oxford University Press Inc., New York

© The various contributors listed on pp. ix-x, 1988

First published 1988
First published in paperback 1992 (with corrections)
Reprinted 1993

British Library Cataloguing in Publication Data
Consciousness in contemporary science.
1. Man. Consciousness
I. Marcel, A.J. (Anthony J.)
II. Bisiach, E. (Edoardo)
153
ISBN 0–19–852237–1 (Pbk)

Library of Congress Cataloging in Publication Data
Consciousness in contemporary science.
Includes bibliographies and index.
1. Consciousness—Congresses. I. Marcel, A.J.
(Anthony J.) II. Bisiach, E. [DNLM: 1. Consciousness.
2. Philosphy. 3. Science. WL 705 C755]
BF311.C653 1988 153 88-5175
ISBN 0–19–852237–1 (Pbk)

Printed in Great Britain by
The Ipswich Book Company,
Ipswich, Suffolk

ACKNOWLEDGEMENTS

We wish to thank Professor Lawrence Weiskrantz, Professor Guglielmo Scarlato, Dr Alan Baddeley, and especially Dr Tim Shallice for advice and encouragement from the time when the original meeting was an idea to the time when it became a book. We would also like to acknowledge our debt to the Consiglio Nazionale delle Ricerche and to the Municipality of Como, which covered expenses for the organization of the meeting. Professor Giulio Casati and the staff of the Centro Scientifico 'Alessandro Volta' of the Villa Olmo were of great assistance in preparing and running the meeting.

Concerning the transformation of the meeting into a book, we are grateful to all the participants for their careful and constructive commentaries on other people's papers. Alan Baddeley kindly allowed us to use facilities of the Medical Research Council's Applied Psychology Unit in Cambridge for editing the manuscript. During this process Sharon Gamble's typing and secretarial help were invaluable. The indexing is very largely due to Rachel Bellamy, who also made many helpful suggestions at the final stages. Lastly, the Oxford University Press are to be thanked for advising and consenting in this publication.

A.J.M. has a large and affectionate debt to E.B. for initiating and organizing the original meeting, and for inviting him to take part in the enterprise of editing the book. E.B.'s own contribution, as well as his tolerance and support, were inestimable.

Finally, E.B. wishes to thank A.J.M. warmly, not only for having accepted the burdensome role of co-editor, but also for having vigorously and creatively carried out such an amount of editorial work, so that he more than anybody else deserves credit for the appearance of the book.

CONTENTS

CONTRIBUTORS

D. ALAN ALLPORT, Department of Experimental Psychology, University of Oxford, South Parks Road, Oxford OX1 3UD, UK.

EDOARDO BISIACH, Istituto di Clinica Neurologica, Università di Milano, Via Francesco Sforza 35, 20122 Milano, Italy.

PATRICIA SMITH CHURCHLAND, Department of Philosophy, University of California San Diego, La Jolla, CA 92093, USA.

DANIEL C. DENNETT, Department of Philosophy, Tufts University, Medford, MA 02155, USA.

MATTHEW HUGH ERDELYI, Department of Psychology, Brooklyn College, City University of New York, Bedford Avenue and Avenue H, Brooklyn, NY 11210, USA.

MICHAEL S. GAZZANIGA, Department of Neurology, Dartmouth Medical School, Pike House, Hanover, NH 03756, USA.

RICHARD L. GREGORY, Brain and Perception Laboratory, Department of Anatomy, The Medical School, University Walk, Bristol BS8 1TD, UK.

PHILIP N. JOHNSON-LAIRD, Department of Psychology, Princeton University, Green Hall, Princeton, NJ 08544, USA.

MARCEL KINSBOURNE, Behavioral Neurology Unit, Sargent College, Boston University, 635 Commonwealth Avenue, Boston, MA 02215, USA.

ANTHONY J. MARCEL, Medical Research Council, Applied Psychology Unit, 15 Chaucer Road, Cambridge, CB2 2EF, UK.

KEITH OATLEY, Centre for Applied Cognitive Science, Ontario Institute for Studies in Education, 252 Bloor Street West, Toronto, Ontario M5S 1V6, Canada.

TIM SHALLICE, Psychology Department, University College London, Gower Street, London WC1E 6BT.

CARLO UMILTÀ, Dipartimento di Psicologia Generale, Università di Padova, Piazza Capitaniato 3, 35139 Padova, Italy.

ROBERT VAN GULICK, Department of Philosophy, Syracuse University, Syracuse, NY 13244, USA.

LAWRENCE WEISKRANTZ, Department of Experimental Psychology, University of Oxford, South Parks Road, Oxford OX1 3UD, UK.

KATHLEEN V. WILKES, St Hilda's College, University of Oxford, Oxford OX4 1DY, UK.

1

A cautious welcome:
an introduction and guide to the book

Anthony J. Marcel and Edoardo Bisiach

Difficult returns

When someone returns after an absence, their home-coming is not always straightforward nor their acceptance always universal. Three stories illustrate different aspects of this.

In sixteenth-century France, a young peasant called Martin Guerre, not long married, quite suddenly left home. Some years later a man came to the village claiming to be Martin. The wife, tentative at first, accepted him and came to love him, as did many of the villagers. However, family disputes led to a judicial examination of whether the pretender was in fact who he claimed to be. Despite the fact that the man was appreciated by many for his various qualities, the official issue of his identity, for administrative and financial purposes, could not be ignored.

In the biblical story of the prodigal son, different issues are at stake. For the father, the pleasure at the return of the prodigal to the fold outweighed the satisfaction derived from the constancy of the son who had stayed. Naturally, the dutiful son resented the welcome given to his prodigal brother and the lack of appreciation of his own less glamorous deeds.

Our third story concerns the arrival in Mexico of Cortes. It was taken by many Aztecs to be the return of Quetzalcoatl. But it is said that some dared to voice the suspicion that Quetzalcoatl could not return since he had never even existed, was just a myth. The very idea of a serpent with feathers showed what an incoherent myth it was. However, yet others, especially some thoughtful Spaniards (anticipating Pirandello?), suggested that if something is believed to be the case, well, in some sense it *is* the case.

As we shall see below, the attitudes in these three stories seem to apply to the reception given to the return of consciousness to the fold of psychology, though in this case our own predecessors had banished the rascal. As in our first story, there are those who either like or dislike the character, irrespective of his[1] name, while others are concerned that the name and the appropriate identity should fit. Sometimes this is for legislative reasons, and sometimes because people do not want to be fooled. As in our second story, when people know only too well who the home-comer is, some positively welcome him with all his drawbacks, either glad to have him back or seeing that he can be set some useful jobs. Others are suspicious of his slipperiness and prodigality, either thinking that they can do perfectly well without him or willing to accept him only if he gives up some of his personality, the bit that to others made him charming. As in our third story, there are those who want back something that they have never known and are duped by a newcomer. If they are not duped, they either reject the interloper, or accept him as hard reality or as the best that can be had. Some of these may think there was never any substance to what they had yearned for. Others suggest that thinking makes it so. There really are a lot of different people at the reception. Who is knocking at the door?

The problems

In recent years, after an absence of over half a century, there has been a growing revival of interest in consciousness among psychologists, philosophers, neuroscientists, social scientists, and clinicians in North America and Western Europe. Of course it has never been entirely absent, being part of the theoretical vocabulary of the psychoanalytic tradition, part of the approach of Vygotsky and Luria, and a focus of interest for philosophers and social scientists outside the natural science framework. It is however only within the last fifteen years or so that mainstream students in the disciplines mentioned above have had increasing recourse to the term. It is significant that Mandler in 1975 entitled a paper 'Consciousness: respectable, useful and probably necessary'. He was responding to the predominant assumption that consciousness is none of these.

The current interest has been both in consciousness as a proper topic of study and in its explanatory use. Although the reasons for this revival of interest have been diverse, some can reasonably be identified. First, in the reaction against behaviourism, psychologists have been more confident in their use of people's experience, realizing not only that it helps

---

[1] In the text, where pronominal reference is made to unspecified individuals, masculine pronouns (*he, his*) are used in most cases. Readers should take such instances to imply both masculine and feminine forms.

them generate explanations, but that much of what they study is indeed that experience (in perception, emotion, memory, and thought). Certain aspects of consciousness such as imagery, dreaming and the stream of thought have become fields of study in their own right. Second, certain phenomena in normal people (attention, slips of action, perception without awareness) and in clinical syndromes (blindsight, amnesia, neglect, split brains, multiple personality) have encouraged many investigators to use the concept of consciousness as a descriptive and explanatory term. Indeed philosophers of science accord special status to some of these latter phenomena, as data which 'denormalize' a previously widely held explanatory framework.

However, both the use of the term and the status of the concept remain unclear and inconsistent. Natsoulas (1978) distinguished seven conceptual uses of the term consciousness, and it is not difficult to find additional usages. One reason for this is that, like many terms in psychology, the word pre-exists scientific terminology in natural language usage. This is much less so in physics. Hence there is much more agreement on the concept of a quark. However, if one examines the history of physics, at several times of paradigm change concepts were unstable. Not only do meanings change over time, but people do not use language by consulting a dictionary (even if they did they would find the term 'consciousness' to be polysemous). Further, scientists, certainly psychologists, tend to use terms according to what they are trying to explain. This difference in usage can be seen in this book (although most of the authors make an attempt to specify what they are referring to). Sometimes it can be depressingly amusing to see scientists legislating on what is meant by a word. It is such polysemy that leads Wilkes (this volume) to scepticism over the scientific status of the term 'consciousness'. However, whether the various referents of a term form a 'natural kind' and whether each of those referents is a valid topic for science are separate issues.

A particular problem is the domain or level of discourse in which the concept is being located. Sometimes the term appears to be used in a functionalist way, so that it is equivalent to concepts such as attention, short-term memory, representation, control, or what people can assert verbally (concepts which themselves are often unclear). At other times it is used to refer to phenomenological concepts, such as subjective experience, qualia, the contents of awareness, intentionality, or personal unity. (Phenomenology is not being used here in its technical, e.g. Husserlian, sense.) Furthermore, the very relationships between descriptions and explanations in phenomenological terms and in functional (e.g. information-processing) terms remain problematic and are in need of scrutiny. The problem of level of discourse is even more obvious when it concerns neuroscientists. They often have occasion to explain one level in terms of another and even use terms from the two levels interchangeably. Legislation is seldom of use to science, but we appear to need some clarification.

The issue of the status of consciousness is largely what has attracted the scientific censor's blue pencil, and is still a matter of dispute among philosophers and psychologists. As regards phenomenal experience, the censor's argument has been not so much that of radical behaviourism, that phenomenal experience is mentalist, but rather that its essential privacy and 'subjectivity' debars it from science. Yet clearly, psychologists and sociologists use reports of such experience not only to help frame hypotheses but also as data. In the former case, perhaps this is just an example of the scientist ultimately relying on at least some tools and heuristics outside the formal practice. But protocol analysts such as Ericsson and Simon (1984) have argued that introspection is a perfectly legitimate and valid tool that ought to be recognized as such. In the latter case, when phenomenal experience is used as data, the problem of status is more acute. Many psychologists treat such reports as informationally equivalent (Simon 1978) to some functionalist internal representation. But it is not clear that this is legitimate, especially if the report is a translation or if it is a description expressing only part of the representation. Additionally, if the report is to be treated as conveying a person's putative phenomenal experience, and if it turns out that isomorphic equivalence with the functionalist counterpart breaks down, then the problem of validity raises its head. It has been suggested by Dennett (1982) that these problems are not real, and even that phenomenal experience, as we conceive of it (though even that is unclear), does not exist (see Dennett, this volume).

These problems can be illustrated by considering how we regard our answers to certain types of question. If we are asked for our telephone number, we usually respond quite quickly with the digits and we have little experience of the information until we say it. The same is true if we are asked if we know any buildings in a well-known place (e.g. Trafalgar Square), especially when a quick reply is required. But if the request is for more detailed information, such as the spatial arrangement of the buildings, or as full a list as possible, or if the request is for spatial information, such as whether the 'toe' of the map of Italy faces east or west, then our experience of our answer is usually rather different. We may experience an image before we answer, we may even feel that we cannot answer *unless* we can experience the image, and we may feel that our answer is a 'description' of what we experience. Now how should the psychologist regard the relation of the answer to the knowledge which it conveys? Is the answer just a response to a specific stimulus? Is it intentional[2] (i.e. is the utterance 'about' something), and if so is it the verbal expression of an informationally equivalent representation of knowledge, which is independent of phenomenology? Or is the answer in some real sense a description? If it is a description, what is it a description of: of what the question referred to (the map), of knowledge, or of the phenomenal experience?

The answers to such questions and the appropriate level of discourse depend on one another. If one wants to treat the answer as a description, then one cannot treat what is described *purely* as a brain state, though it corresponds to one. But is one forced to go beyond the functionalist level to the phenomenological? Essentially the same issues arise when trying to decide on the status of phenomenal reports.

The meeting and the book

This book arises out of a meeting which was convened to make explicit the issues mentioned above and to discuss them. The meeting was held at the Villa Olmo, on the shore of Lake Como, in April 1985[3]. The motivations for the meeting determined and constrained what this book contains, and, indeed, what it omits. The contents are both wide and narrow. They are wide in the sense of the considerations, data, and approaches represented and discussed by the contributors: information-processing approaches to perception, action, and attention; neuropsychological approaches to blindsight, neglect, amnesia, and split-brain syndromes; social and psychodynamic approaches to meaning, memory, the self-concept, and learning; computational approaches; philosophy of science and philosophy of mind.

[2] A problem arises with the spelling of the term *intention/al/ity*. There are in fact three terms.
(a) The term which refers to 'the sum of attributes or objects comprehended in a concept or set', and is the opposite of its extension (the range or enumeration of objects), is *intension* and is always spelt with an *s*.
(b) The term which refers to a goal or purpose that is explicitly represented is *intention* and is always spelt with a *t*.
(c) The term which refers to content, reference, or indication – what something is about – is sometimes spelt with a *t* (*intention*) and sometimes with an *s* (*intension*). The spelling convention is partly a North American versus British matter.
In this book the first usage does not occur. Both of the latter usages are spelt with a *t* – (*intention/al/ity*). In individual chapters authors have made it quite clear in which sense they are using the term. In the subject index entries are listed under two headings (i) 'Intention – aboutness', and (ii) 'Intention – goal'.

[3] Two of the contributors (R V G and K V W) were not present at the Villa Olmo. Their chapters were originally delivered as papers at a symposium at Bielefeld, F.R.G., organized by Philip Smith and Peter Bieri, to whom we are most grateful for consenting to the inclusion of revised versions as chapters in this book.

There are two senses in which the contents are narrow. The issues dealt with are a limited set, of the kind introduced above, which have been raised by the juncture of certain disciplines at a particular time. In addition, the participants were selected from several overlapping areas (cognitive psychology, neurology, philosophy of mind and of science). These areas overlap in their interest in a set of themes and phenomena, which makes the enterprise more focused. It also contributes to the extent to which the participants could speak a common language.

Psychology of various kinds, neurology, neurophysiology, artificial intelligence, and philosophy are interacting at the moment in a way that they did not before. Clinical psychologists are looking to cognitive science for theoretical frameworks, and cognitive psychologists are more interested than previously in understanding emotional disorders. Psychologists and philosophers of mind are finding that the psychological dissociations shown by neurological patients provide test cases for their arguments. Neurologists concerned with brain damage have been adopting to a much greater extent the investigative techniques of psychology and are joining with psychologists to find functionalist models both to characterize the disorders they deal with and to guide their clinical inquiries. Artificial intelligence not only provides functionalist models and theories, and tests their cohesion and workability, but has given psychology a new range of theory and vocabulary. But it too is more influenced by both normal and pathological psychology.

The language in which these disciplines have all been able to converse is that of functionalism. Functionalism is that discourse which focuses on the functions performed by systems and the functional relationships of their components. It deals with them in abstract terms which are indifferent to what is being dealt with by the system and to the particular instantiation of that system. Since it deals in the manipulation of abstract symbols it is the language that characterizes computation. It is also clearly of importance to all the sciences to be able to have a functionalist level of theory. Perhaps the hub of this book, though in some places tacit, is whether consciousness provides a problem for functionalism. Without dwelling too much on this at the moment, the contributors adopt or discuss what is probably the whole range of views on this relationship. Some people think that the absence of consciousness in functionalist models is a feature, i.e. that it represents a substantive theoretical claim, such as that consciousness is not relevant to behaviour. Others think that such absence is not a feature of functionalism, just that it is beyond the scope of 'scientific' psychology. Some believe that there are quite adequate functionalist characterizations;

others believe that we need to tighten up our referential criteria before we can decide. There is also the claim that such characterizations as exist omit the central aspect of consciousness, phenomenal experience, and that psychology is incomplete without it. If consciousness has causal status, if it has an effect, then it ought to be dealt with by functionalist models.

Since these themes were the motivation and agenda of the meeting, and in view of the constraints of overlap and common language, it was inevitable that the participants are not representative of all current approaches to consciousness. Thus there are several phenomena, aspects, and views of consciousness that are not dealt with or represented in this book. Approaches such as the hermeneutic, phenomenological, evolutionary, developmental, and clinical are to be found represented and discussed in other recent collections; for example, the volumes edited by Pope and Singer (1978), by Underwood and Stevens (1979, 1981; Underwood 1982), and by Horowitz (1988). Such approaches clearly have implications for those outlined in this collection. Indeed reference to them is made by several of the present authors. But full representation would have presented difficulties for a small workshop meeting, where the main objective was to clarify commonalities and differences within cognitive science. Indeed the focusing of issues and the common language of contributors is what makes this volume different from other edited books on consciousness.

There is an unfortunate yet easily understandable paradox in the relation of the meeting to this book which stems from it. The purpose of the meeting was discussion, though that had to be engendered by pre-circulated papers. There most certainly *was* discussion, amounting in the formal sessions alone to some 254 pages of typed transcript, excluding what could not be heard and what could not be discreetly transcribed. Since this discussion was one of the main goals we tried very seriously to find a way of including it in the book in some form. But it could not be easily divided up and fitted neatly at the end of each contribution. What we settled on was an attempt to get the contributors to revise their presentations in the light of and in answer to the discussion of their original papers, and to encourage them to cross-reference where appropriate and where possible.

The outcome of this process is that the book has two structures. There are those topics explicitly addressed by each author. There are also the central but more implicit themes which give the book a unity and focus and which produce the major differences of stance and approach.

Themes

The main themes that emerge concern (a) what is to be explained, (b) how to explain it, and (c) what level of discourse or theory is necessary, is most useful, and is scientifically legitimate. These issues arise out of the focus of two complementary types of inquiry: one can address oneself to the question of consciousness or to problems posed by behavioural phenomena.

When the focus of explanation is consciousness, for some of the contributors it is phenomenal experience which constitutes the prime candidate for explanation (M K, A J M, K O, T S, L W)[4]. For others, aspects of the control of behaviour constitute the central problem (P N J-L, C U), while others focus on self-knowledge or our lack of it (M H E, M S G). Different usages of the term consciousness are treated on the one hand as different aspects of a single entity, all requiring related explanation (P N J-L, C U), and on the other hand as suggesting that there is no single coherent phenomenon or topic to be explained (A A, K V W) or that certain usages are beyond scientific explanation (E B). Is it possible that our use of language is so mistaken and incoherent that it not only leads to confusion (A A, K V W), but that in fact there is really nothing to be explained (D C D)? As one of the contributors remarked, it is somewhat easier to take a critical stance to consciousness than to provide effective and rigorous definitions. There is some tension here between those who want the topic specified and those who maintain that definitions should be the theoretical goal rather than the starting point.

Complementary to focusing on consciousness for explanation, is to take behaviour as one's starting point and to ask what is the appropriate way of accounting for it. Indeed what motivated the meeting was that many investigators have recently sought to account for normal and pathological behaviour in terms of consciousness. Several of the authors in this book explain the reasons for this (A J M, L W, M H E, T S). But even if one accepts such reasons [see Holender (1986) for dissent], the question remains as to the appropriate level of discourse. While many of the contributors are committed to functionalist psychology (E B, P N J-L, T S), several emphasize brain-state accounts (P S C, R L G, M K), while others feel that the personal level cannot be forsaken and locate at least some explanation in the social domain (M H E, A J M, K O). As regards the issue of intentionality, although some doubts are raised, it is argued that it can be treated within functionalism (R V G), but the relation of its nature to phenomenal experience remains an open question.

---

[4] Relevant chapters are indicated by the authors' initials.

This divergence is mirrored by the positions taken with regard to the scientific status of the concepts. On one side, it is argued that phenomenological concepts, so far as they relate to something that cannot be completely conveyed by reports, have no place in science (EB); that is, what some conscious state feels like is not itself a representation of the kind that could have informational equivalence with a report. By contrast it is proposed that not only are phenomenological concepts legitimate, but if we want adequate and full accounts then present scientific boundaries may have to be violated (AJM). The most prevalent position is to seek translation of the phenomenological—either to the functionalist level (TS, PNJ-L, CU) or to the neural level (MK, RLG) or to some combination (MSG). A more radical alternative that can be considered is to abandon the phenomenological (DCD) or reduce it to progressively lower levels of neurobiology (PSC). Finally, for several authors (AA, KVW), scientific status depends on more precise terminology than we now have.

Organization of chapters

Since most of the contributions do not deal exclusively with one topic, they have not been explicitly placed in sections. However the ordering of the chapters reflects the main issues addressed by the authors. These can be adequately captured as follows: the status of different aspects of consciousness; criteria for using the concept and identifying instances; the basis of consciousnes in functional brain organization; the relationship between different levels of theoretical discourse; functions of consciousness.

The opening chapters concern the status of the concept of consciousness in its different senses, and the positions adopted set the reference points for what follows. In the first chapter, *Kathleen Wilkes* deals with the adequacy of the term 'consciousness' as referring to something to be explained. She adopts two strategies for this. First, she suggests that if there is an explanandum, one ought to find that it is picked out by language. Second, given that in English the term exists, she asks whether it refers to something which for scientific purposes is unitary or coherent. On the first issue, she concludes that there is little equivalence to the terms 'mind' and 'consciousness' in the lexicons of classical Greek, Chinese, Croatian, or even in the English of a few centuries ago. On the second issue, after listing the referents of 'consciousness' in psychology, she suggests that the relationship between them is that of an arbitrary set rather than that of a 'natural kind' (where the constituents are systematically related). Whether the individual referents are themselves legitimate objects of enquiry she sees as a separate question.

One of these referents of the term 'consciousness' is phenomenal experience, whose qualitative properties are termed qualia. In the following chapter *Daniel Dennett* examines the origin and coherence of a certain candidate notion of qualia. By exposing the groundlessness of qualia-based insights, he undermines the notion of qualia as absolute (non-relational), atomic properties of phenomenal experience. Whether such notions of the properties of phenomenal experience are the same as those countenanced by others is a moot point. That is, most psychologists would argue that they do not assume that qualities of percepts and sensations are free of inference, attitude or memory. However if one considers the tacit assumptions underlying much classical psychophysics, Dennett's discussion has some force. If techniques of scaling qualities of sensation provide reliable results, do they tell us about what they are supposed to be measuring, about the measuring devices or about the assumptions and language of the subjects?

Another of the referents of consciousness indicated by Wilkes is the set of propositional attitudes: mental terms which have content or refer (thinking *about* x or *that* y). This intentionality of representations or processes has often been taken to be the mark of the mental, and artificial intelligence has sometimes been attacked with the claim that intentionality depends on phenomenal experience. *Robert Van Gulick's* chapter offers a clear and careful exploration of what it is to have an intentional state and the relationship of intentionality to subjective experience. He proposes, within a functionalist framework, that *intrinsic* intentionality depends on a system understanding that the symbols it processes have the semantic content that they have. Semantic self-understanding varies in degree, but is hypothesized to be greatest in the case of symbols which enter into subjective experience. It is deliberately left an open question whether conscious subjective experience is a necessary condition for a high degree of semantic self-understanding (which might provide a problem for functionalism). This issue has implications both for the relationship of artificial intelligence, psychology and neuroscience (see Marcel's chapter), and also for how psychologists of perception and language treat their domains of study. Relatively little attention is paid in mainstream psychology to the representational aspects of information and how this is dealt with in processing mechanisms.

The next two chapters also accept the subjective qualitative aspect of phenomenal experience as valid, but take radically different positions on its place in psychology. *Edoardo Bisiach* specifies three types of usage of the term consciousness and asks in what terms science can study consciousness. He maintains that it cannot be studied as first-person phenomenal experience; but that it can be studied within functionalism. In this context it is to be taken mainly as equivalent to monitoring. This function, however, is not unitary: it can be split. Nor can it be defined in objective time. Bisiach is therefore sceptical about a causal role for consciousness, if understood as phenomenal experience, though he concludes that we can make progress on its causal role if consciousness is taken in functionalist terms.

By contrast, *Anthony Marcel* attempts to deal with phenomenal experience without losing sight of its subjective aspects. He argues for the centrality, legitimacy, and necessity in psychology of consciousness as phenomenal experience. He attempts to show not only that certain aspects of phenomenal experience can be known (indirectly and relatively), but also that as such, in its first-person sense, it has causal status. He suggests that since it is necessary to psychology, either functionalism will have to deal with it or will be inadequate.

The following three chapters concentrate in more detail on the criteria for identifying consciousness and for wanting such a concept.

Alan Allport tries to establish what it is that is being referred to by consciousness and how we should recognize an instance of it. He engages in a rigorous search for such criteria. Since he suggests that different criteria can dissociate, he questions whether there is any unitary phenomenon. The criteria that he considers appear to yield answers that are consistent with each other only when behaviour is integrated and coherent. He concludes that consciousness is related to behavioural integration and that studying behavioural integration is therefore the way to study consciousness.

Lawrence Weiskrantz and *Matthew Erdelyi* present two alternatives to Allport's approach to criteria. In the first of these, rather than starting from theoretical definitions, Weiskrantz explains how recourse to consciousness as an explanatory characterization is called for by empirical data, specifically in pathologies of vision and memory. His main distinction is between objectively adequate behaviour (when induced) and the ability to comment on such functions or the acknowledgement of awareness. He suggests that the main (perhaps the only) impairment in certain neurological disorders is one of conscious awareness, and that this has disabling consequences. As some other contributors do, he characterizes the capacity as one of monitoring. He gives such a capacity a causal role in donating an adaptive flexibility. Finally he confronts a question left untouched by others—how one might know if non-verbal species also have consciousness.

The second alternative to Allport's approach is that of Erdelyi. He also presents criteria, finding them, like Weiskrantz, to be demanded by phenomena. But he advocates that as opposed to direct criteria we might use an oblique strategy, specifically that of focusing on what is *not* conscious in order to justify the concept of consciousness. Further, he suggests that recovery of unconscious information through hypermnesia and psychoanalytic insight shows a possible causal role for consciousness. It is causal, according to Erdelyi, in that through conscious intention and conceptualization consciousness 'in part determines what is accessible to itself'.

A particular point of interest arises from Erdelyi's suggestion that an extremely common instance of the nonconscious is to be found in the *latent* as opposed to *manifest content* of our experience and actions. In the sense of what can and cannot be acknowledged, this can also be applied to phenomenal experience. However, the main point is that by this criterion, instances, of consciousness are defined by their content.

Two contributors are specifically interested in functional neurological models as the brain basis of consciousness. *Michael Gazzaniga* asks whether there is anything about brain organization that tells us about consciousness. The answer given is twofold, and comes from an examination of data from split-brain patients. First, he proposes that those activities of the brain which underlie conscious experience are distributed over different functional modules. Second, he argues that data from patients with a sectioned corpus callosum indicate that a property of the left, but not right, cerebral hemisphere is interpretation of the results of such modular activities. Split-brain phenomena such as those studied by Gazzaniga are intriguing and provocative (especially regarding the question of the unitariness of consciousness), although their interpretation is both complicated and contentious.

Marcel Kinsbourne attempts to specify phenomenological characteristics of conscious experience in order to provide a basis for them in aspects of brain activity. Twelve such correspondences are proposed and illustrated. The fact that, as acknowledged by the author, these propositions are controversial makes the arguments for them particularly interesting. Kinsbourne's main theme is that mental operations that are neurally insulated are automatic and outside of awareness, while those that are neurally integrated with others contribute to awareness. Awareness, seen as neural interconnectivity, gives flexibility to behavioural control. Gazzaniga's and Kinsbourne's characterizations of the functional organization of the brain are quite different, one emphasizing modularity, the other integrality.

The concern with correspondences between different levels of description in Kinsbourne's chapter links him with those contributors who focus on the question of the appropriate level of discourse onto which to map consciousness and how to achieve such mappings. *Richard Gregory* and *Patricia Churchland* argue for different methods of reducing consciousness to brain language, while *Tim Shallice* believes that the functionalist level is needed to mediate.

Gregory's recourse to the brain is for him a possible solution to the old problem of whether it is conceivable that we could know another person's private phenomenal experience. He makes the provocative suggestion that mind language is reducible to brain language and that the representation of a brain state will permit reconstruction of phenomenal experience. Gregory provides a most interesting personal account of changes in phenomenal experience under progressive anaesthesia. What is interesting about this tracing of an altering state of consciousness is that we get some glimpse of what it is like to lose particular phenomenal experiences. Gregory implies that if one developed the necessary techniques one could relate these phenomena of consciousness to particular functional brain mechanisms.

As a philosopher of science committed to reductionism, Churchland asks whether we can understand in neurobiological terms what it is for an organism to be conscious. Her starting point is with mental phenomena as understood in folk psychology. She takes these to refer primarily to consciousness of sensations and of other mental episodes, to planning, and to keeping track of oneself and the environment. Like Wilkes and Allport, she raises the issue of whether there is a unitary phenomenon to be dealt with, emphasizing how various denormalizing data may undermine the often supposed unitariness of consciousness. She alights on the sleep–dream–wakefulness cycle as central, since it refers to global states of awareness and since it is an aspect of consciousness likely to provide good co-evolution of psychological and neurobiological studies. She concludes that if consciousness does not reduce smoothly to neurobiology, then, rather than implying dualism, it may mean abandoning what is a folk psychology concept.

It is uncertain whether or not Churchland's conception of consciousness preserves the primary concern of other authors, phenomenal experience. Indeed, what motivates Shallice is the paradox of advances in mechanistic theorizing in psychology being based partly on techniques and data involving phenomenal experience. His solution is to use the functionalist level of information processing to make a bridge with phenomenology. His chapter attempts the arduous task of detailing phenomenological properties of consciousness and mapping them onto information-processing concepts. Shallice discusses the problems which this creates and suggests possible solutions.

Several of the earlier contributors, especially those concerned with the question of causal status, discuss potential functions of consciousness. However, the last three chapters are more explicitly addressed to the issue of the roles it plays.

Both *Carlo Umiltà* and *Philip Johnson-Laird* treat consciousness at a functionalist level and see its main role as one of control and organization of parallel non-conscious operations. Umiltà first illustrates the conversion of common-sense accounts of behaviour to information-processing accounts by developing three interpretations of a historical incident in terms of attentional mechanisms affecting awareness. This produces several propositions which relate characteristics of awareness to control mechanisms. After placing such correspondences in the framework of the current theoretical literature, Umiltà discusses how his own experiments shed light on how conscious states are involved in the initiation and control of action. It is interesting that, in quite different ways, Allport, Umiltà, and Johnson-Laird all choose to make consciousness tractable by opting to treat it in terms of the cohesion and control of behaviour.

Johnson-Laird's chapter serves to illustrate computational modelling. He proposes that the appropriate level of discourse for a theory of consciousness is that of computation, locating explanation in the software rather than the hardware of the brain. By focusing on the functional role of consciousness, he attempts to account for four of its aspects: awarness, self-awareness, intentions, and control. This last aspect is taken to correspond, in a computational model, to the regulation of the output of parallel processes by an operating system with access to a model of the self. Recursive embedding is what is claimed to account for awareness and self-awareness. Whether an instantiation of this model would produce subjective qualities of experience (or what we could agree to be their equivalent) is a fascinating question.

Keith Oatley's treatment of the role of consciousness, while sharing the computational approach, is quite different. He distinguishes four aspects of consciousness, two involuntary and two voluntary. The latter are built on the former and serve the functions of developing plans and goals based on a model of the self. Like Marcel, Oatley emphasizes the social construction of consciousness, providing persuasive illustrations. His bringing together of the domain of emotions with insights from the artificial intelligence metaphor is an exciting enterprise.

Reading, editing, and re-reading the contents of this book raised the question of what conclusions we are left with. At least the two editors now understand how they and other contributors agree and disagree. We acknowledge the continued existence of many and large disagreements. However, the discussion and debate across a range of disciplines and viewpoints has made several philosophical issues which are relevant to cognitive sciences much more manifest and concrete. In addition, conceptual issues in psychology and neuropsychology in several cases have been more deeply analysed than usual, and in other cases have been spotlighted by evident differences in approach. For us this is where the main value of the book lies.

References

Dennett, D. C. (1982). How to study human consciousness empirically. *Synthese*, **53**, 159–80.

Ericsson, K. A. and Simon, H. A. (1984). *Protocol analysis: verbal reports as data*. MIT Press, Cambridge, MA.

Holender, D. (1986). Semantic activation without conscious identification. *Behavioral and Brain Sciences*, **9**, 1–6.

Horowitz, M. (ed.) (1988). *Psychodynamics and cognition*. University of Chicago Press.

Mandler, G. (1975). Consciousness: respectable, useful and probably necessary. In *Information processing and cognition: the Loyola symposium* (ed. R. Solso), pp. 229–54. Erlbaum, Hillsdale, NJ.

Natsoulas, T. (1978). Consciousness. *American Psychologist*, **33**, 906–914.

Pope, K. S. and Singer, J. L. (1978). *The stream of consciousness*. Plenum Press, New York.

Simon, H. A. (1978). On the forms of mental representation. In *Minnesota studies in the philosophy of science, vol. 9. Perception and cognition: issues in the foundations of psychology*. (ed. C. W. Savage), pp. 3–18. University of Minnesota Press.

Underwood, G. and Stevens, R. (eds.) (1979). *Aspects of consciousness, Vol. 1. Psychological issues*. Academic Press, London.

Underwood, G. and Stevens, R. (eds.) (1981). *Aspects of consciousness, Vol. 2. Structural issues*. Academic Press, London.

Underwood, G. (ed.) (1982). *Aspects of consciousness, Vol. 3. Awareness and self-awareness*. Academic Press, London.

2

——, yìshì, duh, um, and consciousness

Kathleen V. Wilkes

Introduction

Two intriguing facts. First, the terms 'mind' and 'conscious(ness)' are notoriously difficult to translate into some other languages. Second, in English (and other European languages) one of these terms—'conscious' and its cognates—is in its present range of senses scarcely three centuries old. Yet, apparently, these terms are central to contemporary psychology: is not psychology the study of the mind (even though some might add 'and behaviour'), and is not its greatest unsolved problem the understanding of consciousness? For some, no doubt, this suggests a failure in those other languages and in European self-understanding before the seventeenth century. For the sceptic, it suggests a hard look at the contemporary English terms.

A word on the strategy. It is important to emphasize that the translation problem, and the relative youth (in English) of one of the terms in question, only *suggest* such an investigation. For the linguistic data must be inconclusive, in two respects at least. The first is that, after all, it is boringly and trivially true that every language contains and lacks terms that other languages lack or contain; every language has terms that are more or less hard to translate; and every language enriches itself by acquiring new terms throughout its history. I accept this fully. Language differences of this sort become significant only when the terms in question are (purportedly) those that pick out *explananda* which are central for scientific investigation. We would surely be sceptical about a theory in physics that lacked the notions of 'force', 'energy', or 'mass'; is the notion of 'conscious(ness)' anything like these? To put it mildly, the problem of whether 'conscious(ness)'—and, albeit to a lesser extent, 'mind'— must be central to present and future theory deserves discussion; i.e. whether either picks out genuine or central *explananda* for the science. It is therefore prima facie interesting that other languages, and English before the seventeenth century, appear to lack either or both of these terms, or anything that corresponds more than roughly to them; in other words, that what strikes some of us so forcefully seems to have left little impression on others.

The second reason why linguistic data must be inconclusive is just as simple: not all the central terms in scientific theories derive from ordinary observation (consider 'ether', 'electron', 'tectonic plates', 'the hippocampus', 'genes'). Since that is so, the mere fact that the everyday language of any culture may lack such terms obviously fails to prove that their scientific theories should not acknowledge them. The term 'conscious(ness)' could be like this, referring to something which common-sense psychology in other cultures, and in our own before the seventeenth century, may have failed to spot, but which none the less any adequate theory needs to highlight. I fully acknowledge this possibility too. All the same, the supporters of the centrality of consciousness incessantly appeal to the *obviousness* of consciousness—'we all know' we are conscious. That is, 'consciousness' should be something easily recognizable, more like 'tiger' than 'electron'; we know that we have it—now, how should we describe and explain it? Insofar as 'consciousness' (and indeed 'mind') are assumed to be thus 'obvious', then it is clearly interesting to find cultures, and our own culture until recently, that happily dispense with one or both notions.

So I shall offer the linguistic evidence as something to provoke thought, no more. Arguments to suggest that it leads us in the right direction follow after.

Why examine both these notions—'mind' and 'consciousness'? There are two reasons. First, it was (I shall claim) a very specific characterization of 'mind' that gave us 'consciousness', in all its present glory, in the first place. The rise of this (narrow) concept of mind, and of consciousness, can be explained together. Second, the term 'mind' no longer troubles psychologists—and rightly so. I want to suggest that the term 'consciousness' should not trouble them either, and for essentially the same reasons. Thus although 'consciousness' is the primary target, 'mind' is a useful stalking-horse.

Of other languages, I shall look at three: ancient Greek, Chinese, and Croatian.[1] Thus there is a good spread in time, space, culture, and language types. In ancient Greek there is nothing corresponding to either 'mind' or 'consciousness' (hence the '——' in the title of this

---

[1] Greek I know reasonably well, the other two scarcely at all, but I have been well-advised by native speakers (and ignorance has some advantages, since the native speakers have to work harder to explain where the problems lie). My particular thanks to Qiu Renzong, Zhang Liping, Xiao Xiaoxin, Srđan Lelas, Nenad Miščević, Ivo Mimica, and Matjaž Potrč. Incidentally, what is suggested about Croatian probably applies equally to Serbian and Slovenian. But my advisors were (almost) all Croatian, so it seemed wiser to restrict the claims.

chapter). In Chinese, there are considerable problems in capturing 'conscious(ness)'. And in Croatian, 'mind' poses interesting difficulties.

One essential point of clarification, however. I am not denying that there are specific contexts in which the English terms are translated perfectly by Greek, Chinese, or Croatian terms: by *psyche, sophia, nous, metanoia,* or *aisthesis,* or by *'yìshì'*, or by 'duh' or 'um'. The substantial point is rather that there is no *generally* adequate translation for 'conscious(ness)' and/or for 'mind'. So the integrity of the English notions—as far as *systematic* (scientific) study of human psychology is concerned (this is an essential qualification; see the caveats above about the inconclusiveness of the linguistic data)—requires examination.

If we now turn to the history of the English term 'conscious(ness)', we see that it arrives late in its present (range of) sense(s). The *Oxford English Dictionary* (1971) finds the first use of 'conscious' to mean 'inwardly sensible or aware', as used in a sermon given by Archbishop Ussher in 1620, who said '[I was] so conscious vnto myself of my great weaknesse'; but even then Hobbes, as late as 1651, continued to use the term with its etymologically based sense of 'shared knowledge': 'When two or more men know of one and the same fact, they are said to be "conscious" of it one to another; which is as much as to know it together' (Hobbes 1946, Vol. I, p. vii). Some early uses of 'conscience' —which also retained in the seventeenth century the idea of *sharing* knowledge—were extended to overlap with some of today's uses of 'conscious', as the immediate continuation of Hobbes' quotation makes clear:

And because such are fittest witnesses of the facts of one another, or of a third: it was, and ever will be, reputed a very evil act, for any man to speak against his 'conscience': or to corrupt or force another so to do: insomuch that the plea of conscience has been always hearkened unto very diligently in all times. *Afterwards, men made use of the same word metaphorically, for the knowledge of their own secret facts, and secret thoughts; and therefore it is rhetorically said, that the conscience is a thousand witnesses* (Hobbes 1946, Vol. I, p. 37; my emphasis. Note the 'metaphorically' here.)

Earlier than that, the middle-English term 'inwit' had some overlap with today's term 'consciousness' too. But the use of the term 'conscious' itself, with a recognizably modern meaning, awaited the first quarter of the seventeenth century. The term 'conscious*ness*' did not appear until 1678; 'self-consciousness' not until 1690. (To the extent that the appropriate French and German terms are indeed equivalent, they appeared at roughly the same times, in French perhaps a few years later.)

In the next section I discuss ancient Greek, but then sketch briefly the seventeenth-century revolution which transformed the understanding of the mind, and which saw consciousness elevated to the pedestal it occupies today. The following two sections mention various difficulties for the translation of 'conscious(ness)' and 'mind' into Chinese and Croatian. Then I turn to contemporary psychology.

Ancient Greek: the contrast with Cartesianism

We have in classical Greece a great and glorious psychological understanding. It is not necessary to argue for this: the works of Euripides and Aristophanes alone prove the point. Yet there is no term that *even roughly* translates either 'mind' or 'consciousness'.

Instead of anything like 'mind' we find the term *psyche*—typically, but misleadingly, translated 'soul'; I shall use the transliteration. This changed considerably in sense between Homer and Aristotle, but in its final Aristotelian version it is of enormous (contemporary) interest: because I think it is of *contemporary* interest, I shall explain it in rough outline. Throughout its career, the term *psyche* kept one highly significant connotation: to be alive was to have *psyche*, to have *psyche* was to be alive. Hence grass and amoebas have *psyche*. They have it just as much as humans do (having a *psyche* is an all-or-nothing matter), even though they have it in a less sophisticated form. In its Aristotelian development, the *psyche* was a 'form', and forms are 'what it is to be' something. (*Every* single object is in-formed matter, en-mattered form; neither form nor 'bare' matter can exist separately in the natural world.) But the *psyche* was the most interesting subclass of forms, labelling forms of *living* things. The *psyche*—spelling out what it is to be living thing *X*—described the characteristic and defining capacities of *X*. Rudimentary life forms—grass, say—just had metabolism, nutrition, and growth. A little higher—at the level of amoebas, or flies—we find all that, plus some capacity for movement and perception. Higher still—with dogs, perhaps, or pigs—we find more diverse capacities for locomotion, and more varied and richer perceptual capacity. Above that, we are told, there is increasingly flexible sensori-motor control, and some capacity for imagination. It is evidently difficult to offer clear examples of non-human animals in this category; but maybe we could try highly encephalized animals like apes or dolphins. Laboratory and field studies of both species indicate that some problems may be solved by a measure of 'previewing'—running through, 'in the mind's eye', the strategies to adopt. How else, one might ask, did Köhler's ape Sultan work out how to reach his bananas (Köhler 1925)? (But nothing in this area is clear; see Macphail 1982, 1986.) Finally, with the human species, we find all the capacities characterizing lower levels of the pyramid, but also a rational faculty, for practical and theoretical reason. So 'what it is to be' a human rather than any other animal requires essential reference to his unique faculties for reason; but all the other capacities are presupposed, and there is constant feedback between all levels of this pyramid. The 'matter' of every animal, which the *psyche* in-forms, is flesh and bones, which is the physical realization of the *psyche*.

We shall return to the *psyche*. But first let us look at our second concept: what about the Greek 'failure' to have any term that systematically translates 'conscious(ness)'? Several authors note this, and are unhappy about it; *vide* Hamlyn (1968; p. xiii):

... there is almost total neglect of any problem arising from psycho–physical dualism and the facts of consciousness. Such problems do not seem to arise for [Aristotle]. The reason appears to be that concepts like that of consciousness do not figure in his conceptual scheme at all; they play no part in his analysis of perception, thought etc. (Nor do they play any significant role in Greek thought in general.) It is this perhaps that gives his definition of the soul itself a certain inadequacy for the modern reader.

Hamlyn is right that Aristotle, and Greek thought in general, ignore 'facts of consciousness' *per se*. But since in this paper I am explicitly questioning whether 'facts of consciousness' merit systematic study, I shall not linger long in defence of Aristotle. Clearly the onus of proof must be Hamlyn's—to spell out what such facts Aristotle neglects (and which Euripides presumably neglects too in his analysis of Medea's psychology). But we might note briefly what Aristotle does and does not talk about, to see the difficulty of charging him with any clear sin of omission; he seems to be able to talk about almost all the phenomena that we group as 'conscious', and what he admittedly ignores may perhaps deserve neglect.

First, what he does discuss: he writes about sleeping and waking, and dreaming with its *aisthemata* (images). He examines the five senses and their integration, and sensori-motor control. He discusses pains, emotions—thoroughly 'conscious', 'phenomenal', and 'subjective' phenomena, in our terminology. He has a long and intricate analysis of the imagination. Above all he has a great deal to say about practical and theoretical ratiocination and deliberation: these are the most characteristically *human* capacities.

What are the omissions? He is not, it is true, particularly interested in 'introspective self-access', 'privacy', etc. He does indeed comment that we usually perceive *that* we perceive, and are rarely wrong about it; but before we excitedly dub him a closet Cartesian, talking about epistemological privilege in our access to our mental states, we should bear in mind that he sensibly goes on to note that we usually also perceive that we are walking. Another 'omission' is of *experiences* of seeing, hearing, etc: phenomenal sensations, qualia, sense data. Some might deny that he omits these entirely—for, given that (a) he allows that we often perceive that we perceive, and (b) he does describe the *aisthemata* of dreams, is he not getting close to acknowledging something like 'phenomenal qualities'? That too would be a misleading inference: no term in Greek translates 'sense data', 'qualia', or 'raw feels', and Aristotle never talks about sensory experiences in the plural. (The plural form *aisthemata* occurs only in his treatment of dreaming, never in connection with the senses.) Certainly Aristotle does not consider any 'phenomenal qualities' of vision or hearing to be *explananda* as such. For this to be an 'omission', though, we must establish that there are indeed such things as qualia. This should not be assumed (see Dennett 1988, this volume); a little later we shall have further reason to question the integrity of the postulate of qualia. Aristotle certainly has the linguistic resources for expressing '*X* thinks that he sees *O*'; this is significant, since many today would argue that the content of '*X* has a sense datum of *O*' is exhausted by '*X* seems to see (thinks he sees) *O*', a locution that avoids the count-noun terms 'quale' or 'sense datum'. The burden of proof, again, rests on the other side—to say just why 'visual experience' cannot be fully examined without postulating 'visual experiences'. (For further discussion, see Matson 1966.)

So it is at least unclear where, or if, Aristotle is at fault. But to see why his position might even be superior, we must move on.

For now we can link all this to the English terms 'mind' and 'consciousness'. Aristotle's psychobiological term *psyche*—with its 'neglect' of 'the facts of consciousness'—was overthrown by Descartes in his *Discourse on the Method* of 1637 (published in French), although the revolution is most clearly seen in the second of his *Meditations* (published in Latin in 1641, in French in 1642). As Descartes' line of argument in the *Meditations* makes clear, the *psyche* had stood the test of time up to that point[2]; even though the Church Fathers, needing to inject into Aristotle's monism a dualistic *psyche*-substance which would survive bodily death, had added something Descartes called 'a wind, a flame, or an ether . . . spread throughout my grosser parts'. We see Descartes in a mere two pages making the transition from *psyche* to mind:

---

[2] I am not of course suggesting that Descartes' revolution just appeared out of the blue. Foreshadowing of some strands can easily be seen in the work of earlier writers (e.g. Augustine, and some Hellenistic or Stoic writers). It seems clear, though, that Descartes was the father of the Cartesian revolution in the sense that he brought the various strands together and argued explicitly for them and explicitly against more traditional views.

In the first place, then, I considered myself as having a face, hands, arms, and all that system of members composed of bones and flesh as seen in a corpse which I designated by the name of body. In addition to this I considered that I was nourished, that I walked, that I felt, and that I thought, and I referred all these actions to the soul: but I did not stop to consider what the soul was, or if I did stop, I imagined that it was something extremely rare and subtle like a wind, a flame, or an ether, which was spread throughout my grosser parts (Haldane and Ross 1967, Vol. I, p. 151).

This is almost pure Aristotle, listing first the 'matter' (body), and then the various *psyche*-capacities in correct hierarchical order. Only 'almost', though, because of the extra etherial soul-stuff, and because Descartes has forgotten the imagination. But by the time we have turned *one* page:

But what then am I? A thing which thinks. What is a thing which thinks? It is a thing which doubts, understands, [conceives], affirms, denies, wills, refuses, which also imagines and feels (Haldane and Ross 1967, Vol. I, p. 153.).

We get a formal definition of 'thinking' in the *'Arguments demonstrating the existence of God'* (Haldane and Ross 1967, Vol. II, p. 52):

Thought is a word that covers everything that exists in us in such a way that we are immediately conscious of it. Thus all the operations of will, intellect, imagination, and of the senses are thoughts.[3]

And to cap it all, in a letter to Mersenne, Descartes writes:

As to [the proposition] . . . *that nothing can be in me, that is, in my mind, of which I am not conscious,* I have proved it in the *Meditations*, and it follows from the fact that the soul is distinct from the body and that its essence is to think (Kenny 1970, p. 90).

(The French texts usually run *'avoir* connaissance de [mes pensées]', and the Latin *'esse conscius';* both are standardly translated as 'to be conscious'.)

Here we see very explicitly both the birth of the 'conscious mind' and its dramatic separation from anything bodily. Yet, as Rorty (most clearly in 1970, but see also 1980) has shown, *epistemological* considerations

---

[3] Note that 'sensation', explicitly thrown out by Descartes' observation that if he has no body he cannot be seeing anything, is smuggled back in by the argument that even if he is deceived *that* he is seeing, he cannot be deceived that he *seems* to see; and this, he claims, is 'the proper sense' of sensation words. So, welcome to sense data, qualia, sensory experiences (note the plural); it is no accident that ancient Greek has no terms to translate any of these smoothly. We have a transformed sense of 'see', which is emphatically not the everyday use: it has been equated with 'think I see'. (Note too that Aristotle could ask Descartes to put 'locomotion' back by precisely the same argument: I can be wrong about whether I am walking, but not about whether I seem to walk.)

had made this necessary (he has said—in conversation—that epistemology gives rise to ontology as sin gives birth to death). To defeat the sceptic, the epistemological foundationalist must find 'indubitable' foundations upon which—if possible—to build. Such incorrigible foundations are discovered only in first-person, present-tense, psychological statements concerning the individual current contents of the introspective gaze: a gaze focusing on objects as heterogeneous as 'pains' and metaphysical 'thoughts about being'. The mind becomes a private inner stage (*vide* Hume's analogy which, precisely, compares the mind to an internal theatre (1748; see 1965, p. 253)) in which everything 'mental' passes chaotically before an unblinking inner eye.

Now, although few of us are foundationalists, and although most of us are justifiably sceptical about the incorrigibility of introspection, and although epistemology is not clearly of direct relevance to the study of the mind anyway, we have lived with a descendant of the Cartesian *mens* ever since. The British empiricists jumped on 'mind' as referring to 'the conscious mind'—still (in part) because they were trying to defend a foundationalist epistemology. For Locke (1690; see 1959, p. 138), cognition seemed altogether impossible without consciousness: 'it is altogether as intelligible to say that a body is extended without parts, as that anything thinks without being conscious of it'. His definition of personal identity is of course famous, but the clauses following—here emphasized in italics—are less often quoted:

[A person] is a thinking intelligent being, that has reason and reflection, and can consider itself as itself, the same thinking thing, in different times and places; *which it does only by that consciousness which is inseparable from thinking, and, as it seems to me, essential to it; it being impossible for any one to perceive without perceiving that he does perceive.* . . . *For, since consciousness always accompanies thinking, and it is that which makes every one to be what he calls self* . . . (Locke 1959, pp. 448–9).

Hume makes the foundationalist–epistemological claim economically: he asserts, bluntly, that 'the perceptions of the mind are perfectly known' (1965, p. 366) and 'consciousness never deceives' (1963, p. 66).

That the notion of 'mind' is not luminously clear scarcely needs saying.[4] That its (Cartesian) restriction to *conscious* phenomena has proved both a brake and a hurdle for the development of psychology is well known. Scholars worked within a broadly Humean associationist framework for centuries; Wundt had to consign to his *Völkerpsychologie* 'those mental products which are . . . inexplicable in terms merely of individual consciousness' (1916, p. 3). The 'Cartesian revolution' is

---

[4] Squires (1971) entertainingly brings out the literal incoherence of our idioms describing the mind.

sometimes called the 'Cartesian catastrophe'. Thorndike, although complaining about more than Descartes' restriction to *conscious* mentality, and the all-or-nothing mind–body dichotomy, would find many supporters today for his complaint voiced in 1898:

[Descartes'] physiological theories have all been sloughed off by science long ago. No one ever quotes him as an authority in morphology or physiology. . . . Yet his theory of the nature of the mind is still upheld by not a few, and the differences between his doctrines of imagination, memory, and of the emotions, and those of many present-day psychological books, are comparatively unimportant (Thorndike 1898, p. 1).

There are modern echoes of this. Hayes (unpublished manuscript, 1986) comments on 'the Cartesian model of mind' as follows:

Why should psychologists be reluctant to accept that humans can learn in this way? [i.e., that task-relevant information need not be represented in a form that is open to conscious inspection, or that people can learn to perform a task without an initial declarative encoding—K V W.] An obvious candidate is an allegiance (explicitly or implicitly held) to the Cartesian model of mind. According to this view, a system which learns without the products of its learning being available to processes which can inspect, report, and if necessary modify them is essentially out of control (pp. 4–5).

Now, before Descartes nobody doubted, and nobody worried about, the obvious fact that the mind went beyond consciousness. It is easy to show, from Heraclitus on, that our self-knowledge was seen as fundamentally partial and limited; think of Plotinus, St Augustine ('the mind is not large enough to contain itself'), Dante, Shakespeare, St Teresa of Avila, Montaigne, Boehme—who would doubt that these are 'introspective' writers?—and countless more. Similarly, *some* sorts of dualism (drawing the dividing line at various points between something roughly 'bodily' and something more mental, spiritual, intellectual, or whatever) can be found throughout recorded history: consider Plato, or the Bible. The pre-Cartesian idea of the 'mind' was assuredly vague—like most common-sense notions—and may have few suitably translatable terms in other languages; but nobody had hitherto seen fit to make *consciousness* the crucial dividing line between two radically different sorts of things: mind and body, mental and physical.

Once 'the mind' was seen pre-eminently as the *conscious* mind, those objecting to this restriction—and there were many—needed to contrast their own positions; but now they had to redefine them in terms of the 'accepted' post-Cartesian mental–physical schism. A simple observation illustrates this. Before Descartes and the British empiricists, talk of 'the unconscious' had not been needed; it was taken for granted that human psychology had unexplored chasms—that the Heraclitean search into the *psyche*, and the Greek injunction to 'know thyself', were lifelong and difficult ambitions, not given free by infallible conscious introspection. No 'conscious/non-conscious' distinction had seemed necessary. But once 'consciousness' had been highlighed as *the* defining characteristic of mind, then the goal-posts were shifted: anyone wanting to combat the Cartesian revolution by stressing the role of non-conscious factors had to devise his own vocabulary. And so, reacting to the post-Cartesian stress on consciousness, we find the term 'unconscious(ness)' appearing: in English first in 1751, although rarely until the early nineteenth century; as 'unbewusstsein' and 'bewusstlos' in German in 1776; and as 'inconscient' in French in 1850—and then primarily in translations of German texts. (A cautionary note is needed here. Today, talk of 'the unconscious' tends to lead the unwary or casual reader to think of the *Freudian* unconscious. For clarity, therefore —having pointed out the relatively late, but pre-Freudian, date of the introduction of the terms, I shall in future talk not of 'unconscious' but of 'non-conscious' mental phenomena.[5]

I am claiming that the 'Cartesian catastrophe' had two main effects. First, it forced a schism between 'conscious' and 'non-conscious', compelling virtually everyone thereafter to assess the role of each, and to cast their own theories in terms dictated by the dichotomy. Second, psychology and philosophy were now stuck with two separate realms; and the task for centuries was how to relate them, how to bring them back together again. 'The mind', in other words, was hived-off from the body. We can see the loss this entails by contrasting 'mind' once again with the supplanted term *psyche*.

The clearest merit of the *psyche* lies in its denial of a mind–body dichotomy[6]: in its insistence upon the unity of the biological sciences (where 'biology' includes 'psychology'). This is highly significant. Significant, because it takes for granted that the study of 'sensation', 'emotion', 'desire', etc—just like the study of 'digestion', and for the same reasons—presupposes reference to the bodily 'matter' of these

---

[5] Anyone who inclines to believe that Freud invented the non-conscious mind would find Whyte (1960) a useful corrective.

[6] It is of course notorious that Aristotle is not an unqualified 'monist', whatever I seem to suggest here. There is the puzzling discussion of the separable 'active intellect' in *De Anima III*, ch. 5 (Hamlyn 1968, p. 60), which I do not pretend to understand—and I am not sure that anyone else does either. Still, two comments are worth making. First, it is not a 'separation' of anything like a *mind*; the active intellect is something which stands to thinking as light stands to seeing: a most mysterious metaphysical notion. Second, this passage seems to me to link up with his equally puzzling comments here and there about the prime mover as being 'pure actuality'; and where semi-theological reverberations appear, they help explain (albeit they do not excuse) many philosophical inconsistencies.

competences. It requires us to examine sensori-motor control, rather than sensation or locomotion separately (the *psyche* puts perception and locomotion at the same, intertwined levels of the hierarchy). It insists that there is no theoretically interesting schism between cognition and behaviour. Study of the *psyche* would not permit the isolation of 'cognition', all too frequently found in contemporary cognitive science; nor the belief of many post-Cartesian philosophers and psychologists that 'perception' can be understood without 'locomotion'; nor, more generally, that 'psychological' capacities can be understood without 'physiological' or 'behavioural' ones.

Although it is difficult to exaggerate the significance of the mind–body split on subsequent philosophy and psychology, it would be foolish to ignore the substantial resistance to it. Scheerer (1984, 1987) has traced the history of motor theories of cognition and perception, and the 'muscle sense concept' which developed in the eighteenth century. Unsurprisingly, the path-breakers in this area were typically physiologically trained—as were so many of the early German psychologists—or were in fields we might today think of as 'physiological psychology'. Even here, though, some of the work was contaminated by Cartesian dualism, in the following sense: the objective was sometimes seen as how to explain how two quite different sorts of things—e.g. eye movements and visual sensations—interacted, rather than as a denial of any theoretically significant difference. This *accepted*, as it should not have done, the conscious/non-conscious, phenomenal/physiological distinction and the existence of such things as visual sensations. (Remember that no Greek term adequately translates 'visual sensations'—note the plural.) Similarly, most philosophers writing about 'the mind–body problem' still see their task—however staunchly physicalistic they may be—as attempting to get two importantly different sorts of things together. Watson, who as the founding father of behaviourism was no friend of Descartes, explicitly accepted the mental/physical dichotomy, inasmuch as he junked everything on one side of it.

There are encouraging signs today that the dichotomy—in other words 'the mind'—is losing its grip. Consider the very label 'neuropsychology'. Neuropsychologists studying the hippocampus, or the amygdala, see no need to defend their ascription of 'mental' terms, such as 'classifying', 'analysing', 'comparing', to these brain masses. Nor, of course, can we handle sophisticated computers without such ascriptions. It is becoming increasingly clear that simulations of 'mentality' will never proceed beyond a shallow and uninteresting level unless and until locomoting robots rather than 'bedridden' computers are used —systems with sensory transducers, able to 'learn' by picking up information from the environment and causally interacting with objects (for a discussion, see Gunderson 1985). Instead of asking how *two* levels—mental and physical—interrelate and interact, there is, increasingly, the admission that there are *dozens* of levels: from abstract ratiocination to cell membranes and beyond. This is of course unsurprising. '*The*' mental or psychological distinction seemed fairly clear (intuitively) when it was fixed by the Cartesian criteria of incorrigibility and conscious introspective access. When the non-conscious was re-admitted into the mind, then these intuitive considerations were supplemented by 'intentionality' (roughly, the idea that mental states are 'about' things or states of affairs which might or might not exist or obtain). No conjunctive or disjunctive deployment of these two— highly dissimilar—properties of 'the mind' has ever satisfied those looking for a principled basis for the mind–body, mental–physical dichotomy.

Consider next a second very obvious merit of the *psyche*. It directs attention to the natural (indeed, natural-kind) capacities of organisms. For, as we have seen, Aristotelian forms (the *psyche* among them) highlight 'what it is to be' an axe, a cat, or a human: systematic characteristic properties. Emphasis on the mind, especially the *conscious* mind, tends by contrast to highlight events, or qualia, or 'ideas'— countable things—rather than capacities; to highlight the individual items that introspection purportedly catches, illuminated on the inner stage of the mental theatre. Psychologists (although maybe not philosophers) are today returning to study capacities rather than events or mental *items*; but we need only remember Titchener (1898, p. 7): 'Mind, then, as the sum of thoughts and feelings and the rest, is a sum of processes. The objects of the "science of mind" are the processes of mind; the objects of "mental science" are mental processes'; and '[t]hese simple processes are called *mental elements*. They are very numerous: there are probably some 50,000 of them' (p. 21). Further, the long and remarkable history of associationist psychology, initiated by Hume but pursued vigorously well into this century, shows how long it took to shift the focus from mental items to mental capacities.

Chinese

Let us look next at Chinese. Since our term 'mind' has today reverted to something much broader and woollier than the Cartesian *conscious* mind, the translator has few problems with this (although he would have had problems with the Cartesian mind). He can offer '*Xīn*' (literally, 'heart'), or '*Jīng Shén*' (literally 'spirit/God'); and if translating the phrase 'the mind–body problem', 'mind–body' will probably be '*Xīn-*

Shēn' (note the accents here: this second 'Shen' is a different character from the first). We should be happy enough with this. Certainly both 'heart' and 'spirit/God' have connotations 'mind' lacks, and vice versa. But it seemed to me—because of the comparative ease of translation —that the vagueness of the contemporary English term is adequately enough reflected by the equally unspecific Chinese term(s). Indeed, in English too we occasionally find 'spirit' or 'soul' as acceptable substitutes for 'mind', and not only in religious writings; and our idioms often put the seat of mental functioning where Aristotle thought it was, in the heart. We have already seen that what counts as 'the' spiritual, intellectual, or mental side of dualism varies considerably throughout ages and across cultures.

The term 'conscious(ness)' presents greater difficulty. '*Yìshì*' comes closest. Its two components, '*yì*' and '*shì*', originally had much the same meaning: knowing, or remembering. Furthermore, in ancient Chinese, '*yì*' seems to have been closer to *tacit, implicit* knowledge than to 'conscious', front-of-mind knowledge. Qiu (1984, p. 2) gives an intriguing example:

In one of the Chinese classics *Da Dai Li* it was said that 'Wu Wang (the Military King) asked if the Tao of Huang Di and Zhuang Xu existed in *yì*, or could be seen?'
Obviously the Tao of ruling the country by Huang Di and Zhuang Xu was supposed to be a kind of tacit, implicit knowledge.

Today, according to Qiu, '*yì*' includes (a) meaning or implication; (b) wish, desire, or intention; (c) conception, idea; (d) anticipation or expectation. '*Shì*' is primarily knowing, or knowledge. Both the roots of the combination, *yì–shì*, are thus highly cognitive.

Qiu characterizes three contemporary uses of the combination. The first is 'an opposite concept of matter. . . . It covers the various processes or products of psychology and thinking such as sensation, perception, representation, concepts, judgment, inference, abstraction, generalization and even feeling, emotion, will and personality' (Qiu 1984, p. 1). Second, as a psychological term, it is 'that at the highest level of psychological development which is inseparably connected with language and is supposed to be characteristic of [the] human being' (Qiu 1984, p. 1). The third is 'being aware of', 'realizing', etc. (Qiu 1984, p. 2).

To put it mildly, the scope of '*yìshì*' is highly *unclear*. The first of the three senses would allow us to include under '*yìshì*' much that we would *not* in English call 'conscious'; indeed, it seems closer to 'psychological', and includes much more than is suggested by the highly intellectual—cognitive—meanings of the two components, '*yì*' and '*shì*'. Indeed, Qiu suspects that this meaning is a relatively new one. The second sense, by contrast, *excludes* much we would probably wish to include: most would want to argue for non-linguistic, but none the less conscious, states both in humans and in animals. The third sense highlighted does indeed tally with one of our uses, 'being conscious/ aware *of*', but, by tallying with only one such use, gives a narrower meaning than the English term.

If the English term 'conscious' does *not* group a set of mental phenomena that has the coherence and integrity required for 'conscious' to be a useful theoretical term, it is scarcely surprising if Chinese has no roughly equivalent term that *is* consistent. Meanwhile, however, note that although both the '*yì*' and '*shì*' components of '*yìshì*' refer to *knowing*, there seems little concern with incorrigibility, immediacy, the inner eye, private mental items on an internal stage, knowledge by acquaintance rather than by description—in fact little concern with the features of consciousness that so attracted Descartes, inasmuch as they seemed to provide a basis for combatting scepticism.

Croatian

Here we find the other side of the coin. 'Svijest' and its derivatives will do for 'consciousness' and its collateral terms. The problem arises with 'mind'. 'Um' was a popular candidate. One defender, a professional psychologist, was certain that this was really an exact translation, and hence that 'uman' translated as 'mental'. But two philosophers preferred 'duh', while conceding that 'duh' was not really all that close; and 'duševni' as the adjective. Opposition to 'um' stemmed from the fact that it seemed too intellectual, focusing rather on rationality, wisdom, controlled emotions. It is not exclusively 'rational': 'razum' is the direct translation for analytical reason, and 'um' is a bit more synthetic than that, with a suggestion of a value judgment; but 'um' is still closer to the Greek *sophia* than to the English 'mind'. Thus, no more than *sophia*, it cannot readily serve to capture the irrational, uncontrolled, irresponsible thoughts and feelings that 'mind' allows for. 'Duh', however, smacked rather too much of 'spiritual' to translate the term 'mind' smoothly. Interestingly, a (Slovenian) psychology textbook of 1924 (Veber 1924) freely employed derivatives of both terms: an alternative to 'psihologija' (psychology) was 'duševslovje', and 'telo in *duša*' was the phrase used to express the contrast of mind with body ('telo'); but it equally talked of the 'psihologija *um*skega doživljanja'.

('Um', and 'duh', seemed to be the only serious contenders; but it illustrates well the chaos of the *English* term to note that in a medium-sized English–Croatian dictionary there were no fewer than ten terms offered alongside 'mind'.)

It was evidently interesting to see, given all this, how the science, psychology, now characterizes its subject matter. As in English, Croatian wisely has terms independent of either 'um' or 'duh'; e.g. 'psihologia', with the adjectival 'psihološki', meaning 'psychological'. What was studied? Here the most popular candidate was 'ličnost'—approximately translated as 'personality'. (However, the psychologist mentioned above would have felt equally happy to describe psychology as studying 'um'. I had the impression, however, that he was—wrongly—thinking that I was *criticizing* Croatian for having no exact equivalent to 'mind', and saw himself as defending the language by insisting that 'um' was such an equivalent.)

One important point to note here is that both 'um' and 'duh' seem to have resisted the links with 'svijest' which 'mind', because of the Cartesian and Empiricist tradition, has with 'consciousness'. 'Um' is certainly closer than 'mind' to Aristotle's view of what was distinctively human. It might therefore be interesting for sociologists of science to see whether contemporary Croatian psychologists are quite as bothered by 'the' problem of 'svijest' as are their English-language colleagues with 'consciousness'. (Sadly, however, the irreversible imperialism of English as the language of science would probably contaminate such data.) On the other hand, of course, it might be argued that any—new—preoccupation with 'consciousness' would be a step in the right direction. But that is, essentially, our central question.

I suggested earlier (see pp. 21–27)—when comparing 'mind' unfavourably with *'psyche'*—that the importance of mind has (a) been exaggerated and (b) is losing its grip. Croatian seems to me to indicate that this particular concept is dispensable; to argue otherwise would require one to indict Croatian common-sense psychology as in some respects lacking. Here the argumentative boot needs to be socked firmly onto the foot of those who claim that it is inadequate. This is a point to which we shall return.

Science, philosophy, and common sense

One thing to be clear about. Whatever our judgment about the coherence of terms like 'mind', or 'conscious', no proposal for linguistic legislation or reform will succeed. Little impresses the everyday speaker —certainly no philosophical arguments will shift him. Despite all the anguish of the purists, people persist in 'misusing' terms like 'hopefully', or equate 'uninterested' with 'disinterested'; the French are having no success in eliminating 'Franglais'. We all carry on, blithely disregarding grammatical, or political, or philosophical preoccupations. In science, though, one indispensable ambition is to secure clear and unambiguous concepts. Psychology has very few such; this is both a symptom of, and a reason for, its underdeveloped state.

The term 'mind' perhaps does not matter much for experimental psychology *now*—since it is no longer regarded as coextensive with 'the conscious mind'. We have seen already that 'the' mind–body, mental–physical dichotomy is increasingly being supplanted by the idea that there are numerous levels of description; indeed, each of the terms 'psychological', 'neuropsychological', and 'neurophysiological' cover several such levels. Croatian shows well that psychology can proceed with alternative terms, without even worrying (until pressed by an inquisitive philosopher) which of 'duh' or 'um' better captures whatever it is we want 'mind' to capture. In other words, Croatian presumably does not single out 'the' mental/psychological *per se*, except inasmuch as it has the bland term 'psihološki' ('psychological') to describe the scope of the science. Few psychologists are any longer misled into worrying about 'mind' as such. Insofar as it has any place in science, it is as a (dummy) umbrella term covering 'the sum total of our mental or psychological states, events, and processes'—and that is pretty unhelpful, unless we know what we mean by 'mental' and 'psychological'. Or we can say that 'mind' is the general term that picks out the subject matter of all non-behaviourist psychologies—and then we have to try to say when some area of concern is, or is not, 'psychology' (again, what about neuropsychology?) In any case, psychologists will go on doing whatever they were doing in the first place: labelling of this sort has its justification only in bureaucratic convenience.[7] However, I want to keep 'mind' in play, because I shall be suggesting that 'conscious(ness)' might—indeed should—find itself in the same boat.

For the use of the term 'consciousness' is currently more serious. Again, there is no problem with everyday usage. The ordinary speaker has no 'theory' about consciousness[8], but uses the term faultlessly. He

---

[7] It may matter rather more for philosophy, though. That part of philosophy which studies our everyday psychology—let us call it 'philosophy of mind'—cannot do better than give as broad and vague a definition as the one here: the sum total of mental phenomena. So the existence of the concept of mind reinforces and is reinforced by the dualism that claims, *contra* Aristotle, that there is a difference of kind between 'perceiving' and 'walking'; that there is indeed a single mind–body (mental–physical) distinction.

[8] Recently it has been argued—particularly by both Churchlands (P. M. Churchland 1984, P. S. Churchland 1986)—that common-sense psychology incorporates a 'theory'. I disagree fundamentally with this assumption: common sense and science have such dramatically different aims that it seems absurd to compare them *as theories*. All the same, anyone reading this chapter should realize that I am taking for granted, and not (here) arguing for, a substantial difference between 'common-sense psychology' and 'scientific psychology'.

uses it inconsistently, perhaps; but ordinary language is, literally construed, inconsistent through and through. (We talk of the sun rising,
knowing what we say to be *literally* false; we do something for someone's sake, while remaining resolutely anti-realist about 'sakes'; we
might describe someone as 'possessed', or another as being 'Oedipal'
while deploring demon theories and Freudian theories alike; we give
someone a piece of our mind, have something weighing on our mind,
or at the back of it, without reifying 'the mind'.) Our use of the
term 'conscious' will differ from context to context, but that is normal;
so is it with thousands of everyday terms. The mere fact that ordinary
language talks about X's does not mean that X's exist (think of sakes);
nor, even if X's do exist, need they be matters of scientific investigation: think of briefcases, fences, watering cans. As Dupré (1981) has
noted, lay interests often diverge from scientific ones. It is more
important for the man in the street to distinguish garlic from lilies
than to acknowledge their similarities, and 'the gourmet puts more
emphasis on the distinction between garlic and onions than is
implicit in taxonomy' (p. 83). So the layman's inconsistent and 'un-
scientific' use of terms like 'conscious'—and indeed 'mind'—can be
left alone.

But there are perhaps major difficulties with the notion of 'consciousness' for science. One reason for suggesting this derives from the
arguments above: ancient Greek managed without the term, and
Chinese seems to group things rather differently. English managed
without it until the seventeenth century, and then needed it in conjunction with a suspect—redefined—concept of 'mind', which Croatian
clearly lacks, to solve specifically epistemological problems. These, as I
have already emphasized strongly, are only hints. They prove nothing;
some cultures have only two colour terms, and what implications can
we draw from that? Maybe Euripides would have described Medea even
better if he had had a Greek term for 'consciousness'. We need fuller
arguments.

Natural kinds

What does science describe and explain? One standard answer here is
that science studies 'natural kinds': the joints into which Nature is
carved. Natural kinds, in short, are systematically fruitful *explananda*
and *explanantia*, where the members of the kind are held together and
governed by law(s) (and sometimes by symmetry principles, or descriptions of structural isomorphism; but for simplicity I shall ignore these in
what follows). Thus 'gold', but not 'briefcases'; 'tigers', but not 'fences';

'habituation', but (probably) not 'prejudices'. 'Memory' is a good example of a term whose status is thoroughly disputed.[9]

There are two essential points to make about natural kinds. First, everyday (non-scientific) language is riotously rich, flexible, and contextually nuance-ridden. The scientific conceptual apparatus, by contrast, tries for economy, precision (*inflexibility*), and context neutrality. Thus, if science adopts everyday terms for its own purposes—as it does—it must try to *discover* the appropriate (systematically fruitful, natural-kind) ones, weeding out the innumerable everyday terms like 'briefcase' which resist systematic exploration. Further, as Mandler and Kessen (1959) emphasized, once everyday terms are adopted, they must then be adapted: 'baked in the theoretical kiln' to eliminate their context dependency, vagueness, and imprecision. Contrast the everyday, and the scientific, use of 'force', 'mass', 'energy', '(electron) spin', and 'intelligence' to see the point here. So, clearly, if 'conscious(ness)' is a natural-kind *explanandum*, we can take it for granted that any science which adopts it will certainly want and need to adapt—redefine—it to some extent: maybe to a considerable extent.

Second, some natural-kind terms will be more or less 'unitary' in the following sense: they will group a set of individual items, properties, processes which are such that virtually all the laws or principles governing the kind include virtually all the members of that kind. 'Tiger', 'gold', and 'water' would be examples. At the other extreme a kind term might split into sub classes where, although *some* laws may hold true of all instances falling under that kind, other laws will apply only to the constituent sub classes; I shall call these 'cluster' natural kinds. 'Metal' would be one example; also 'acid', 'heat', and 'fish'. But note that there is a difference between 'kinds' and mere 'sets'. A set can be wholly arbitrary, such as the set consisting of 'all the words with "g" as the fourth letter'. A kind must have more coherence than that if it is to be a 'natural joint'. What sort of coherence? Well, even though most of the interesting laws may concern only the subclasses, there might be some laws at least which are interestingly—non-trivially—true of all

---

[9] That 'memory' is not now regarded as a likely candidate for natural-kind status probably does not need stressing; the number of distinctions people have seen fit to draw (short-term, long-term, and 'working' memory; procedural vs. declarative; semantic vs. episodic, iconic, non-cognitive, somatic, etc.) illustrate the diversity of the phenomenon. Alternatively one could consider the bewildering variety of types of amnesia, to make the same point. Amnesia can be anterograde, retrograde, or both; some amnesics can remember skills (like the Tower of Hanoi puzzle) while being unable to remember any *facts*—such as the fact that this puzzle has been seen before; some memory failures seem due to an inability to store information, others to a failure to retrieve it; some diseases (e.g. the Korsakoff syndrome) might spare some remote memories, whereas others (Huntington's, Alzheimer's) do not; the list could continue long (see Butters and Miliotis 1985.)

the subclasses; or, even if this is not so, the laws that concern the subclasses may have significant structural analogy or isomorphism.[10] (Evidently we shall need to say something about what makes a scientifically *interesting* law.)

Thus whenever we ask whether we have a scientific *explanandum*, there will be two intertwined issues: (a) whether the term in question —e.g. 'learning' or 'intelligence'—picks out a natural kind at all, and, if so, what redefinition and modification the everyday term will need; and (b) the extent to which, if it is a natural kind, it is unitary or cluster. Before returning to consciousness, it may be helpful to illustrate the point with slightly easier examples.

Consider learning. Dickinson (1980), for example, thinks (*contra* many other learning theorists) that it is at least possible that there is a 'general learning process'; in my terminology, he would regard at least some kinds of learning as forming a fairly unitary natural kind. The matter, though, is far from clear. Many others regard 'learning' as a label that groups sets of species-, task-, or niche-specific abilities; compare Kling (1971, p. 553), who suggests that we should consider 'learning' as 'a heading for a set of chapters in a textbook'—a classificatory term for a range of interrelated psychological and physiological issues, where presumably the most interesting laws would be species- or task-specific. If Kling is right, 'learning' might either be no natural kind at all, or it might be a cluster kind.

Then look at 'recognition':

I believe in fact that there is no single faculty of 'recognition' but that the term covers the totality of all the associations aroused by any object. Phrased another way, we 'manifest recognition' by responding appropriately; to the extent that any appropriate response occurs, we have shown 'recognition'. But this view abolishes the notion of a unitary step of 'recognition'; instead, there are multiple parallel processes of appropriate response to a stimulus (Geschwind 1974, p. 167).

If we follow Geschwind, 'recognition' might be a natural kind; but if so, it is at the cluster end of the spectrum.

What of 'intelligence'? There is a discussion of this, highly relevant to my theme, in Menzel and Juno (1985, pp. 155–6):

Marmoset intelligence, as we see it, is whatever marmosets do . . . the major ingredients of most definitions of intelligence . . . amount largely to roundabout definitions of *human* intelligence . . . judgments regarding intelligence might best be viewed as folk taxonomy rather than scientific taxonomy, and . . . no taxonomist should trust any external 'field marker' implicitly. . . . Species do not have any platonic 'essence'; and neither does intelligence. At the same time,

---

[10] I am grateful to Barbara von Eckardt for suggesting this way of putting the point.

folk taxonomy is probably not completely mistaken . . . to say that *questions* about animal intelligence are 'outdated' or 'scientifically useless' would be short-sighted.

This last quotation highlights all the points I want to make. First, Menzel and Juno reinforce the point that science adopts everyday terms but then has to adapt them for scientific purposes. Common sense ('folk taxonomy') may *point* to a genuine natural kind; but not all such terms prove in the end to point reliably: not all, that is, are suitable for either adoption or adaptation. Second, they stress the legitimacy of asking to what degree 'intelligence' is a natural kind at all; and, if it is, to what extent it is what I have termed 'unitary' or 'cluster'. (With this example, of course, one answer seems just obvious: whether or not 'intelligence' picks out a natural kind, it will not pick out a unitary one: there is certainly no 'essence' to it.)

Returning to 'consciousness', we can ask: is it a natural kind, suitable for 'adaptation' after its 'adoption' from the everyday language? If it is, then—like most terms in psychology—it is likely to be closer to the 'cluster' end of the spectrum. I have discussed this (in different terms) in a recent paper (Wilkes 1984), from which it should be clear that I regard 'conscious phenomena' rather as a *set* than as a ('cluster') kind. I shall not therefore spend more than a paragraph or two summarizing the discussion of the four disparate bunches of phenomena that I think 'conscious(ness)' includes.

First, someone awake is conscious, someone asleep is not. That may seem clear enough, but note the many grey areas. We are not sure what to say of someone dreaming, in a fugue, under hypnosis, undergoing epileptic automatism. Second, consider bodily sensations: pains, itches, butterflies in the stomach, pins and needles. There is a popular belief that feeling pain (to take the paradigmatic example from this set) is a way of being conscious; that there can be no such things as non-conscious pains. The strange behaviour of some patients under analgesics of the morphine group, or under hypnotic anaesthesia, make this less clear than we might hope, though. Moreover, clear cases of pain seem to be on a continuum, with the unnoticed strains and tensions that lead us—waking and sleeping—to shift head position, uncross our legs, and so forth. Those insensitive to pain are also unable to make these minute postural adjustments in response to minor muscular overwork; this seems to me to suggest that such non-conscious stimuli might be included, in a scientific taxonomy, under the general heading of 'pain' even though they are not responses to consciously felt sensations: even though their 'esse' is not 'percipi'.

My third category needs a little more introduction, for many (con)fuse it with the sensations of the second: it is sensory experience, the

operation of sight, hearing, taste, smell, and touch. This was assimilated
to sensations like pain by the Cartesian revolution, since that revolution exploited mental *items* as the objects of incorrigible report. But
seeing and hearing are not prima facie entity-like, and—if we remember
the hints from ancient Greek—before Descartes there was no use and no
place for the *plural* of 'experience' in connection with the senses, and
hence no need either for philosophical terms of art like 'qualia', 'sensa',
or 'raw feels' to describe their operation. I have already mentioned that
many today want to get rid of count nouns like 'sense data', 'visual
sensations', or 'qualia' in favour of non-count terms like 'seeming to
see'. Quite apart from the fact that sensory experience is not 'jointed', as
pains and itches can be ('pain' is a genuine count noun, unlike 'seeing'),
there are other differences: it is rare that sense perception can *per se* be
called 'intense', 'pleasant/unpleasant', 'located', 'lasting five minutes',
'stabbing/throbbing', 'better than before', all of which apply readily and
easily to bodily sensations; and when such adjectives do apply to
sensory experience, the reasons for their application are very unlike the
reasons for applying them to sensations proper. Moreover, even though
the slogan *esse est percipi* ('to be is to be perceived') seems plausible to
some in connection with bodily sensations, it is—as we shall see
—much less so where the five senses (to which we should add kinaesthesia) are concerned. No doubt there is a continuum here, as the addition
of kinaesthesia suggests; but differences of degree can be substantial. I
have defended the distinction here more fully in my 1984 paper.

To resume, then, normal perception is usually called 'conscious', by
contrast to subliminal perception or 'blindsight'. Here again there are
grey areas: what should we say about the visual experience of the skilled
driver who puts his driving on 'automatic control' when engaging in
lively conversation, and who can remember practically nothing of the
road over which he drove, or the lorry he braked to avoid?[11] What should
we say about Anton's syndrome—the mirror-image of 'blindsight':
those with cortical blindness who none the less assert that they can see
(see Anton 1899; Benson and Greenberg 1969)?

And finally, fourth, there is the ascription of the propositional attitudes—deliberating, pondering, desiring, believing. (A subset of these
would include *self*-conscious states: when what one is thinking or
deliberating about is oneself or one's own mental states.) Some of these
seem clearly 'conscious', specifically occurrent thoughts—Proustian
memory floods, or the thought in words running through the head: I'm
running out of toothpaste; must remember to buy some. Others seem

---

[11] The question whether the car-driver driving smoothly while conversing was 'conscious' or 'unconscious' divided psychologists at a recent conference in Bielefeld approximately fifty–fifty.

'tacit', or non-, pre-, or sub-conscious (take your pick), such as our ascription 'he thought there was another step' when we see someone stumble. Prosopagnosics (people who cannot recognize faces, however familiar—even their own faces in a mirror) presumably are not conscious of a thought 'that's John' when shown a photograph; yet they may react strongly to it if, say, John is a close friend who has recently died. So at some non-conscious level, we have to assume, recognition is achieved. There are many grey areas here: in particular, what to say about the beliefs that we unhesitatingly ascribe to people which they might never have thought about *explicitly*, but which are trivial consequences of other propositions they believe; e.g. 'few people eat daffodils'.

How can we decide whether, given this heterogeneity, 'consciousness' is any sort of natural kind? Much evidently revolves around the 'usefulness', and the 'interest', of laws; this must be a matter of degree. It *might* so happen—although it is unlikely—that 'aquatic animal' is a more useful natural kind, throws up more interesting laws, than 'fish'. It seems likely that 'memory' is not after all any sort of natural kind, nor 'intelligence'. The problems are considerable, and require us to have some grip on what count as *interesting* or informative laws. Put another way, to be 'interesting' the laws that relate the various subclasses of the natural kind should not have their content exhausted by a conjunction of the subordinate laws describing each subclass. If I understand her correctly, Cartwright (1983) thinks that many 'super laws' in physics are uninteresting for just this reason: she says, e.g., 'what the unified laws [super laws] dictate should happen, happens *because* of the combined action of laws from separate domains, like the law of gravity and Coulomb's law' (p. 71). If we want to understand some phenomenon, we will be best advised to look at the contributory laws: 'super laws' may add little or nothing. But here I admit that I am issuing promissory notes, because of constraints of space; in general, to decide on what counts as an 'interesting' law is a fascinating and important problem I cannot discuss now. Full treatment would require a look at all the 'values of scientific theories', such as their scope, range, accuracy, simplicity or economy, predictive power, capacity for extension to new domains, internal logical consistency, consistency with theories in related areas, and so on and so forth; and would then need a consideration how these values can be applied to individual laws, or to fragments of theories. Here I must reluctantly leave the idea of 'an interesting law' to intuition.

Back to consciousness. If it is a scientifically fruitful natural kind, then the multiplicity of phenomena called 'conscious' (a) should be profitable *explananda* for scientific study, and (b) should show systematic and *interesting* relationships between them all.

Summary and conclusion

The chances that 'conscious(ness)' is a scientifically fruitful *explanandum* look slim. I shall not here rehearse the arguments of my 1984 paper, which were explicitly designed to highlight the gross heterogeneity in phenomena called 'conscious'. Building on that paper, to say that the term 'conscious' is woolly needs no argument. That alone would not rule it out from science; so is the term 'representation'. Newton-Smith (1981) points out that there are several theoretical notions, such as 'field', and 'space–time', which are admittedly vague when introduced, but which allow for—make possible—precisely the experiments which serve or served to refine them. Maybe (let us hope) the term 'representation', so much (ab)used in cognitive science, will prove to be as legitimate a theoretical notion as 'field', and will eventually acquire a precise sense. But we must remember that the strategy of science is to adopt a term from the ordinary language and then to adapt it: to redefine it to fit it for its assigned role. There are thousand upon thousand of ordinary language psychological terms, of which 'conscious' is but one—it is one that does not even exist in other languages with the same range and scope; before 'adopting' it, there should be some reason to think that doing so will serve a genuine theoretical need.

There is no such need. To sum up: (a) no two people agree on what to include under 'conscious'; (b) the roots of the term's popularity lie in epistemology, not in observation or experiment (so that 'consciousness' should be the nightmare of the philosopher, not of the psychologist); (c) any conscious/non-conscious dichotomy leaves far too much—and most of the really interesting phenomena—unclassifiably in the middle, thereby encouraging us to downplay them; (d) it is improbable that something bunching together pains, and thoughts about mathematics, is going to be a reliable pointer to a legitimate natural kind; (e) dropping it, *just like dropping 'mind'*, would leave nothing out (i.e. Hamlyn's 'problems of consciousness' (see p. 20 of this volume) are either pseudo-problems, or can be better answered by a theoretical apparatus that ignores 'consciousness *per se*'); (f) dropping 'consciousness' would leave us free to consider whether, and to what extent, each of the four kinds of conscious phenomena listed above (and there might well be more) is a natural kind *explanandum* and therefore deserving of explanation in its own right, without contamination from *allegedly* associated areas of investigation. (Last century we were obsessed by 'the will'. Probably few today think of this as a 'useful' natural kind, whether unitary or cluster.)

Essentially, I am trying to say two distinguishable things. First, that in all the contexts in which it tends to be deployed, the term 'conscious' and its cognates are, for *scientific* purposes, both unhelpful and unnecessary. The assorted domains of research, so crudely *indicated* by the ordinary language term, can and should be carved up into taxonomies that cross-classify those which emphasis on 'consciousness' would suggest. Second, that we have little if any reason to suppose that these various domains have anything interesting in common: that is, consciousness will not be a (cluster) natural kind.

In sum, just as psychologists do not study 'mind' *per se*, so they need not bother with consciousness. Just as not bothering with 'mind' *per se* leaves nothing out, so not bothering with 'consciousness' would not restrict research. Back—or rather forwards—to Aristotle.

Acknowledgements

Earlier drafts of this paper were read at Bielefeld and Liverpool, and I am particularly grateful to Peter Bieri, Barbara von Eckardt, Tim Shallice, George Mandler, and Eckart Scheerer (albeit among others) for their helpful comments. I have a large debt to Tony Marcel for his detailed suggestions (at Bielefeld and subsequently). I am aware that I have probably not converted the doubters with this revised version.

References

Anton, G. (1899). Ueber die Selbstwahrnehmung der Herderkrankungen des Gehirns durch den Kranken bei Rindenblindheit und Rindentaubheit. *Archiv für Psychiatrie und Nervenkrankheiten*, **32**, 86–127.

Benson, D. F. and Greenberg, J. P. (1969). Visual form agnosia. *Archives of Neurology*, **20**, 82–9.

Butters, N. and Miliotis, P. (1985). Amnesic disorders. In *Clinical neuropsychology*, (2nd edn), (eds K. M. Heilman and E. Valenstein), pp. 403–51. Oxford University Press.

Cartwright, N. (1983). *How the laws of physics lie*. Clarendon Press, Oxford.

Churchland, P. M. (1984). *Matter and consciousness*. MIT Press, Cambridge, MA.

Churchland, P. S. (1986). *Neurophilosophy*. MIT Press, Cambridge, MA.

Dennett, D. C. (1988). Quining qualia. In *Consciousness in contemporary science*, (ed. A. Marcel and E. Bisiach), pp. 42–77. Oxford University Press.

Dickinson, A. (1980). *Contemporary animal learning theory*. Cambridge University Press.

Dupré, D. (1981). Natural kinds and biological taxa. *Philosophical Review*, **90**, 66–90.

Geschwind, N. (1974). Disconnexion syndromes in animals and man. In *Selected papers on language and the brain: Boston studies in the philosophy of science*, (ed. N. Geschwind), pp. 105–236. Dordrecht, Holland.

Gunderson, K. (1985). *Mentality and machines*, (2nd edn). Croom Helm, London.

Haldane, S. and Ross, G. R. T. (ed.) (1967). *The philosophical works of Descartes*, 2 volumes, (trans. S. Haldane and G. R. T. Ross). Cambridge University Press.

Hamlyn, D. W. (1968). *Aristotle's De Anima Books II, III.* Clarendon Press, Oxford.

Hayes, N. A. (1986). Consciousness and modes of learning. Unpublished manuscript. Department of Experimental Psychology, Oxford.

Hobbes, T. (1946). *Leviathan.* George Routledge, London.

Hume, D. (1963). In *Enquiries concerning the human understanding*, (ed. L. A. Selby-Bigge), Clarendon Press, Oxford.

Hume, D. (1965). *A treatise of human nature*, (ed. L. A. Selby-Bigge), Oxford University Press.

Kenny, A. J. P. (ed.), (1970). *Descartes: philosophical letters*, (trans A. J. P. Kenny), Clarendon Press, Oxford.

Kling, J. W. (1971). Learning: an introductory survey. In *Experimental psychology*, (ed. J. W. Kling and L. A. Riggs), pp. 515–62, Holt, Rinehart and Winston, New York.

Köhler, W. (1925). *The mentality of apes.* Routledge and Kegan Paul, London.

Locke, J. (1959). *An essay concerning human understanding*, (ed. A. C. Fraser), Dover Publications, New York.

Macphail, E. M. (1982). *Brain and intelligence in vertebrates.* Clarendon Press, Oxford.

Macphail, E. M. (1986). Vertebrate intelligence: the null hypothesis. In *Animal intelligence*, (ed. L. Weiskrantz), pp. 37–50. Clarendon Press, Oxford.

Mandler, G. and Kessen, W. (1959). *The language of psychology.* New York.

Matson, W. (1966). Why isn't the mind–body problem ancient? In *Mind, matter and method: essays in philosophy and science in honor of Herbert Feigl*, (ed. P. Feyerabend and G. Maxwell, pp. 92–102. University of Minnesota Press, Minneapolis.

Menzel, E. W. Jr. and Juno, C. (1985). Social foraging in marmoset monkeys. In *Animal intelligence*, (ed. L. Weiskrantz), pp. 145–57, Clarendon Press, Oxford.

Newton-Smith, W. H. (1981). *The rationality of science.* Routledge and Kegan Paul, London.

Oxford English Dictionary, Compact Edition (1971). Volume 1, p. 212. Oxford University Press.

Qiu Renzong (1984). The vocabulary of 'consciousness' in Chinese language. Unpublished manuscript, Chinese Academy of Social Science, Beijing.

Rorty, R. (1970). Cartesian epistemology and changes in ontology. In *Contemporary American philosophy*, (ed. J. E. Smith), pp. 273–92. George Allen and Unwin, London.

Rorty, R. (1980). *Philosophy and the mirror of nature.* Basil Blackwell, Oxford.

Scheerer, E. (1984). Motor theories of cognitive structure: a historical review. In

Cognition and motor processes, (ed. W. Prinz and A. F. Sanders), pp. 77–98. Springer Verlag, Berlin.

Scheerer, E. (1987). Muscle sense and innervation feelings: a chapter in the history of perception and action. In *Issues in perception and action*, (ed. H. Heuer and A. F. Sanders), pp. 171–94. Erlbaum Hillsdale, N J.

Squires, R. (1971). On one's mind. *Philosophical Quarterly*, **20**, 347–66.

Thorndike, E. L. (1898). The psychology of Descartes. In *The Columbia University Seminar in philosophy, 1897–8*. Unpublished manuscript. Thorndike Manuscripts, Columbia University, New York.

Titchener, E. B. (1898). *A primer of psychology*. Norwood M A.

Veber, F. (1924). *Očrt psihologije*. Zvezna Tiskarna in Knijigama, Ljubljana.

Whyte, L. L. (1960). *The unconscious before Freud*. Basic Books, New York.

Wilkes, K. V. (1984). Is consciousness important? *British Journal for the Philosophy of Science*, **35**, 223–43.

Wundt, W. M. (1916). *Elements of folk psychology*. George Allen and Unwin, London. Trans. of *Völkerpsychologie: Eine Untersuchung der Entwicklungsgesetze von Sprache, Mythus und Sitte*.

3

Quining qualia

Daniel C. Dennett

Corralling the quicksilver

'Qualia' is an unfamiliar term for something that could not be more familiar to each of us: the *ways things seem to us*. As is so often the case with philosophical jargon, it is easier to give examples than to give a definition of the term. Look at a glass of milk at sunset; *the way it looks to you*—the particular, personal, subjective visual quality of the glass of milk is the *quale* of your visual experience at the moment. The *way the milk tastes to you then* is another, gustatory *quale*, and *how it sounds to you* as you swallow is an auditory *quale*. These various 'properties of conscious experience' are prime examples of *qualia*. Nothing, it seems, could you know more intimately than your own qualia; let the entire universe be some vast illusion, some mere figment of Descartes' evil demon, and yet what the figment is *made of* (for you) will be the *qualia* of your hallucinatory experiences. Descartes claimed to doubt everything that could be doubted, but he never doubted that his conscious experiences had qualia, the properties by which he knew or apprehended them.

The verb 'to quine' is even more esoteric. It comes from *The Philosophical Lexicon* (Dennett 1987a), a satirical dictionary of eponyms: 'quine, v. To deny resolutely the existence or importance of something real or significant'. At first blush it would be hard to imagine a more quixotic quest than trying to convince people that there are no such properties as qualia; hence the ironic title of this chapter. But I am not kidding.

My goal is subversive. I am out to overthrow an idea that, in one form or another, is 'obvious' to most people—to scientists, philosophers, lay people. My quarry is frustratingly elusive; no sooner does it retreat in the face of one argument than 'it' reappears, apparently innocent of all charges, in a new guise.

Which idea of qualia am I trying to extirpate? Everything real has properties, and since I do not deny the reality of conscious experience, I grant that conscious experience has properties. I grant moreover that each person's states of consciousness have properties in virtue of which those states have the experiential content that they do. That is to say, whenever someone experiences something as being one way rather than another, this is true in virtue of some property of something happening in them at the time, but these properties are so unlike the properties traditionally imputed to consciousness that it would be grossly misleading to call any of them the long-sought qualia. Qualia are supposed to be *special* properties, in some hard-to-define way. My claim—which can only come into focus as we proceed—is that conscious experience has *no* properties that are special in *any* of the ways qualia have been supposed to be special.

The standard reaction to this claim is the complacent acknowledgement that while some people may indeed have succumbed to one confusion or fanaticism or another, one's own appeal to a modest, innocent notion of properties of subjective experience is surely safe. It is just that presumption of innocence I want to overthrow. I want to shift the burden of proof, so that anyone who wants to appeal to private, subjective properties has to prove first that in so doing they are *not* making a mistake. This status of *guilty until proven innocent* is neither unprecedented nor indefensible (so long as we restrict ourselves to concepts). Today, no biologist would dream of supposing that it was quite all right to appeal to some innocent concept of *élan vital*. Of course one *could* use the term to mean something in good standing; one could use *élan vital* as one's name for DNA, for instance, but this would be foolish nomenclature, considering the deserved suspicion with which the term is nowadays burdened. I want to make it just as uncomfortable for anyone to talk of qualia—or 'raw feels' or 'phenomenal properties' or 'subjective and intrinsic properties' or 'the qualitative character' of experience—with the standard presumption that they, and everyone else, knows what on earth they are talking about.[1]

What are qualia, *exactly*? This obstreperous query is dismissed by one author ('only half in jest') by invoking Louis Armstrong's legendary reply when asked what jazz was: 'If you got to ask, you ain't never gonna get to know' (Block 1978, p. 281). This amusing tactic perfectly illustrates the presumption that is my target. If I succeed in my task, this move, which passes muster in most circles today, will look as quaint and insupportable as a jocular appeal to the ludicrousness of a living thing—a living thing, mind you!—doubting the existence of *élan vital*.

---

[1] A representative sample of the most recent literature on qualia would include Block 1980; Shoemaker 1981, 1982; Davis 1982; White 1985; Armstrong and Malcolm 1984; Churchland 1985; and Conee 1985.

My claim, then, is not just that the various technical or theoretical concepts of qualia are vague or equivocal, but that the source concept, the 'pre-theoretical' notion of which the former are presumed to be refinements, is so thoroughly confused that, even if we undertook to salvage some 'lowest common denominator' from the theoreticians' proposals, any acceptable version would have to be so radically unlike the ill-formed notions that are commonly appealed to that it would be tactically obtuse—not to say Pickwickian—to cling to the term. Far better, tactically, to declare that there simply are no qualia at all.[2]

Rigorous arguments only work on well-defined materials, and, since my goal is to destroy our faith in the pre-theoretical or 'intuitive' concept, the right tools for my task are intuition pumps, not formal arguments. What follows is a series of fifteen intuition pumps, posed in a sequence designed to flush out—and then flush away—the offending intuitions. In the next section, I will use the first two intuition pumps to focus attention on the traditional notion. It will be the burden of the rest of the paper to convince you that these two pumps, for all their effectiveness, mislead us and should be discarded. In the following section, the next four intuition pumps create and refine a 'paradox' lurking in the tradition. This is not a formal paradox, but only a very powerful argument pitted against some almost irresistibly attractive ideas. In the next section, six more intuition pumps are arrayed in order to dissipate the attractiveness of those ideas, and the following section drives this point home by showing how hapless those ideas prove to be when confronted with some real cases of anomalous experience. This will leave something of a vacuum, and in the final section three more intuition pumps are used to introduce and motivate some suitable replacements for the banished notions.

The special properties of qualia

Intuition pump 1: watching you eat cauliflower. I see you tucking eagerly into a helping of steaming cauliflower, the merest whiff of which makes me feel faintly nauseated, and I find myself wondering how you

---

[2] The difference between 'eliminative materialism'—of which my position on qualia is an instance—and a 'reductive' materialism that takes on the burden of identifying the problematic item in terms of the foundational materialistic theory is thus often best seen not so much as a doctrinal issue but as a tactical issue: how might we most gracefully or effectively enlighten the confused in this instance? See my discussion of 'fatigues' in the Introduction to *Brainstorms* (Dennett 1978a) and, earlier, my discussion of what the enlightened ought to say about the metaphysical status of *sakes* and *voices* in *Content and consciousness* (Dennett 1969, chapter 1).

could possibly relish *that taste*, and then it occurs to me that, to you, cauliflower probably tastes (must taste?) different. A plausible hypothesis, it seems, especially since I know that the very same food often tastes different to me at different times. For instance, my first sip of breakfast orange juice tastes much sweeter than my second sip if I interpose a bit of pancakes and maple syrup, but after a swallow or two of coffee, the orange juice goes back to tasting (roughly? exactly?) the way it did with the first sip. Surely we want to say (or think about) such things, and surely we are not wildly wrong when we do, so . . . surely it is quite OK to talk of *the way the juice tastes to Dennett at time t*, and ask whether it is just the same as or different from *the way the juice tastes to Dennett at time t'* or *the way the juice tastes to Jones at time t.*

This 'conclusion' seems innocent, but right here we have already made the big mistake. The final step presumes that we can isolate the qualia from everything else that is going on—at least in principle or for the sake of argument. What counts as *the way the juice tastes to x* can be distinguished, one supposes, from what is a mere accompaniment, contributory cause, or byproduct of this 'central' way. One dimly imagines taking such cases and stripping them down gradually to the essentials, leaving their common residuum, the way things look, sound, feel, taste, smell to various individuals at various times, independently of how they are subsequently disposed to behave or believe. The mistake is not in supposing that we can in practice ever or always perform this act of purification with certainty, but the more fundamental mistake of supposing that there is such a residual property to take seriously, however uncertain our actual attempts at isolation of instances might be.

The examples that seduce us are abundant in every modality. I cannot imagine, will never know, could never know, it seems, how Bach sounded to Glenn Gould. (I can barely recover in my memory the way Bach sounded to me when I was a child.) And I cannot know, it seems, what it is like to be a bat (Nagel 1974), or whether you see what I see, colourwise, when we look up at a clear 'blue' sky. The homely cases convince us of the reality of these special properties—those subjective tastes, looks, aromas, sounds—that we then apparently isolate for definition by this philosophical distillation.

The specialness of these properties is hard to pin down, but can be seen at work in *intuition pump 2: the wine-tasting machine.* Could Gallo Brothers replace their human wine-tasters with a machine? A computer-based 'expert system' for quality control and classification is probably within the bounds of existing technology. We now know enough about the relevant chemistry to make the transducers that would replace taste buds and olfactory organs (delicate colour vision would perhaps be more problematic), and we can imagine using the output of such transducers as the raw material—the 'sense data' in effect—for elaborate evaluations, descriptions, classifications. Pour the sample in the funnel and, in a few minutes or hours, the system would type out a chemical assay, along with commentary: 'a flamboyant and velvety Pinot, though lacking in stamina'—or words to such effect. Such a machine might well perform better than human wine-tasters on all reasonable tests of accuracy and consistency the wine-makers could devise[3], but *surely* no matter how 'sensitive' and 'discriminating' such a system becomes, it will never have, and enjoy, what *we* do when we taste a wine: the qualia of conscious experience. Whatever informational, dispositional, functional properties its internal states have, none of them will be special in the way qualia are. If you share that intuition, you believe that there are qualia in the sense I am targeting for demolition.

What is special about qualia? Traditional analyses suggest some fascinating second-order properties of these properties. First, since one *cannot say* to another, no matter how eloquent one is and no matter how co-operative and imaginative one's audience is, exactly what way one is currently seeing, tasting, smelling, and so forth, qualia are *ineffable*—in fact the paradigm cases of ineffable items. According to tradition, at least part of the reason why qualia are ineffable is that they are *intrinsic* properties—which seems to imply *inter alia* that they are somehow atomic and unanalysable. Since they are 'simple' or 'homogeneous' there is nothing to get hold of when trying to describe such a property to one unacquainted with the particular instance in question.

Moreover, verbal comparisons are not the only cross-checks ruled out. *Any* objective, physiological, or 'merely behavioral' test—such as those passed by the imaginary wine-tasting system—would of necessity miss the target (one can plausible argue), so all interpersonal comparisons of these ways of appearing are (apparently) systematically impossible. In other words, qualia are essentially *private* properties. And, finally, since they *are* properties of *my experiences* (they are not chopped liver, and they are not properties of, say, my cerebral blood flow—or haven't you been paying attention?), qualia are essentially directly accessible to the consciousness of their experiencer (whatever that means), or qualia are properties of one's experience with which one is intimately or directly acquainted (whatever that means), or 'immediate phenomenological qualities' (Block 1978) (whatever that

---

[3] The plausibility of this concession depends less on a high regard for the technology than on a proper scepticism about human powers, now documented in a fascinating study by Lehrer (1983).

means). They are, after all, the very properties the appreciation of which permits us to identify our conscious states. So, to summarize the tradition, qualia are supposed to be properties of a subject's mental states that are

(1) ineffable

(2) intrinsic

(3) private

(4) directly or immediately apprehensible in consciousness.

Thus are qualia introduced onto the philosophical stage. They have seemed to be very significant properties to some theorists because they have seemed to provide an insurmountable and unavoidable stumbling block to functionalism or, more broadly, to materialism or, more broadly still, to any purely 'third-person' objective viewpoint or approach to the world (Nagel 1986). Theorists of the contrary persuasion have patiently and ingeniously knocked down all the arguments, and said most of the right things, but they have made a tactical error, I am claiming, of saying in one way or another: 'We theorists can handle *those qualia* you talk about just fine; we will show that you are just slightly in error about the nature of qualia'. What they ought to have said is: 'What qualia?'

My challenge strikes some theorists as outrageous or misguided because they think they have a much blander and hence less vulnerable notion of qualia to begin with. They think I am setting up and knocking down a straw man, and ask, in effect: 'Who said qualia are ineffable, intrinsic, private, directly apprehensible ways things seem to one?' Since my suggested fourfold essence of qualia may strike many readers as tendentious, it may be instructive to consider, briefly, an apparently milder alternative: qualia are simply 'the qualitative or phenomenal features of sense experience[s], in virtue of having which they resemble and differ from each other, qualitatively, in the ways they do' (Shoemaker, 1982, p. 367). Surely I do not mean to deny *those* features.

I reply: it all depends on what 'qualitative or phenomenal' comes to. Shoemaker contrasts *qualitative* similarity and difference with 'intentional' similarity and difference—similarity and difference of the properties an experience represents or is 'of'. That is clear enough, but what then of 'phenomenal'? Among the non-intentional (and hence qualitative?) properties of my visual states are their physiological properties. Might these very properties be the qualia Shoemaker speaks of? It is supposed to be obvious, I take it, that these sorts of features are ruled out, because they are not 'accessible to introspection' (S. Shoemaker, personal communication). These are features of my visual *state*, perhaps, but not of my visual *experience*. They are not *phenomenal* properties.

But then another non-intentional similarity some of my visual states share is that they tend to make me think about going to bed. I think this feature of them *is* accessible to introspection—on any ordinary, pre-theoretical construal. Is that a phenomenal property or not? The term 'phenomenal' means nothing obvious and untendentious to me, and looks suspiciously like a gesture in the direction leading back to ineffable, private, directly apprehensible ways things seem to one.[4]

I suspect, in fact, that many are unwilling to take my radical challenge seriously, largely because they want so much for qualia to be acknowledged. Qualia seem to many people to be the last ditch defence of the inwardness and elusiveness of our minds, a bulwark against creeping mechanism. They are sure there must be *some* sound path from the homely cases to the redoubtable category of the philosophers, since otherwise their last bastion of specialness will be stormed by science.

This special status for these presumed properties has a long and eminent tradition. I believe it was Einstein who once advised us that science could not give us the *taste* of the soup. Could such a wise man have been wrong? Yes, if he is taken to have been trying to remind us of the qualia that hide forever from objective science in the subjective inner sancta of our minds. There are no such things. Another wise man said so—Wittgenstein (1958, especially pp. 91–100). Actually, what he said was:

> The thing in the box has no place in the language-game at all; not even as a *something*; for the box might even be empty.—No, one can 'divide through' by the thing in the box; it cancels out, whatever it is (p.100);

and then he went on to hedge his bets by saying 'It is not a *something*, but not a *nothing* either! The conclusion was only that a nothing would serve just as well as a something about which nothing could be said' (p. 102). Both Einstein's and Wittgenstein's remarks are endlessly amenable to exegesis, but, rather than undertaking to referee this War of the Titans, I choose to take what may well be a more radical stand than

[4] Shoemaker (1984, p. 356) seems to be moving reluctantly towards agreement with this conclusion: 'So unless we can find some grounds on which we can deny the possibility of the sort of situation envisaged . . . we must apparently choose between rejecting the functionalist account of qualitative similarity and rejecting the standard conception of qualia. I would prefer not to have to make this choice; but if I am forced to make it, I reject the standard conception of qualia'.

Wittgenstein's.[5] Qualia are not even 'something about which nothing can be said'; 'qualia' is a philosophers' term which fosters[6] nothing but confusion, and refers in the end to no properties or features at all.

The traditional paradox regained

Qualia have not always been in good odour among philosophers. Although many have thought, along with Descartes and Locke, that it made sense to talk about private, ineffable properties of minds, others have argued that this is strictly nonsense—however naturally it trips off the tongue. It is worth recalling how qualia were presumably rehabilitated as properties to be taken seriously in the wake of Wittgensteinian and verificationist attacks on them as pseudo-hypotheses. The original version of *intuition pump 3: the inverted spectrum* (Locke 1690; see 1959 edn.) is a speculation about two people: how do I know that you and I see the same subjective colour when we look at something? Since we both learned colour words by being shown public coloured objects, our verbal behaviour will match *even if we experience entirely different subjective colours*. The intuition that this hypothesis is systematically unconfirmable (and undisconfirmable, of course) has always been quite robust, but some people have always been tempted to think technology could (in principle) bridge the gap.

Suppose, in *intuition pump 4: the Brainstorm machine*, there were some neuroscientific apparatus that fits on your head and feeds your visual experience into my brain (as in the movie, *Brainstorm*, which is not to be confused with the book, *Brainstorms*). With eyes closed I accurately report everything you are looking at, except that I marvel at how the sky is yellow, the grass red, and so forth. Would this not confirm, empirically, that our qualia were different? But suppose the

---

[5] Shoemaker (1982) attributes a view to Wittgenstein (acknowledging that 'it is none too clear' that this is actually what Wittgenstein held) which is very close to the view I defend here. But to Shoemaker, 'it would seem offhand that Wittgenstein was mistaken' (p. 360), a claim Shoemaker supports with a far from offhand thought experiment—which Shoemaker misanalyses if the present paper is correct. (There is no good reason, contrary to Shoemaker's declaration, to believe that his subject's *experience* is systematically different from what it was before the inversion.) Smart (1959) expresses guarded and partial approval of Wittgenstein's hard line, but cannot see his way clear to as uncompromising an eliminativism as I maintain here.

[6] In 1979, I read an earlier version of this paper in Oxford, with a commentary by John Foster, who defended qualia to the last breath, which was: 'qualia should not be quined but fostered!' Symmetry demands, of course, the following definition for the most recent edition of *The philosophical lexicon* (Dennett 1987a): 'foster, v. To acclaim resolutely the existence or importance of something chimerical or insignificant'.

technician then pulls the plug on the connecting cable, inverts it 180 degrees, and reinserts it in the socket. Now I report the sky is blue, the grass green, and so forth. Which is the 'right' orientation of the plug? Designing and building such a device would require that its 'fidelity' be tuned or calibrated by the normalization of the two subjects' reports—so we would be right back at our evidential starting point. The moral of this intuition pump is that no intersubjective comparison of qualia is possible, even with perfect technology.

So matters stood until someone dreamt up the presumably improved version of the thought experiment: the *intra*personal inverted spectrum. The idea seems to have occurred to several people independently (Gert 1965; Putnam 1965; Taylor 1966; Shoemaker 1969, 1975; Lycan 1973). Probably Block and Fodor (1972) have it in mind when they say 'It seems to us that the standard verificationist counterarguments against the view that the "inverted spectrum" hypothesis is conceptually incoherent are not persuasive' (p. 172). In this version, *intuition pump 5: the neurosurgical prank*, the experiences to be compared are all in one mind. You wake up one morning to find that the grass has turned red, the sky yellow, and so forth. No one else notices any colour anomalies in the world, so the problem must be in you. You are entitled, it seems, to conclude that you have undergone visual colour qualia inversion (and we later discover, if you like, just how the evil neurophysiologists tampered with your neurons to accomplish this).

Here it seems at first—and indeed for quite a while—that qualia are acceptable properties after all, because propositions about them can be justifiably asserted, empirically verified, and even explained. After all, in the imagined case, we can tell a tale in which we confirm a detailed neurophysiological account of the precise etiology of the dramatic change you undergo. It is tempting to suppose, then, that neuro-physiological evidence, incorporated into a robust and ramifying theory, would have all the resolving power we could ever need for determining whether or not someone's qualia have actually shifted.

But this is a mistake. It will take some patient exploration to reveal the mistake in depth, but the conclusion can be reached—if not secured—quickly with the help of *intuition pump 6: alternative neurosurgery*. There are (at least) two different ways the evil neurosurgeon might create the inversion effect described in intuition pump 5:

1. Invert one of the 'early' qualia-producing channels, e.g. in the optic nerve, so that all relevant neural events 'downstream' are the 'opposite' of their original and normal values. *Ex hypothesi* this inverts your qualia.

2. Leave all those early pathways intact and simply invert certain memory-access links—whatever it is that accomplishes your tacit (and even unconscious) comparison of today's hues with those of yore. *Ex hypothesi* this does *not* invert your qualia at all, but just your memory-anchored dispositions to react to them.

On waking up and finding your visual world highly anomalous, you should exclaim 'Egad! *Something* has happened! Either my qualia have been inverted or my memory-linked qualia reactions have been inverted. I wonder which!'

The intrapersonal, inverted spectrum thought experiment was widely supposed to be an improvement, since it moved the needed comparison into one subject's head. But now we can see that this is an illusion, since the link to earlier experiences, the link via memory, is analogous to the imaginary cable that might link two subjects in the original version.

This point is routinely—one might say traditionally—missed by the constructors of 'intrasubjective, inverted spectrum' thought experiments, who suppose that the subject's *noticing the difference*—surely a vivid experience of discovery by the subject—would have to be an instance of (directly? incorrigibly?) recognizing the difference *as a shift in qualia*. But as my example shows, we could achieve the same startling effect in a subject without tampering with his presumed qualia at all. Since *ex hypothesi* the two different surgical invasions can produce exactly the same introspective effects, while only one operation inverts the qualia, nothing in the subject's experience can favour one of the hypotheses over the other. So unless he seeks outside help, the state of his own qualia must be as unknowable to him as the state of anyone else's qualia: hardly the privileged access or immediate acquaintance or direct apprehension the friends of qualia had supposed 'phenomenal features' to enjoy!

The outcome of this series of thought experiments is an intensification of the 'verificationist' argument against qualia. *If* there are qualia, they are even less accessible to our ken than we had thought. Not only are the classical intersubjective comparisons impossible (as the *Brainstorm* machine shows), but we cannot tell in our own cases whether our qualia have been inverted—at least not by introspection. It is surely tempting at this point—especially to non-philosophers—to decide that this paradoxical result must be an artefact of some philosophical misanalysis or other, the sort of thing that might well happen if you took a perfectly good pre-theoretical notion—our everyday notion of qualia —and illicitly stretched it beyond the breaking point. The philosophers have made a mess; let them clean it up; meanwhile we others can get back to work, relying as always on our sober and unmetaphysical acquaintance with qualia.

Overcoming this ubiquitous temptation is the task of the next section, which will seek to establish the unsalvageable incoherence of the hunches that lead to the paradox by looking more closely at their sources and their motivation.

Making mistakes about qualia

The idea that people might be mistaken about their own qualia is at the heart of the ongoing confusion and must be explored in more detail, and with somewhat more realistic examples if we are to see the delicate role it plays.

Intuition pump 7: Chase and Sanborn. Once upon a time there were two coffee-tasters, Mr Chase and Mr Sanborn, who worked for Maxwell House.[7] Along with half a dozen other coffee-tasters, their job was to ensure that the taste of Maxwell House coffee stayed constant, year after year. One day, about six years after Chase had come to work for Maxwell House, he confessed to Sanborn:

> I hate to admit it, but I'm not enjoying this work anymore. When I came to Maxwell House six years ago, I thought Maxwell House coffee was the best-tasting coffee in the world. I was proud to have a share in the responsibility for preserving that flavour over the years. And we've done our job well; the coffee tastes just the same today as it tasted when I arrived. But, you know, I no longer like it! My tastes have changed. I've become a more sophisticated coffee drinker. I no longer like *that taste* at all.

Sanborn greeted this revelation with considerable interest. 'It's funny you should mention it,' he replied, 'for something rather similar has happened to me.' He went on:

> When I arrived here, shortly before you did, I, like you, thought Maxwell House coffee was tops in flavour. And now I, like you, really don't care for the coffee we're making. But *my* tastes haven't changed; my ... *tasters* have changed. That is, I think something has gone wrong with my taste buds or some other part of my taste-analyzing perceptual machinery. Maxwell House coffee doesn't taste to me the way it used to taste; if only it did, I'd still love it, for I still think *that taste* is the best taste in coffee. Now I'm not saying we haven't done our job well. You other tasters all agree that the taste is the same, and I must admit that on a day-to-day basis I can detect no change either. So it must be my problem alone. I guess I'm no longer cut out for this work.

---

[7] This example first appeared in print in my reflections on Smullyan in *The Mind's I* (Hofstadter and Dennett 1981, pp. 427–8).

Chase and Sanborn are alike in one way at least: they both used to like Maxwell House coffee, and now neither likes it. But they claim to be different in another way. Maxwell House tastes to Chase just the way it always did, but not so for Sanborn. But can we take their protestations at face value? Must we? Might one or both of them simply be wrong? Might their predicaments be importantly the same and their apparent disagreement more a difference in manner of expression than in experiential or psychological state? Since both of them make claims that depend on the reliability of their memories, is there any way to check on this reliability?

My reason for introducing two characters in the example is not to set up an interpersonal comparison between how the coffee tastes to Chase and how it tastes to Sanborn, but just to exhibit, side by side, two poles between which cases of intrapersonal experiential shift can wander. Such cases of intrapersonal experiential shift, and the possibility of adaptation to them, or interference with memory in them, have often been discussed in the literature on qualia, but without sufficient attention to the details, in my opinion. Let us look at Chase first. If we fall in for the nonce with the received manner of speaking, it appears at first that there are the following possibilities:

(a) Chase's coffee-taste qualia have stayed constant, while his reactive attitudes to those qualia, devolving on his canons of aesthetic judgment, etc., have shifted—which is what he seems, in his informal, casual way, to be asserting.

(b) Chase is simply wrong about the constancy of his qualia; they have shifted gradually and imperceptibly over the years, while his standards of taste have not budged—in spite of his delusions about having become more sophisticated. He is in the state Sanborn claims to be in, but just lacks Sanborn's self-knowledge.

(c) Chase is in some predicament intermediate between (a) and (b); his qualia have shifted some *and* his standards of judgment have also slipped.

Sanborn's case seems amenable to three counterpart versions:

(a) Sanborn is right; his qualia have shifted, due to some sort of derangement in his perceptual machinery, but his standards have indeed remained constant.

(b) Sanborn's standards have shifted unbeknownst to him. He is thus misremembering his past experiences, in what we might call a

nostalgia effect. Think of the familiar experience of returning to
some object from your childhood (a classroom desk, a tree-house)
and finding it much smaller than you remember it to have been.
Presumably as you grew larger your internal standard for what was
large grew with you somehow, but your memories (which are stored
as fractions or multiples of that standard) did not compensate, and
hence, when you consult your memory, it returns a distorted judgment. Sanborn's nostalgia-tinged memory of good old Maxwell
House is similarly distorted. (There are obviously many different
ways this impressionistic sketch of a memory mechanism could be
implemented, and there is considerable experimental work in cognitive psychology that suggests how different hypotheses about such
mechanisms could be tested.)

(c) As before, Sanborn's state is some combination of (a) and (b).

I think that everyone writing about qualia today would agree that
there are all these possibilities for Chase and Sanborn. I know of no one
these days who is tempted to defend the high line on infallibility or
incorrigibility that would declare that alternative (a) is—and must
be—the truth in each case, since people just cannot be wrong about such
private, subjective matters.[8]

Since quandaries are about to arise, however, it might be wise to
review in outline why the attractiveness of the infallibilist position is
only superficial, so it will not recover its erstwhile allure when the going
gets tough. First, in the wake of Wittgenstein (1958) and Malcolm (1956,
1959), we have seen that one way to buy such infallibility is to acquiesce
in the complete evaporation of content (Dennett 1976). 'Imagine someone saying: "But I know how tall I am!" and laying his hand on top of his
head to prove it' (Wittgenstein 1958, p. 96). By diminishing one's claim
until there is nothing left to be right or wrong about, one can achieve a
certain empty invincibility, but that will not do in this case. One of the
things we want Chase to be right about (if he is right) is that he is not in
Sanborn's predicament, so if the claim is to be viewed as infallible it can
hardly be because it declines to assert anything.

There is a strong temptation, I have found, to respond to my claims in
this chapter more or less as follows: 'But after all is said and done, there
is still something I know in a special way: I know *how it is with me right
now*'. But if absolutely nothing follows from this presumed knowledge

---

[8] Kripke (1982, p. 40) comes close when he asks rhetorically: 'Do I not know, directly,
and *with a fair degree of certainty*, that I mean plus [by the function I call "plus"]?' [my
emphasis]. Kripke does not tell us what is implied by 'a fair degree of certainty', but
presumably he means by this remark to declare his allegiance to what Millikan (1984)
attacks under the name of 'meaning rationalism'.

—nothing, for instance, that would shed any light on the different psychological claims that might be true of Chase or Sanborn—what is the point of asserting that one has it? Perhaps people just want to reaffirm their sense of proprietorship over their own conscious states.

The infallibilist line on qualia treats them as properties of one's experience one cannot in principle misdiscover, and this is a mysterious doctrine (at least as mysterious as papal infallibility) unless we shift the emphasis a little and treat qualia as *logical constructs* out of subjects' qualia judgements: a subject's experience has the quale F if and only if the subject judges his experience to have quale F. We can then treat such judgings as constitutive acts, in effect, bringing the quale into existence by the same sort of licence as novelists have to determine the hair colour of their characters by fiat. We do not ask how Dostoevski knows that Raskolnikov's hair is light brown.

There is a limited use for such interpretations of subjects' protocols, I have argued (Dennett 1978*a*, 1979, especially pp. 109–10, 1982), but they will not help the defenders of qualia here. Logical constructs out of judgments must be viewed as akin to theorists' fictions, and the friends of qualia want the existence of a particular quale in any particular case to be an empirical fact in good standing, not a theorist's useful interpretive fiction, else it will not loom as a challenge to functionalism or materialism or third-person objective science.

It seems easy enough, then, to dream up empirical tests that would tend to confirm Chase and Sanborn's different tales, but if passing such tests could support their authority (that is to say, their reliability), failing the tests would have to undermine it. The price you pay for the possibility of empirically confirming your assertions is the outside chance of being discredited. The friends of qualia are prepared, today, to pay that price, but perhaps only because they have not reckoned how the bargain they have struck will subvert the concept they want to defend.

Consider how we could shed light on the question of where the truth lies in the particular cases of Chase and Sanborn, even if we might not be able to settle the matter definitively. It is obvious that there might be telling objective support for one extreme version or another of their stories. Thus if Chase is unable to re-identify coffees, teas, and wines in blind tastings in which only minutes intervene between first and second sips, his claim to *know* that Maxwell House tastes just the same to him now as it did six years ago will be seriously undercut. Alternatively, if he does excellently in blind tastings, and exhibits considerable knowledge about the canons of coffee style (if such there be), his claim to have become a more sophisticated taster will be supported. Exploitation of the standard principles of inductive testing—basically Mill's method of differences—can go a long way toward indicating what sort of change has occurred in Chase or Sanborn—a change near the brute perceptual processing end of the spectrum or a change near the ultimate reactive judgment end of the spectrum. And as Shoemaker (1982) and others have noted, physiological measures, suitably interpreted in some larger theoretical framework, could also weight the scales in favour of one extreme or the other. For instance, the well-studied phenomenon of induced illusory boundaries (see Fig. 3.1) has often been claimed to be a particularly 'cognitive' illusion, dependent on 'top–down' processes and hence, presumably, near the reactive judgment end of the spectrum, but recent experimental work (Von der Heydt *et al.* 1984) has revealed that 'edge detector' neurons *relatively* low in the visual pathways—in area 18 of the visual cortex—are as responsive to illusory edges as to real light–dark boundaries on the retina, suggesting (but not quite proving, since these might somehow still be 'descending effects') that illusory contours are not imposed from on high, but generated quite early in visual processing. One can imagine discovering a similarly 'early' anomaly in the pathways leading from taste buds to judgment in Sanborn, for instance, tending to confirm his claim that he has suffered some change in his basic perceptual—as opposed to judgmental— machinery.

But let us not overestimate the resolving power of such empirical testing. The space in each case between the two poles represented by possibility (a) and possibility (b) would be occupied by phenomena that were the product, somehow, of two factors in varying proportion: roughly, dispositions to generate or produce qualia and dispositions to react to the qualia once they are produced. (That is how our intuitive picture of qualia would envisage it.) Qualia are supposed to affect our

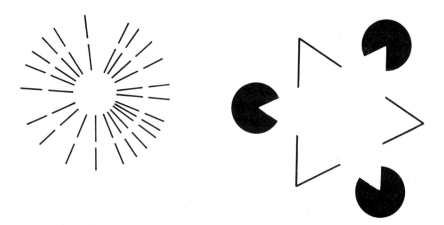

Fig. 3.1.   Induced illusory contours

action or behaviour only via the intermediary of our judgments about them, so any behavioural test, such as a discrimination or memory test, since it takes acts based on judgments as its primary data, can give us direct evidence only about the *resultant* of our two factors. In extreme cases we can have indirect evidence to suggest that one factor has varied a great deal, the other factor hardly at all, and we can test the hypothesis further by checking the relative sensitivity of the subject to variations in the conditions that presumably alter the two component factors. But such indirect testing cannot be expected to resolve the issue when the effects are relatively small—when, for instance, our rival hypotheses are Chase's preferred hypothesis (a) and the minor variant to the effect that his qualia have shifted *a little* and his standards *less than he thinks*. This will be true even when we include in our data any unintended or unconscious behavioural effects, for their import will be ambiguous. (Would a longer response latency in Chase today be indicative of a process of 'attempted qualia renormalization' or 'extended aesthetic evaluation?')

The limited evidential power of neurophysiology comes out particularly clearly if we imagine a case of adaptation. Suppose, in *intuition pump 8: the gradual post-operative recovery*, that we have somehow 'surgically inverted' Chase's taste bud connections in the standard imaginary way: post-operatively, sugar tastes salty, salt tastes sour, etc. But suppose further—and this is as realistic a supposition as its denial —that Chase has subsequently compensated—as revealed by his behaviour. He now *says* that the sugary substance we place on his tongue is sweet, and no longer favours gravy on his ice-cream. Let us suppose the compensation is so thorough that on all behavioural and verbal tests his performance is indistinguishable from that of normal subjects—and from his own pre-surgical performance.

If all the internal compensatory adjustment has been accomplished early in the process—intuitively, pre-qualia—then his qualia today are restored to just as they were (relative to external sources of stimulation) before the surgery. If on the other hand some or all of the internal compensatory adjustment is post-qualia, then his qualia have not been renormalized *even if he thinks they have*. But the physiological facts will not in themselves shed any light on where in the stream of physiological process twixt tasting and telling to draw the line at which the putative qualia appear as properties of that phase of the process. The qualia are the 'immediate or phenomenal' properties, of course, but this description will not serve to locate the right phase in the physiological stream, for, echoing intuition pump 6, there will always be at least two possible ways of interpreting the neurophysiological theory, however it comes out. Suppose our physiological theory tells us (in as much detail

as you like) that the compensatory effect in him has been achieved by an
adjustment in the memory-accessing process that is required for our
victim to compare today's hues to those of yore. There are *still* two
stories that might be told:

(I) Chase's current qualia are still abnormal, but thanks to the revision
    in his memory-accessing process, he has in effect adjusted his
    memories of how things used to taste, so he no longer notices any
    anomaly.

(II) The memory-comparison step occurs just prior to the qualia phase
     in taste perception; thanks to the revision, it now *yields* the same
     old qualia for the same stimulation.

In (I) the qualia contribute to the input; in effect, to the memory
comparator. In (II) they are part of the output of the memory comparator.
These seem to be two substantially different hypotheses, but the
physiological evidence, no matter how well developed, will not tell us
on which side of memory to put the qualia. Chase's introspective
evidence will not settle the issue between (I) and (II) either, since *ex
hypothesi* those stories are not reliably distinguishable by him. Remember that it was to confirm or disconfirm Chase's opinion that we turned
to the neurophysiological evidence in the first place. We can hardly use
his opinion in the end to settle the matter between our rival neurophysiological theories. Chase may think that he thinks his experiences are
the same as before *because* they really are (and he remembers accurately
how it used to be), but he must admit that he has no introspective
resources for distinguishing that possibility from alternative (I), on
which he thinks things are as they used to be *because* his memory of
how they used to be has been distorted by his new compensatory habits.

Faced with their subject's systematic neutrality, the physiologists
may have their own reasons for preferring (I) to (II), or vice versa, for they
may have *appropriated* the term 'qualia' to their own theoretical ends,
to denote some family of detectable properties that strike them as
playing an important role in their neurophysiological theory of perceptual recognition and memory. Chase or Sanborn might complain—in
the company of more than a few philosophical spokesmen—that these
properties the neurophysiologists choose to call 'qualia' are not the
qualia they are speaking of. The scientists' retort is: 'If we cannot
distinguish (I) from (II), we certainly cannot support either of your
claims. If you want our support, you must relinquish your concept of
qualia.'

What is striking about this is not just that the empirical methods would fall short of distinguishing what seem to be such different claims about qualia, but that they would fall short *in spite of being better evidence than the subject's own introspective convictions.* For the subject's own judgments, like the behaviours or actions that express them, are the resultant of our two postulated factors, and cannot discern the component proportions any better than external behavioural tests can. Indeed, a subject's 'introspective' convictions will generally be *worse* evidence than what outside observers can gather. For if our subject is—as most are—a 'naive subject', unacquainted with statistical data about his own case or similar cases, his immediate, frank judgments are, evidentially, like any naive observer's perceptual judgments about factors in the outside world. Chase's intuitive judgments about his qualia constancy are no better off, epistemically, than his intuitive judgments about, say, lighting intensity constancy or room temperature constancy—or his own body temperature constancy. Moving to a condition inside his body does not change the intimacy of the epistemic relation in any special way. Is Chase running a fever or just feeling feverish? Unless he has taken steps to calibrate and cross-check his own performance, his opinion that his fever-perception apparatus is undisturbed is no better than a hunch. Similarly, Chase may have a strongly held opinion about the degree to which his taste-perceiving apparatus has maintained its integrity, and the degree to which his judgment has evolved through sophistication, but, pending the results of the sort of laborious third-person testing just imagined, he would be a fool to claim to know—especially to know directly or immediately—that his was a pure case (a), closer to (a) than to (b), or a case near (b).

Chase is on quite firm ground, epistemically, when he reports that *the relation* between his coffee-sipping activity and his judging activity has changed. Recall that this is the factor that Chase and Sanborn have in common: they used to like Maxwell House; now they do not. But unless he carries out on himself the sorts of tests others might carry out on him, his convictions about what has stayed constant (or nearly so) and what has shifted *must be sheer guessing.*

But then qualia—supposing for the time being that we know what we are talking about—must lose one of their 'essential' second-order properties: far from being directly or immediately apprehensible properties of our experience, they are properties whose changes or constancies are either entirely beyond our ken, or inferrable (at best) from 'third-person' examinations of our behavioural and physiological reaction patterns (if Chase and Sanborn acquiesce in the neurophysiologists' sense of the term). On this view, Chase and Sanborn should be viewed not as introspectors capable of a privileged view of these properties, but as autopsychologists, theorists whose convictions about the properties of their own nervous systems are based not only on their 'immediate' or current experiential convictions, but also on their appreciation of the import of events they remember from the recent past.

There are, as we shall see, good reasons for neurophysiologists and other 'objective, third-person' theorists to single out such a class of properties to study. But they are not qualia, for the simple reason that one's epistemic relation to them is *exactly* the same as one's epistemic relation to such external, but readily—if fallibly—detectable, properties as room temperature or weight. The idea that one should consult an outside expert, and perform elaborate behavioural tests on oneself to confirm what qualia one had, surely takes us too far away from our original idea of qualia as properties with which we have a particularly intimate acquaintance.

So perhaps we have taken a wrong turning. The doctrine that led to this embarrassing result was the doctrine that sharply distinguished qualia from their (normal) effects on reactions. Consider Chase again. He claims that coffee tastes 'just the same' as it always did, but he admits—nay insists—that his reaction to 'that taste' is not what it used to be. That is, he pretends to be able to divorce his apprehension (or recollection) of the quale—the taste, in ordinary parlance—from his different reactions to the taste. But this apprehension or recollection is itself a reaction to the presumed quale, so some sleight of hand is being perpetrated—innocently no doubt—by Chase. So suppose instead that Chase had insisted that precisely *because* his reaction was now different, the taste had changed for him. (When he told his wife his original tale, she said 'Don't be silly! Once you add the dislike you change the experience!'—and the more he thought about it, the more he decided she was right.)

Intuition pump 9: the experienced beer drinker. It is familiarly said that beer, for example, is an acquired taste; one gradually trains onself —or just comes—to enjoy that flavour. What flavour? The flavour of the first sip? No one could like *that* flavour, an experienced beer drinker might retort:

Beer tastes different to the experienced beer drinker. If beer went on tasting to me the way the first sip tasted, I would never have gone on drinking beer! Or to put the same point the other way around, if my first sip of beer had tasted to me the way my most recent sip just tasted, I would never have had to acquire the taste in the first place! I would have loved the first sip as much as the one I just enjoyed.

If we let this speech pass, we must admit that beer is *not* an acquired taste. No one comes to enjoy *the way the first sip tasted*. Instead, prolonged beer drinking leads people to experience a taste they enjoy, but precisely their enjoying the taste guarantees that it is not the taste they first experience.[9]

But this conclusion, if it is accepted, wreaks havoc of a different sort with the traditional philosophical view of qualia. For if it is admitted that one's attitudes towards, or reactions to, experiences are in any way and in any degree constitutive of their experiential qualities, so that a change in reactivity *amounts to* or *guarantees* a change in the property, then those properties, those 'qualitative or phenomenal features', cease to be 'intrinsic' properties and in fact become paradigmatically extrinsic, relational properties.

Properties that 'seem intrinsic' at first often turn out on more careful analysis to be relational. Bennett (1965) is the author of *intuition pump 10: the world-wide eugenics experiment*. He draws our attention to phenol-thio-urea, a substance which tastes very bitter to three-fourths of humanity, and as tasteless as water to the rest. Is it bitter? Since the reactivity to phenol-thio-urea is genetically transmitted, we could make it paradigmatically bitter by performing a large-scale breeding experiment: prevent the people to whom it is tasteless from breeding, and in a few generations phenol would be as bitter as anything to be found in the world. But we could also (in principle) perform the contrary feat of mass 'eugenics' and thereby make phenol paradigmatically tasteless—as tasteless as water—without ever touching phenol. Clearly, public bitterness or tastelessness is not an intrinsic property of phenol-thio-urea but a relational property, since the property is changed by a change in the reference class of normal detectors.

The public versions of perceptual 'qualia' all *seem* intrinsic, in spite of their relationality. They are not alone. Think of the 'felt value' of a dollar (or whatever your native currency is). 'How much is that in *real* money?' the American tourist is reputed to have asked, hoping to translate a foreign price onto the scale of 'intrinsic value' he keeps in his head. As Elster (1985) claims, 'there is a tendency to overlook the implicitly relational character of certain monadic predicates'. Walzer (1985) points out that 'a ten-dollar bill might seem to have a life of its own as a thing of value, but, as Elster suggests, its value implicitly depends on "other people who are prepared to accept money as payment for goods"'. But even as one concedes this, there is still a tendency to reserve something subjective, felt value, as an 'intrinsic' property of that ten-dollar bill. But as we now see, such intrinsic properties cannot be properties to which a subject's access is in any way privileged.

[9] We can save the traditional claim by ignoring presumably private or subjective qualia and talking always of public tastes—such as the public taste of Maxwell House coffee that both Chase and Sanborn agree has remained constant. Individuals can be said to acquire a taste for such a public taste.

Which way should Chase go? Should he take his wife's advice and declare that since he cannot stand the coffee anymore, it no longer tastes the same to him (it used to taste good and now it tastes bad)? Or should he say that really, in a certain sense, it does taste the way it always did, or at least it sort of does—when you subtract the fact that it tastes so bad now, of course?

We have now reached the heart of my case. The fact is that we have to ask Chase which way he wants to go, and there really are two drastically different alternatives available to him *if we force the issue*. Which way would *you* go? Which concept of qualia did you 'always have in the back of your mind', guiding your imagination as you thought about theories? If you acknowledge that the answer is not obvious, and especially if you complain that this forced choice drives apart two aspects that you had supposed united in your pre-theoretic concept, you support my contention that there is no secure foundation in ordinary 'folk psychology' for a concept of qualia. We *normally* think in a confused and potentially incoherent way when we think about the ways things seem to us.

When Chase thinks of 'that taste' he thinks equivocally or vaguely. He harkens back in memory to earlier experiences, but need not try—or be able—to settle whether he is including any or all of his reactions or excluding them from what he intends by 'that taste'. His state then and his state now are different—*that* he can avow with confidence—but he has no 'immediate' resources for making a finer distinction, nor any need to do so.[10]

This suggests that qualia are no more essential to the professional vocabulary of the phenomenologist (or professional coffee-taster) than to the vocabulary of the physiologist (Dennett 1978b). To see this, consider again the example of my dislike of cauliflower. Imagine now, in *intuition pump 11: the cauliflower cure*, that someone offers me a pill to cure my loathing for cauliflower. He promises that after I swallow this pill cauliflower will taste exactly the same to me as it always has, but I will like that taste. 'Hang on', I might reply, 'I think you may have just contradicted yourself'. But in any event I take the pill and it works. I become an instant cauliflower-appreciater, but if I am asked which of the two possible effects (Chase-type or Sanborn-type) the pill has had on me, I will be puzzled, and will find nothing *in my experience* to shed light on the question. Of course I recognize that the taste is (sort of) the same—the pill has not made cauliflower taste like chocolate cake, after all—but at the same time my experience is so different now that I resist

---

[10] 'I am not so wild as to deny that my sensation of red today is like my sensation of red yesterday. I only say that the similarity can *consist* only in the physiological force behind consciousness—which leads me to say, I recognize this feeling the same as the former one, and so does not consist in a community of sensation.'—(Peirce, *Collected Works*, Vol. V, p. 172 fn 2).

saying that cauliflower tastes the way it used to taste. There is in any event no reason to be cowed into supposing that my cauliflower experiences have some intrinsic properties behind, or in addition to, their various dispositional, reaction-provoking properties.

'But in principle there has to be a right answer to the question of how it is, intrinsically, with you now, even if you are unable to say with any confidence' Why? Would one say the same about all other properties of experience? Consider *intuition pump 12: visual field inversion created by wearing inverting spectacles*, a phenomenon which has been empirically studied for years. (G. M. Stratton published the pioneering work in 1896, and J. J. Gibson and I. Kohler were among the principal investigators; for an introductory account, see Gregory 1977.) After wearing inverting spectacles for several days subjects make an astonishingly successful adaptation. Suppose we pressed on them this question: 'Does your adaptation consist in your re-inverting your visual field or in your turning the rest of your mind upside-down in a host of compensations?' If they demur, may we insist that there has to be a right answer, even if they cannot say with any confidence which it is? Such an insistence would lead directly to a new version of the old inverted spectrum thought experiment: 'How do I know whether some people see things upside-down (but are perfectly used to it), while others see things right-side-up?'

Only a very naive view of visual perception could sustain the idea that one's visual field has a property of right-side-upness or upside-downness *independent of one's dispositions to react to it*—'intrinsic right-side-upness' we could call it (see my discussion of the properties of the 'images' processed by the robot, SHAKEY, in Dennett 1982). So not all properties of conscious experience invite or require treatment as 'intrinsic' properties. Is there something distinguishing about a certain subclass of properties (the 'qualitative or phenomenal' subclass, presumably) that forces us to treat them—unlike subjective right-side-upness—as intrinsic properties? If not, such properties have no role to play, in either physiological theories of experience, or in introspective theories.

Some may be inclined to argue this way: I can definitely imagine the experience of 'spectrum inversion' from the inside; after all, I have actually experienced temporary effects of the same type, such as the 'taste-displacement' effect of the maple syrup on the orange juice. What is imaginable, or actual, is possible. Therefore spectrum inversion or displacement (in all sensory modalities) is possible. But such phenomena just *are* the inversion or displacement of qualia, or intrinsic subjective properties. Therefore there must be qualia: intrinsic subjective properties.

This is fallacious. What one imagines and what one says one imagines may be two different things. To imagine visual field inversion, of the sort Stratton and Kohler's subjects experienced, is not necessarily to imagine the absolute inversion of a visual field (even if that is what it 'feels like' to the subjects). Less obviously, you imagining—as vividly as you like—a case of subjective colour-perception displacement is not necessarily you imagining what that phenomenon is typically called by philosophers: an inverted or displaced spectrum *of qualia*. In so far as that term carries the problematic implications scouted here, there is no support for its use arising simply from the vividness or naturalness of the imagined possibility.

If there are no such properties as qualia, does that mean that 'spectrum inversion' is impossible? Yes and no. Spectrum inversion as classically debated is impossible, but something like it is perfectly possible— something that is as like 'qualia inversion' as visual field inversion is like the impossible *absolute* visual image inversion we just dismissed.

Some puzzling real cases

It is not enough to withhold our theoretical allegiances until the sunny day when the philosophers complete the tricky task of purifying the everyday concept of qualia. Unless we take active steps to shed this source concept, and replace it with better ideas, it will continue to cripple our imaginations and systematically distort our attempts to understand the phenomena already encountered.

What we find, if we look at the actual phenomena of anomalies of colour perception, for instance, amply bears out our suspicions about the inadequacy of the traditional notion of qualia. Several varieties of *cerebral achromatopsia* (brain-based impairment of colour vision) have been reported, and while there remains much that is unsettled about their analysis, there is little doubt that the philosophical thought experiments have underestimated or overlooked the possibilities for counterintuitive collections of symptoms, as a few very brief excerpts from case histories will reveal.

Objects to the right of the vertical meridian appeared to be of normal hue, while to the left they were perceived only in shades of gray, though without distortions of form. . . . He was unable to recognize or name any color in any portion of the left field of either eye, including bright reds, blues, greens and yellows. As soon as any portion of the colored object crossed the vertical meridian, he was able to instantly recognize and accurately name its color (Damasio *et al.* 1980).

This patient would seem at first to be unproblematically describable as suffering a shift or loss of colour qualia in the left hemifield, but there is a problem of interpretation here, brought about by another case:

The patient failed in all tasks in which he was required to match the seen color with its spoken name. Thus, the patient failed to give the names of colors and failed to choose a color in response to its name. By contrast, he succeeded on all tasks where the matching was either purely verbal or purely nonverbal. Thus, he could give verbally the names of colors corresponding to named objects and vice versa. He could match seen colors to each other and to pictures of objects and could sort colors without error (Geschwind and Fusillo 1966).

This second patient was quite unaware of any deficit. He 'never replied with a simple "I don't know" to the demand for naming a colour' (Geschwind and Fusillo 1966, p. 140). There is a striking contrast between these two patients: both have impaired ability to name the colours of things in at least part of their visual field, but, whereas the former is acutely aware of his deficit, the latter is not. Does this difference make all the difference about qualia? If so, what on earth should we say about this third patient?

His other main complaint was that 'everything looked black or grey' and this caused him some difficulty in everyday life. . . . He had considerable difficulty recognizing and naming colours. He would, for example, usually describe bright red objects as either red or black, bright green objects as either green, blue or black, and bright blue objects as black. The difficulty appeared to be perceptual and he would make remarks suggesting this; for example when shown a bright red object he said 'a dirty smudgy red, not as red as you would normally see red.' Colours of lesser saturation or brightness were described in such terms as 'grey' 'off-white' or 'black,' but if told to guess at the colour, he would be correct on about 50 per cent of occasions, being notably less successful with blues and greens than reds (Meadows 1974).

This man's awareness of his deficit is problematic to say the least. It contrasts rather sharply with yet another case:

One morning in November 1977, upon awakening, she noted that although she was able to see details of objects and people, colors appeared 'drained out' and 'not true.' She had no other complaint . . . her vision was good, 20/20 in each eye. . . . The difficulty in color perception persisted, and she had to seek the advice of her husband to choose what to wear. Eight weeks later she noted that she could no longer recognize the faces of her husband and daughter . . . [So in] addition to achromatopsia, the patient had prosopagnosia, but her linguistic and cognitive performances were otherwise unaffected. The patient was able to tell her story cogently and to have remarkable insight about her defects (Damasio *et al.* 1980).

As Meadows notes, 'Some patients thus complain that their vision for colours is defective while others have no spontaneous complaint but show striking abnormalities on testing'.

What should one say in these cases? When no complaint is volunteered but the patient shows an impairment in colour vision, is this a sign that his qualia are unaffected? ('His capacities to discriminate are terribly impaired, but, luckily for him, his inner life is untouched by this merely public loss.') We could line up the qualia this way, but equally we could claim that the patient has simply not noticed the perhaps gradual draining away or inversion or merging of his qualia revealed by his poor performance. ('So slowly did his inner life lose its complexity and variety that he never noticed how impoverished it had become.') What if our last patient described her complaint just as she did above, but performed normally on testing? One hypothesis would be that her qualia had indeed, as she suggested, become washed out. Another would be that in the light of her sterling performance on the colour discrimination tests, her qualia were fine; she was suffering from some hysterical or depressive anomaly, a sort of colour-vision hypochondria that makes her complain about a loss of colour perception. Or perhaps one could claim that her qualia were untouched; her disorder was purely verbal: an anomalous understanding of the words she uses to describe her experience. (Other startlingly specific, colour–word disorders have been reported in the literature.)

The traditional concept leads us to overlook genuine possibilities. Once we have learned of the curious deficit reported by Geschwind and Fusillo (1966), for instance, we realize that our first patient was never tested to see if he could still sort colours seen on the left or pass other non-naming, non-verbal, colour-blindness tests. Those tests are by no means superfluous. Perhaps he would have passed them; perhaps, *in spite of what he says*, his qualia are as intact for the left field as for the right—if we take the capacity to pass such tests as 'criterial'. Perhaps his problem is 'purely verbal'. If your reaction to this hypothesis is that this is impossible, that must mean you are making his verbal, reporting behaviour sovereign in settling the issue—but then you must rule out a priori the possibility of the condition I described as colour-vision hypochondria.

There is no prospect of *finding* the answers to these brain-teasers in our everyday usage or the intuitions it arouses, but it is of course open to the philosopher to *create* an edifice of theory defending a particular set of interlocking proposals. The problem is that although normally a certain family of stimulus and bodily conditions yields a certain family of effects, any particular effect can be disconnected, and our intuitions do not tell us which effects are 'essential' to quale identity or qualia constancy (cf. Dennett 1978*a*, chapter 11.). It seems fairly obvious to me that none of the real problems of interpretation that face us in these curious cases are advanced by any analysis of how the concept of *qualia* is to be applied—unless we wish to propose a novel, technical sense for which the traditional term might be appropriated. But that would be at least a tactical error: the intuitions that surround and *purport* to anchor the current understanding of the term are revealed to be in utter disarray when confronted with these cases.

My informal sampling shows that some philosophers have strong opinions about each case and how it should be described in terms of qualia, but they find they are in strident (and ultimately comic) disagreement with other philosophers about how these 'obvious' descriptions should go. Other philosophers discover that they really do not know what to say—not because there are not enough facts presented in the descriptions of the cases, but because it begins to dawn on them that they have not really known what they were talking about over the years.

Filling the vacuum

If qualia are such a bad idea, why have they seemed to be such a good idea? Why does it seem as if there are these intrinsic, ineffable, private, 'qualitative' properties in our experience? A review of the presumptive second-order properties of the properties of our conscious experiences will permit us to diagnose their attractiveness and find suitable substitutes (for a similar exercise, see Kitcher 1979).

Consider 'intrinsic' first. It is far from clear what an intrinsic property would be. Although the term has had a certain vogue in philosophy, and often seems to secure an important contrast, there has never been an accepted definition of the second-order property of intrinsicality. If even such a brilliant theory-monger as David Lewis can try and fail, by his own admission, to define the extrinsic/intrinsic distinction coherently, we can begin to wonder if the concept deserves our further attention after all. In fact Lewis (1983) begins his survey of versions of the distinction by listing as one option: 'We could Quine the lot, give over the entire family as unintelligible and dispensable', but he dismisses the suggestion immediately: 'That would be absurd' (p. 197). In the end, however, his effort to salvage the accounts of Chisholm (1976) and Kim (1982) are stymied, and he conjectures that 'if we still want to break in we had best try another window' (p. 200).

Even if we are as loath as Lewis is to abandon the distinction, should we not be suspicious of the following curious fact? If challenged to explain the idea of an intrinsic property to a neophyte, many people would hit on the following sort of example: consider Tom's ball; it has many properties, such as its being made of rubber from India, its belonging to Tom, its having spent the last week in the closet, and its redness. All but the last of these are clearly *relational* or *extrinsic* properties of the ball. Its redness, however, is an intrinsic property. Except that this is not so. Ever since Boyle and Locke we have known better. Redness—public redness—is a quintessentially relational property, as many thought experiments about 'secondary qualities' show. [One of the first was Berkeley's (1713) pail of lukewarm water, and one of the best is Bennett's (1965) phenol-thio-urea.] The seductive step, on learning that public redness (like public bitterness, etc.) is a relational property after all, is to cling to intrinsicality ('*something* has to be intrinsic') and move it into the subject's head. It is often thought, in fact, that if we take a Lockean, relational position on objective bitterness, redness, etc., we *must* complete our account of the relations in question by appeal to non-relational, intrinsic properties. If what it is to be objectively bitter is to produce a certain effect in the members of the class of normal observers, we must be able to specify that effect and distinguish it from the effect produced by objective sourness and so forth.

What else could distinguish this effect but some intrinsic property? Why not another relational or extrinsic property? The relational treatment of monetary value does not require, for its completion, the supposition of items of intrinsic value (value independent of the valuers' dispositions to react behaviourally). The claim that certain perceptual properties are different is, in the absence of any supporting argument, just question-begging. It will not do to say that it is just obvious that they are intrinsic. It may have seemed obvious to some, but the considerations raised by Chase's quandary show that it is far from obvious that any intrinsic property (whatever that comes to) could play the role of anchor for the Lockean, relational treatment of the public perceptual properties.

Why not give up intrinsicality as a second-order property altogether, at least pending resolution of the disarray of philosophical opinion about what intrinsicality might be? Until such time the insistence that qualia are the intrinsic properties of experience is an empty gesture at best; no one could claim that it provides a clear, coherent, understood prerequisite for theory.[11]

---

[11] A heroic (and, to me, baffling) refusal to abandon intrinsicality is Sellars' (1981) contemplation over the years of his famous pink ice cube, which leads him to postulate a revolution in microphysics, restoring objective 'absolute sensory processes' in the face of Boyle and Locke and almost everybody since them (also see my commentary in Dennett 1981).

What, then, of ineffability? Why does it seem that our conscious experiences have ineffable properties? Because they do have *practically* ineffable properties. Suppose, in *intuition pump 13: the osprey cry*, that I have never heard the cry of an osprey, even in a recording, but know roughly, from reading my bird books, what to listen for: 'a series of short, sharp, cheeping whistles, *cheep, cheep* or *chewk chewk*, etc; sounds annoyed' (Peterson 1947) (or words to that effect or better). The verbal description gives me a partial confinement of the logical space of possible bird cries. On its basis I can rule out many bird calls I have heard or might hear, but there is still a broad range of discriminable-by-me possibilities within which the actuality lies hidden from me like a needle in a haystack.

Then one day, armed with both my verbal description and my binoculars, I identify an osprey visually, and then hear its cry. 'So *that's* what it sounds like', I say to myself, ostending—it seems—a particular mental complex of intrinsic, ineffable qualia. I dub the complex '*S*' (*pace* Wittgenstein), rehearse it in short-term memory, check it against the bird book descriptions, and see that, while the verbal descriptions are true, accurate, and even poetically evocative—I decide I could not do better with a thousand words—they still fall short of *capturing* the qualia complex I have called S. In fact, that is why I need the neologisim, '*S*', to refer directly to the ineffable property I cannot pick out by description. My perceptual experience has pin-pointed for me the location of the osprey cry in the logical space of possibilities in a way verbal description could not.

But tempting as this view of matters is, it is overstated. First of all, it is obvious that from a single experience of this sort I do not—cannot —know how to generalize to other osprey calls. Would a cry that differed only in being half an octave higher also be an osprey call? That is an empirical, ornithological question for which my experience provides scant evidence. But moreover—and this is a psychological, not ornithological, matter—I do not and cannot know, from a single such experience, which physical variations and constancies in stimuli would produce an indistinguishable experience in me. Nor can I know whether I would react the same (have the same experience) if I were presented with what was, by all physical measures, a re-stimulation identical to the first. I cannot know the modulating effect, if any, of variations in my body (or psyche).

This inscrutability of projection is surely one of the sources of plausibility of Wittgenstein's scepticism regarding the possibility of a private language:

Wittgenstein emphasizes that ostensive definitions are always in principle capable of being misunderstood, even the ostensive definition of a color word such as 'sepia'. How someone understands the word is exhibited in the way someone goes on, 'the use that he makes of the word defined'. One may go on in the right way given a purely minimal explanation, while on the other hand one may go on in another way no matter how many clarifications are added, since these too can be misunderstood (Kripke 1982, p. 83; see also pp. 40–6).

But what is inscrutable in a single glance, and somewhat ambiguous after limited testing, can come to be justifiably seen as the deliverance of a highly specific, reliable, and projectible property detector, once it has been field-tested under a suitably wide variety of circumstances.

In other words, when first I hear the osprey cry, I may have identified a property detector in myself, but I have no idea (yet) what property my newfound property detector detects. It might seem then that I know nothing new at all—that my novel experience has not improved my epistemic predicament in the slightest. But of course this is not so. I may not be able to describe the property or identify it relative to any readily usable public landmarks (yet), but I am acquainted with it in a modest way: I can refer to the property I detected: it is the property I detected in *that* event. My experience of the osprey cry has given me a new way of thinking about osprey cries (an unavoidably inflated way of saying something very simple) which is practically ineffable both because it has (as yet for me) an untested profile in response to perceptual circumstances, and because it is—as the poverty of the bird book description attests—such a highly informative way of thinking: a deliverance of an informationally very sensitive portion of my nervous system.

In this instance I mean information in the formal information theory sense of the term. Consider (*intuition pump 14: the Jello box*) the old spy trick, most famously encountered in the case of Julius and Ethel Rosenberg, of improving on a password system by tearing something in two (a Jello box, in the Rosenberg's case), and giving half to each of the two parties who must be careful about identifying each other. Why does it work? Because tearing the paper in two produces an edge of such informational complexity that it would be virtually impossible to reproduce by deliberate construction. (Cutting the Jello box along a straight edge with a razor would entirely defeat the purpose.) The particular jagged edge of one piece becomes a *practically* unique pattern-recognition device for its mate; it is an apparatus for detecting the shape property M, where M is uniquely instantiated by its mate. It is of the essence of the trick that we cannot replace our dummy predicate 'M' with a longer, more complex, but accurate and exhaustive description of the property, for, if we could, we could use the description as a recipe or feasible algorithm for producing another instance of M or another M

detector. The only *readily available* way of saying what property *M* is is just to point to our *M* detector and say that *M* is the shape property detected by this thing here.

And that is just what we do when we seem to ostend, with the mental finger of inner intention, a quale or qualia complex in our experience. We refer to a property—a public property of uncharted boundaries—via reference to our personal and idiosyncratic capacity to respond to it. That idiosyncracy is the extent of our privacy. If I wonder whether your blue is my blue, your middle C is my middle C, I can coherently be wondering whether our discrimination profiles over a wide variation in conditions will be approximately the same. And they may not be; people experience the world quite differently. But that is empiricially discoverable by all the usual objective testing procedures.[12]

Peter Bieri has pointed out to me that there is a natural way of exploiting Dretske's (1981) sense of information in a reformulation of my first three second-order properties of qualia: intrinsicality, ineffability, and privacy. (There are problems with Dretske's attempt to harness information theory in this way—see my discussion in Dennett 1987*b*, chapter 8—but they are not relevant to this point.) We could speak of what Bieri would call 'phenomenal information properties' of psychological events. Consider the information—what Dretske would call the *natural meaning*—that a type of internal perceptual event might carry. That it carries that information is an objective (and hence, in a loose sense, intrinsic) matter since it is independent of what information (if any) the subject *takes* the event type to carry. Exactly what information is carried is (practically) ineffable, for the reasons just given. And it is private in the sense just given: proprietary and potentially idiosyncratic.

Consider how Bieri's proposed 'phenomenal information properties' (let us call them *pips*) would apply in the case of Chase and Sanborn. Both Chase and Sanborn ought to wonder whether their pips have changed. Chase's speech shows that he is under the impression that his pips are unchanged (under normal circumstances—all bets are off if he has just eaten horse-radish). He believes that the same objective things in the world—in particular, chemically identical caffeine-rich fluids —give rise to his particular types of taste experiences now as six years ago.

Sanborn is under the impression that his pips are different. He thinks

---

[12] Stich (1983) discusses the implications for psychological theory of incommensurability problems that can arise from such differences in discrimination profiles (see, especially, chapters 4 and 5).

his objective property detectors are deranged. He no longer has confidence that their deliverances today inform him of what they did six
years ago. And what, exactly, did they inform him of then? If Sanborn
were an ordinary person, we would not expect him to have an explicit
answer, since most of us treat our taste detectors as mere M detectors,
detecting whatever it is that they detect. (There are good reasons for this,
analysed by Akins, unpublished thesis, 1987.) But professional coffee-
tasters are probably different. They probably have some pretty good idea
of what kind of chemical-analysis transduction machinery they have in
their mouths and nervous systems.

So far, so good. We could reinterpret Chase's and Sanborn's speeches
as hypotheses about the constancies or changes in the outputs of their
perceptual information-processing apparatus, and just the sort of empirical testing we imagined before would tend to confirm or disconfirm
their opinions thus interpreted. But what would justify calling such an
information-bearing property 'phenomenal'?

Such a pip has, as the testimony of Chase and Sanborn reveals, the
power to provoke in Chase and Sanborn acts of (apparent) re-
identification or recognition. This power is of course a Lockean, dispositional property on a par with the power of bitter things to provoke a
certain reaction in people. It is this power alone, however it might be
realized in the brain, that gives Chase and Sanborn 'access' to the
deliverances of their individual property detectors.

We may 'point inwardly' to one of the deliverances of our idiosyncratic, proprietary property detectors, but when we do, what are we
pointing *at*? What does that deliverance itself *consist of*? Or what are its
consciously apprehensible properties, if not just our banished friends
the qualia? We must be careful here, for if we invoke an inner perceptual
process in which we observe the deliverance with some inner eye and
thereby discern its properties, we will be stepping back into the frying
pan of the view according to which qualia are just ordinary properties of
our inner states.

But nothing requires us to make such an invocation. We do not have to
know how we identify or re-identify or gain access to such internal
response types in order to be able so to identify them. This is a point that
was forcefully made by the pioneer functionalists and materialists, and
has never been rebutted (Farrell 1950; Smart 1959). The properties of
the 'thing experienced' are not to be confused with the properties of the
event that realizes the experiencing. To put the matter vividly, the
physical difference between someone's imagining a purple cow and
imagining a green cow *might* be nothing more than the presence or
absence of a particular zero or one in one of the brain's 'registers'. Such a
brute physical presence is all that it would take to anchor the sorts of

dispositional differences between imagining a purple cow and imagining a green cow that could then flow, causally, from that 'intrinsic' fact. (I doubt that this is what the friends of qualia have had in mind when they have insited that qualia are intrinsic properties.)

Moreover, it is our very inability to expand on, or modify, these brute dispositions so to identify or recognize such states that creates the doctrinal illusion of 'homogeneity' or 'atomicity to analysis' or 'grainlessness' that characterizes the qualia of philosophical tradition.

This putative grainlessness, I hypothesize, is nothing but a sort of functional invariability: it is close kin to what Pylyshyn (1980, 1984) calls *cognitive impenetrability*. Moreover, this functional invariability or impenetrability is not absolute but itself plastic over time. Just as on the efferent side of the nervous system, *basic actions*—in the sense of Danto (1963, 1965) and others (see Goldman 1970)—have been discovered to be variable, and subject under training to decomposition (one can learn with the help of 'biofeedback' to will the firing of a particular motor neuron 'directly'), so what counts for an individual as the simple or atomic properties of experienced items is subject to variation with training.[13]

Consider the results of 'educating' the palate of a wine-taster, or 'ear training' for musicians. What had been 'atomic' or 'unanalysable' becomes noticeably compound and describable; pairs that had been indistinguishable become distinguishable, and when this happens we say *the experience changes*. A swift and striking example of this is illustrated in *intuition pump 15: the guitar string*. Pluck the bass or low E string open and listen carefully to the sound. Does it have describable parts or is it one and whole and ineffably guitarish? Many will opt for the latter way of talking. Now pluck the open string again and carefully bring a finger down lightly over the octave fret to create a high 'harmonic'. Suddenly a *new* sound is heard: 'purer' somehow and of course an octave higher. Some people insist that this is an entirely novel sound, while others will describe the experience by saying 'the bottom fell out of the note'— leaving just the top. But then on a third open plucking one can hear, with surprising distinctness, the harmonic overtone that was isolated in the second plucking. The homogeneity and ineffability of the first experience is gone, replaced by a duality as 'directly apprehensible' and clearly describable as that of any chord.

The difference in experience is striking, but the complexity apprehended on the third plucking was *there* all along (being responded

---

[13] See Churchland (1979, especially chapter 2) for supporting observations on the variability of perceptual properties, and for novel arguments against the use of 'intrinsic properties' as determiners of the meaning of perceptual predicates. See also Churchland (1985) for further arguments and observations in support of the position sketched here.

to or discriminated). After all, it was by the complex pattern of overtones that you were able to recognize the sound as that of a guitar rather than of a lute or harpsichord. In other words, although the subjective experience has changed dramatically, the *pip* has not changed; you are still responding, as before, to a complex property so highly informative that it practically defies verbal description.

There is nothing to stop further refinement of one's capacity to describe this heretofore ineffable complexity. At any time, of course, there is one's current horizon of distinguishability—and that horizon is what sets, if anything does, what we should call the primary or atomic properties of what one consciously experiences (Farrell 1950). But it would be a mistake to transform the fact that inevitably there is a limit to our capacity to describe things we experience into the supposition that there are absolutely indescribable properties in our experience.

So when we look one last time at our original characterization of qualia, as ineffable, intrinsic, private, directly apprehensible properties of experience, we find that there is nothing to fill the bill. In their place are relatively or practically ineffable public properties we can refer to indirectly via reference to our private property detectors—private only in the sense of idiosyncratic. And in so far as we wish to cling to our subjective authority about the occurrence within us of states of certain types or with certain properties, we can have some authority—not infallibility or incorrigibility, but something better than sheer guessing —but only if we restrict ourselves to relational, extrinsic properties like the power of certain internal states of ours to provoke acts of apparent re-identification. So contrary to what seems obvious at first blush, there simply are no qualia at all.

Acknowledgements

The first version of this paper was presented at University College, London, in November 1978, and in various revisions at a dozen other universities in 1979 and 1980. It was never published, but was circulated widely as Tufts University Cognitive Science Working Paper 7, December 1979. A second version was presented at the Universities of Adelaide and Sydney in 1984, and in 1985 to psychology department colloquia at Harvard and Brown under the title 'Properties of conscious experience'. The second version was the basis for my presentation at the workshop from which this book arises, and was circulated in pre-print in 1985, again under the title 'Quining qualia'. The present version, the fourth, is a substantial revision, thanks to the helpful comments of many people, including Kathleen Akins, Ned Block, Alan Cowey, Sydney Shoemaker, Peter Bieri, William Lycan, Paul Churchland, Gilbert Harman, and the participants at Villa Olmo.

References

Akins, K. (1987). Information and organisms: or why Nature doesn't build epistemic engines. Unpublished PhD thesis. University of Michigan.

Armstrong, D. and Malcolm, N. (ed.) (1984). *Consciousness and causality.* Blackwell Scientific Publications, Oxford.

Bennett, J. (1965). Substance, reality and primary qualities. *American Philosophical Quarterly*, **2**, 1–17.

Berkeley, G. (1713). *Three dialogues between Hylas and Philonous.* London.

Block, N. (1978). Troubles with functionalism. In *Perception and cognition: issues in the foundations of psychology*, (ed. C. W. Savage), pp. 261–326. University of Minnesota Press.

Block, N. (1980). Are absent qualia impossible? *The Philosophical Review*, **89**, 257.

Block, N. and Fodor, J. (1972). What psychological states are not. *The Philosophical Review*, **81**, 159–81.

Chisholm, R. (1976). *Person and object.* Open Court Press, La Salle, IL.

Churchland, P. M. (1979). *Scientific realism and the plasticity of mind.* Cambridge University Press.

Churchland, P. M. (1985). Reduction, qualia and the direct inspection of brain states. *Journal of Philosophy*, **LXXXII**, 8–28.

Conee, E. (1985). The possibility of absent qualia. *The Philosophical Review*, **XCIV**, (3), 345–66.

Damasio, A., Yamada, T., Damasio, H., Corbett, J., and McKee, J. (1980). Central achromatopsia: behavioral, anatomic, and physiological aspects. *Neurology*, **30**, 1064–71.

Danto, A. (1963). What we can do. *Journal of Philosophy*, **LX**, 435–45.

Danto, A. (1965). Basic actions. *American Philosophical Quarterly*, **60**, 141–48.

Davis, L. (1982). Functionalism and absent qualia. *Philosophical Studies*, **41**, (2), 231–51.

Dennett, D. C. (1969). *Content and consciousness.* Routledge and Kegan Paul, Andover, Hants.

Dennett, D. C. (1976). Are dreams experiences? *The Philosophical Review*, **LXXIII**, 151–71.

Dennett, D. C. (1978a). *Brainstorms.* MIT Press, Cambridge, MA.

Dennett, D. C. (1978b). Two approaches to mental images. In *Brainstorms*, (ed. D. C. Dennett), pp. 174–89. MIT Press, Cambridge, MA.

Dennett, D. C. (1979). On the absence of phenomenology. In *Body, mind, and method*, (ed. D. F. Gustafson and B. L. Tapscott), pp. 93–114. D. Reidel, Dordrecht.

Dennett, D. C. (1981). Wondering where the yellow went. *The Monist*, **64**, 102–8.

Dennett, D. C. (1982). How to study human consciousness empirically: or nothing comes to mind. *Synthese*, **53**, 159–80.

Dennett, D. C. (1987*a*). *The philosophical lexicon*, (8th edn). Copy available from the American Philosophical Association, University of Delaware, Newark, DE.

Dennett, D. C. (1987*b*). *The intentional stance.* MIT Press, Cambridge, MA.

Dretske, F. (1981). *Knowledge and the flow of information.* MIT Press, Cambridge, MA.

Elster, J. (1985). *Making sense of Marx.* Cambridge University Press.

Farrell, B. (1950). Experience. *Mind*, **59**, 170–98.

Gert, B. (1965). Imagination and verifiability. *Philosophical Studies*, **XVI**, 44–7.

Geschwind, N. and Fusillo, M. (1966). Color-naming defects in association with alexia. *Archives of Neurology*, **15**, 137–46.

Goldman, A. (1970). *A theory of human action.* Prentice Hall, Englewood Cliffs, NJ.

Gregory, R. (1977). *Eye and brain* (3rd edn). Weidenfeld and Nicolson, London.

Hofstadter, D. and Dennett, D. C. (1981). *The mind's I: fantasies and reflections on self and soul.* Basic Books, New York.

Kim, J. (1982). Psychophysical supervenience. *Philosophical Studies*, **41**, 51–70.

Kitcher, P. (1979). Phenomenal qualities. *American Philosophical Quarterly*, **16**, 123–9.

Kripke, S. (1982). *Wittgenstein on rules and private language.* Harvard University Press, Cambridge, MA.

Lehrer, A. (1983). *Wine and conversation.* University of Indiana Press.

Lewis, D. (1983). Extrinsic properties. *Philosophical Studies*, **44**, 197–200.

Locke, J. (1959). *An essay concerning human understanding* (ed. A. C. Fraser). Dover, New York.

Lycan, W. (1973). Inverted spectrum. *Ratio*, **XV**, 315–19.

Malcolm, N. (1956). Dreaming and skepticism. *The Philosophical Review*, **LXV**, 14–37.

Malcolm, N. (1959). *Dreaming.* Routledge and Kegan Paul, Andover, Hants.

Meadows, J. C. (1974). Disturbed perception of colours associated with localized cerebral lesions. *Brain*, **XCVII**, 615–32.

Millikan, R. (1984). *Language, thought and other biological categories.* MIT Press, Cambridge, MA.

Nagel, T. (1974). What is it like to be a bat? *The Philosophical Review*, **LXXXIII**, 435–51.

Nagel, T. (1986). *The view from nowhere.* Oxford University Press.

Peirce, C. (1931–58). *Collected Works.* Vol. V, (ed. C. Hartshorne and P. Weiss). Harvard University Press, Cambridge MA.

Peterson, R. T. (1947). *A field guide to the birds.* Houghton Mifflin, Boston, MA.

Putnam, H. (1965). Brains and behavior. In *Analytical philosophy*, second series, (ed. J. Butler), pp. 1–19. Blackwell Scientific Publications, Oxford.

Pylyshyn, Z. (1980). Computation and cognition: issues in the foundations of cognitive science. *Behavioral and Brain Sciences*, **3**, 111–32.

Pylyshyn, Z. (1984). *Computation and cognition: toward a foundation for cognitive science.* MIT Press, Cambridge, MA.

Sellars, W. (1981). Foundations for a metaphysics of pure process (the Carus Lectures). *The Monist,* **64,** 3–90.

Shoemaker, S. (1969). Time without change. *Journal of Philosophy,* **LXVI,** 363–81.

Shoemaker, S. (1975). Functionalism and qualia. *Philosophical Studies,* **27,** 291–315.

Shoemaker, S. (1981). Absent qualia are impossible—a reply to Block. *The Philosophical Review,* **XC,** 581–99.

Shoemaker, S. (1982). The inverted spectrum. *Journal of Philosophy,* **79,** 357–81.

Shoemaker, S. (1984). *Identity, cause, and mind,* pp. 351–7. Cambridge University Press.

Smart, J. C. (1959). Sensations and brain processes. *The Philosophical Review,* **LXVIII,** 141–56.

Stich, S. (1983). *From folk psychology to cognitive science: the case against belief.* MIT Press, Cambridge, MA.

Taylor, D. M. (1966). The incommunicability of content. *Mind,* **LXXV,** 527–41.

Von Der Heydt, R., Peterhans, E., and Baumgartner, G. (1984). Illusory contours and cortical neuron response. *Science,* **224,** 1260–2.

Walzer, M. (1985). What's left of Marx. *The New York Review of Books,* 21 November, 43–6.

White, S. (1985). Professor Shoemaker and so-called 'qualia' of experience. *Philosophical Studies,* **47,** 369–83.

Wittgenstein, L. (1958). *Philosophical investigations.* Blackwell Scientific Publications, Oxford.

4

Consciousness, intrinsic intentionality, and self-understanding machines

Robert Van Gulick

More than thirty years ago in his famous paper 'Computing machinery and intelligence' Turing (1950) began by considering the question 'Can machines think?'. However, he began with that question only to dismiss it as too vague to be clearly answered.

Instead, he offered a more operationally decidable question based on what he called the imitation game. In that now very familiar game, an interrogator must determine which of two sets of conversational responses to his questions or statements is being produced by a human being and which is being produced by a computer. All communication occurs by teleprinter; a more modern test would use a terminal with keyboard and CRT monitor. This forces the interrogator to base his decision entirely on the nature and content of the conversational replies he receives. If the player is not able to discriminate reliably the computer-generated replies from those produced by the human being, the computer is judged to have passed what has come to be called the Turing test and is to be counted as intelligent and having a mind.

Today the Turing test is generally regarded as providing an insufficient criterion for being intelligent or having a mind, or at least it is so regarded by most philosophers, with Dennett (1985) as probably the most notable exception. Current dissatisfaction with the Turing test can be explained by the fact that the sort of behaviourism that was widely accepted in the 1950s when Turing proposed his test has largely been rejected. Few philosophers today would accept the claim that behavioural criteria alone can suffice to define what it is to have a mind or to be in one or another specific mental state, such as believing that the dollar will fall or fearing that the Rhine will soon overflow. The conditions for being in some such mental state are taken to require not only behaving in a given way or being disposed to behave in that way, but also having that behaviour caused in the right sort of way by appropriate types of internal processes. There is little agreement about just how to specify those further constraints regarding how the behaviour must be produced or what counts as an 'appropriate' internal process. None the less, there appear to be some clear cases in which a system might satisfy all the behavioural or performance criteria associated with having a mind, but still be denied such status on the basis of what we knew about the nature of the internal causal processes producing that behaviour. Block (1981) has provided one such case. He offers an imaginary example of a device that passes the Turing test for any conversation of length *n* by relying only on an enormous, but finite list, of all coherent conversations up to length *n* stored alphabetically in its memory. At any given point in the test the device simply uses a rapid memory search to find a conversation on its list which agrees with its current conversation up through to the last utterance by its conversational partner and prints out the next sentence in that listed conversation. The device is of course wildly impractical, and in practice could never be built given the size of the memory required. But, none the less, as a logically possible case it suffices to make the purely conceptual point that at least some conditions concerning how behaviour is produced must be included in our criteria for mental terms; behavioural and performance criteria alone will not do the job. Though this point is generally accepted, I should note that it is not universally accepted. Dennett (1971, 1981) again stands out as the most notable exception. Given his instrumentalist, intentional-systems theory of mental terms, all that is required to justify attributing beliefs or desires to some system is that such attributions should yield reliable systematic predictions of its behaviour. Dennett defines mental terms without reference to any internal causes of behaviour. Although Dennett's position is distinctively his own, it agrees in its externalist perspective with 1950s-style behaviourism, of which it is a lineal descendant through Dennett's early association with Ryle in the 1960s.[1]

Given the inadequacy of the Turing test, I would like to return to the original question Turing asked but rejected, 'Can machines think?' Can that question be made more precise and decidable in a way that incorporates the contemporary view that having a mind is a matter of how behaviour is produced as well as how one behaves?

At least three concepts need clarification. What is meant by a 'machine'? What is it to 'think'? And what is the modal force of 'can' in asking 'Can machines think?' That is, should 'can' be interpreted so that the question is about whether present machines already think, about whether it is logically or physically possible that someday there will be such machines, or about whether machines of an appropriate sort would necessarily be thinkers?

---

[1] See Ryle 1949. Dennett's acknowledgement of his debt to Ryle is clearest in the preface to his book *Content and Consciousness* (1969).

Following the treatment by Searle (1984), we can give several possible interpretations of the question that yield clear 'yes' or 'no' answers. One could mean by 'a machine' simply 'any device whose operation is at some level of description entirely explained by its physical constitution and strictly physical laws'. Our question then becomes

(1) Is it possible for a system whose operation is at some level entirely explained by appeal to its physical constitution and physical laws to think or have a mind?

The answer to this question would surely be 'yes' regardless of how we interpret 'think' or 'can', for human beings are almost certainly machines in the required sense. I say 'almost certainly' because such a claim might be denied by neo-dualists like Eccles and Popper (1977). The fact that humans are such machines guarantees a positive answer to the question, since one could not plausibly define the term 'think' in a way that classified all humans as non-thinkers. Though such suggestions might be made by eliminativists who wish to reject all ordinary mental concepts (Skinner 1953; Churchland 1981; Rorty 1970), I do not find such proposals plausible and they would amount to rejecting our original question altogether as not worth asking. Thus we can assume that some humans do think, and since no matter how we interpret 'can', actually thinking will entail that one can think, it follows that machines, understood as physically explainable systems, can think. However, the very fact that the question so interpreted has an uncontroversially affirmative answer makes it not very interesting.

Another alternative would be to tie the word 'machine' more closely to the idea of a formal computing device, by defining a machine as 'a system that instantiates (or is a realization of) a formally specified machine table', where a machine table is just an exhaustive description of a finite state input–output device which specifies for any pair of state and input the output to be produced and the succession state into which the device is to move (Putnam 1960). Instantiation of such a formal machine description requires roughly that there be some mapping from the formal states, inputs, and outputs of the abstract machine table onto physical states, inputs, and outputs of the instantiating system, such that under that mapping the relations of temporal sequence among those physical items are isomorphic to the relations of formal succession among the machine table items. If we interpret the word 'machine' in this way our question becomes

(2) Is it possible for a system which is an instantiation of some formally specifiable abstract machine to think or have a mind?

The answer is again obviously and uninterestingly 'yes', for the same reason that was given in regard to (1). Human beings are machines in the relevant sense. Every human being instantiates some abstract machine table; indeed every human being instantiates an infinity of non-equivalent machine tables (Block and Fodor 1972). Thus since human beings *do* think and therefore *can* think, it follows that machines in the relevant sense can think, since being a machine in that sense requires only that one instantiate some abstract machine or other. Of course, it does not follow that humans are thinkers in virtue of being such machines.

On a third interpretation, the answer to the question would clearly be negative. We might interpret 'machines' as referring only to present-day, existing digital computers, interpret 'thinking' as equivalent to 'having subjective experiences with phenomenal properties', and interpret 'can' as making an assertion about the actual capacities of the relevant machines. So we get

(3) Do any present-day digital computers have the capacity to have subjective experiences with phenomenal properties?

There is virtually no reason to believe that existing computers (IBM 360s, PCs, VAXs, or DEC 20s) have subjective experiences with phenomenal properties. To use Nagel's (1974, p. 166) phrase, we have no reason to believe that 'there is anything that it is like to be' an IBM 360. Their designers have made no effort to build any such capacity into them, nor is it likely that they would have had any idea how to do so had they wanted to try. So it is quite implausible to suppose that these machines possess such a capacity as an accidental or coincidental result of having been built for quite different purposes. Of course, this does not prove that present-day computers lack phenomenal states, and it is not impossible that such states might result as an unintended emergent feature of the machine's complex organization. However, in the absence of any positive argument for more than the bare possibility, we can assume that in all likelihood there is nothing that it is like to be a present-day computer. Thus, on this third interpretation, the answer to our question 'Can machines think?' remains obvious and uninteresting. Question (3) differs from (1) and (2) only in receiving an obviously negative as opposed to an obviously positive answer.

However, our survey of these three uninteresting interpretations has not been without value, since it has provided us with some of the distinctions needed to formulate more interesting versions of the question. The next thing we must do to obtain an interesting interpretation is to transform the question into one about sufficient conditions for thinking or having a mind. Narrowing somewhat the formal notion of a machine used in the second interpretation, we might ask

(4) Is there some class of formally specifiable abstract machines such that being an instantiation of any machine in that class is a sufficient condition for thinking or having a mind?

Because it makes a claim about sufficiency, the answer to (4) is not obviously positive as was the answer to (2), nor is it obviously negative. However, it remains less than clear in so far as it leaves unspecified what is meant by thinking or having a mind. We might try to spell this out in either of two ways, each of which draws on one of the two traditional marks of the mental: intentionality and subjective experience. Doing so produces

(5) Is there any class of formally specifiable abstract machines such that being an instantiation of any machine in that class is a sufficient condition for having intentional states; i.e. states with intentional content?

and

(6) Is there any class of formally specifiable abstract machines such that being an instantiation of any machine in that class is a sufficient condition for having subjective experiences with phenomenal properties?

Again, neither of these questions has an immediately obvious answer. However, they are not equally controversial; many more philosophers have been inclined to give an affirmative answer to the former (5) than to the latter (6). Indeed, it is a common strategy among functionalist or computationalist philosophers of mind to divide mental states or predicates into two classes—those that involve intentionality but not subjective or phenomenal properties, and those that do involve subjective or phenomenal properties—and then to restrict their claims about the computational or functional nature of mental states to only those not involving subjective properties (Block 1978). Such a strategy is often made in conjunction with the suggestion that a more biochemical or neurophysiological account will be needed to explain subjective and phenomenal properties. Their position is that purely computational conditions (perhaps supplemented by some further constraints on the environmental character of the instantiating system's inputs and outputs) will suffice to define what it is to have a specific belief or a specific desire, but that such conditions will not suffice to define what it is to feel pain or have an experience of a phenomenally red object. The view is so widely held among computationalist and functionalist philosophers that we can label it as the standard computationalist thesis.

Despite its wide acceptance, the position has not been without its critics. Searle, in particular, has carried on a very public and outspoken debate with the defenders of the standard thesis, arguing that the answer to question (5) is definitely 'no'. His position is that no purely formal or computational specification can provide a set of sufficient conditions for having intentional states (or as he would put it, states with intrinsic intentionality). Searle (1984, p. 36) puts his position bluntly as follows:

The question we wanted to ask is this: 'can a digital computer, as defined, think? That is to say: Is instantiating or implementing the right computer program with the right inputs and outputs sufficient for, or constitutive of thinking?' And to this question, . . . , the answer is clearly 'no'.

Among the arguments Searle gives to support his position is the following:

P_1: Syntax is not sufficient for semantics.

P_2: Computer programs are entirely defined by their formal or syntactical structure.

P_3: Minds have mental contents; specifically they have semantic contents.

Therefore,

C: No computer program by itself is sufficient to give a system a mind.[2]

Searle thus gives negative answers to both (5) and (6). Moreover, his negative answer to (5) results in part from his negative answer to (6). For unlike those holding the standard computationalist view, Searle takes the position, which he believes is clearly supported by a variety of intuitive thought experiments, that no system or organism could have any intentional states at all if it lacked subjective experiences. Although Searle comes close to explicitly committing himself to this claim at only one point in his published writings (1979, p. 92), it seems to be the underlying motivation for his view. I have checked this point with him, and he confirms both that he accepts the claim and that it plays an important role in supporting his intuitions. Since on his view having conscious subjective experiences is a necessary condition for having any

[2] Searle 1984, paraphrase of argument on p. 39.

states with intrinsic intentional content, a negative answer to (6) automatically entails a negative answer to (5). If you cannot provide purely computational conditions which are sufficient for having subjective experiences, and such experiences are necessary for having intentional states, then you obviously cannot provide purely computational conditions which are sufficient for having intentional states.

It is important to be clear about just what Searle is claiming. His position is not that all mental or intentional states must be conscious states (or states with a subjective experiential character). He allows that some mental states may be non-conscious or lack any subjective experiential character, but he maintains that such non-conscious mental states can occur only in an organism which has some other mental states which are conscious and have subjective experiential character. What he denies is that there could be a system which has mental states (or states with intrinsic intentionality) even though it had no conscious subjective states at all. To support this position Searle relies on intuitive thought experiments such as the Chinese room example (1980) in which an English speaker allegedly mimics the computational operations needed to pass a Chinese version of the Turing test without really understanding Chinese. He appears to regard the view as a common-sensical one, which will be obvious to all except those who are already ideologically committed to some suspect philosophical theory of mind, such as behaviourism or computationalism. We can state Searle's claim as

(7) A capacity for having conscious subjective experiences is a necessary condition for having any states with intrinsic intentional content.

Though Searle's position runs counter to the mainstream of current philosophical thinking, he is not alone in holding that subjective experience is an essential component of mentality. Nagel has also argued that existing philosophical theories of mind are inadequate because of their failure to take account of the subjective nature of mind. In his recent book *The view from nowhere* he writes:

The subjective features of conscious mental processes—as opposed to their physical causes and effects—cannot be captured by the purified form of thought suitable for dealing with the physical world that underlies the appearances. Not only raw feels but also intentional mental states—however objective their content—must be capable of manifesting themselves in subjective form to be in the mind at all.

The reductionist program that dominates current work in the philosophy of mind is completely misguided, because it is based on the groundless assumption that a particular conception of objective reality is exhaustive of what there is. Eventually, I believe, current attempts to understand the mind by analogy with man-made computers that can perform superbly some of the same external tasks as conscious beings will be recognized as a gigantic waste of time. The true principles underlying the mind will be discovered, if at all, by a more direct approach (1986, pp. 15–16).

Nagel and Searle both regard a capacity for subjective experience as a necessary condition for having any mental states at all. The reasons that Nagel offers in support of his claim are embedded in a much larger epistemological and metaphysical programme, which could not be easily addressed within the limits of this chapter. Thus I will restrict myself to Searle's defence of their shared conclusion.

Let us return then to (7), which provides a statement of Searle's position. One key term requires explanation; we need to get clear about what Searle means by 'intrinsic intentionality'. Some confusion is created by the fact that he uses the term in contrast with three different non-equivalent terms.

Sometimes the notion of intrinsic intentionality is contrasted with that of derivative intentionality. The latter sort of intentionality is associated with symbols, signs, words, sentences, or utterances. Such items also have intentional content in so far as they are about things, refer to things outside themselves, and can have conditions of satisfaction. However, Searle adopts the common view that such items have their intentional content only because of how they are used by or related to agents with mental states, such as beliefs and desires, which have intrinsic or non-derivative intentionality.

At other times intrinsic intentionality is contrasted with mere 'as-if' intentionality. The intent is to draw a distinction between cases in which it is literally true to say of some system that it has intentional states and cases in which such attributions are merely metaphoric or picturesque ways of talking about systems which in some behavioural respects resemble intentional agents but do not literally have intentional states, as when we say that the household thermostat believes the room temperature is 20 °C or that our automobile wants or is thirsty for more petrol when its fuel warning light comes on. Searle sometimes highlights the contrast between this metaphoric sense of intentionality and intrinsic intentionality by describing it as 'observer-relative intentionality'. The point is to stress that the intentionality is only in the eyes of the beholder (or mind of the attributor), who finds it pragmatically useful to treat some non-intentional system as if it had intentional states. According to Searle, the thermostat itself has no intentional states; it is just that we may sometimes find it helpful to attribute intentional content to some of its physical states in predicting its action.

Those states have a content for us, but they have no content for the thermostat.

The third use of the term 'intrinsic intentionality' is in contrast with extrinsic (or relational) intentionality. Many philosophers, especially those who view mental states as type-individuated by their functional roles, believe that facts about the external environment in which an organism lives are relevant to determining the intentional content of its mental states. On such a view, a particular state has the content that it does because of the role that it plays in mediating the organism's interaction with its environment. If that environment were different from how it is, we might need to attribute different intentional contents to the organism's mental states, even if those states considered in isolation from the external environment were actually just as they are at present. For the functionalist, intentional content is determined at least in part by extrinsic relations to the world outside the organism; thus the content of its mental states is not completely determined by (or supervenient on) its purely intrinsic properties and internal causal structure. For the functionalist there is no contradiction in assuming that two organisms might share all their intrinsic causal properties (even to the point of being molecule-for-molecule duplicates) but differ in the content of their intentional states because of differences between their respective environments. Many examples of such cases have been offered in the philosophical literature over the past twelve years by Putnam (1975), Burge (1979), Stich (1983), and others to show that intentional content is sensitive to external factors about the organism's past history, physical environment, and social context. None the less, part of what is involved in Searle's use of the term 'intrinsic intentionality' is a rejection of the thesis that external relations play a role in the determination of content. According to Searle, the content of an organism's intentional states is entirely determined by its intrinsic causal structure. That is, if two organisms are qualitatively identical with respect to all the properties they possess purely in virtue of their structure, they will also be identical with respect to their intentional states, even if they differ greatly with respect to how they are situated in or related to their external environments. His views on intentional content are thus solipsistic: were my brain to be removed and placed in the skull of another organism living in a wildly different environment, or placed in a vat and artificially stimulated by electrodes, my mental and intentional states would have the same contents they do at present, as long as the intrinsic causal properties of my brain were not altered in the transfer process.

Searle thus uses the term 'intrinsic intentionality' to mark three different distinctions. The first two are relatively unproblematic, but still open to some dispute. With respect to his first distinction, it is clear that there is a difference between the sense in which a belief or other mental state can be said to have content and the sense in which a sentence or word has content. What is controversial is the question of whether or not the intentionality of language is in some sense derived from the intentionality of mental states as Searle suggests. I find Searle's position on this issue plausible because I am inclined to accept a basically Gricean analysis of linguistic meaning (Grice 1957, 1969), one which explains linguistic meaning in terms of the psychological states of agents, in particular in terms of their complex communicative intentions. But the question of whether thought or language is the primary bearer of content has been a matter of dispute among philosophers, as in the Chisholm and Sellars (1958) debate. Today, most parties seem to favour some sort of complex interdependence between the two (Schiffer 1972). However, we need not settle the issue here, since it seems clear that even if one denies that linguistic acts, such as asserting or ordering, owe their intentional content to that of mental states, linguistic symbols, such as words or sentences, have no content in themselves, but do so only on the basis of their relations to other bearers of content, be they mental states or only linguistic acts. Thus there is some agreement about the distinction and the derivative sense in which words and symbols have content that is shared by all parties to the dispute about the relative primacy of thought and language as the bearers of content.

The second distinction is also relatively unproblematic; there does seem to be a clear difference between literal and metaphoric attributions of intentionality. It seems clear that a thermostat does not literally have any beliefs about the temperature of its surroundings. However, there is again some room for disagreement, especially about how to draw the line between those systems which have intentional states in the literal sense and those which do not. For example, one might argue that though a thermostat does not literally have any beliefs, it does literally have some cruder states which have intentional content in a non-metaphoric sense, such as possessing information about the temperature of its surroundings. However, even if there are such disputes about where to draw the boundaries, it is still necessary to maintain the basic distinction between literal and metaphoric attributions of intentionality.

The third distinction is by far the most questionable of the three, since it embodies a substantial and not generally accepted thesis about how intentional content is determined. If environmental and social factors do play a role in determining content, then there is nothing which corresponds to intrinsic intentionality in Searle's third sense.

Fortunately, we can set that issue aside at least temporarily. For though it is relevant to the larger question before us, it seems that only the first two senses of 'intrinsic intentionality' are intended in the claim we have stated as (7). Substituting those two notions into (7) we can restate Searle's claim as

(8) A capacity for having subjective experiences is a necessary condition for having any states with literal non-derivative intentional content.

What reasons, if any, are there for accepting (8) or believing that it is true? Searle himself seems led to it by adopting what he calls a 'first-person perspective on intentionality' (1984). His method is to take his introspective awareness of his own mental states as the primary datum in determining what is essential to having a mind. He is willing to admit that there is a great deal about the operation and physical basis of his mind that cannot be discovered by mere introspective reflection. None the less, he maintains it is the mind as revealed in self-conscious awareness which is the primary object to be explained by our theories of mind. In any field of science researchers are often able to pick out or refer to what is to be explained before they are able to explain it. According to Searle, it is conscious introspection which allows us to fix the referent of our inquiry into the nature of mind. Thus although Searle is thoroughly materialistic and anti-Cartesian in his metaphysics, he is very Cartesian in his methodology and in his view of the defining features of mentality. His approach is individualistic, solipsistic, and introspectionist. Given such a first-person perspective on intentionality, it is hardly surprising that Searle regards a capacity for subjective experience as necessary for having a mind in any sense worthy of the name; for it is the world of subjective experience of which we are aware when we introspect.

However, such an explanation is diagnostic rather than justificatory. We still need to ask what reasons can be given in support of (8) that might persuade someone who did not accept Searle's first-person methodology and introspectivist way of delimiting the domain of the mental.

Consider again the brief argument given above for the conclusion that instantiating a computer program cannot be sufficient for having mental states. The argument turned on two claims: (i) that mental states have semantic or intentional properties and (ii) that computer programs are defined in a purely syntactic way. If confronted with a computing device that processes strings of words in a way which produces output strings which count as conversationally appropriate responses to its input strings (that is, if confronted with a computer capable of passing the Turing test), Searle would deny that there was any reason to believe that the device had any states with literal, non-derivative, intentional content. The symbols that the device processes do have content in so far as we regard them as symbols of English or German, but the content involved is only derivative. Moreover, they have such content only from our perspective, not from the perspective of the machine. Thus it is only 'as-if' or observer-relative intentional content; no intrinsic intentionality is involved at all. Searle would again remind us that you cannot get semantics from mere syntax.

Though I am inclined to accept Searle's negative judgement about the absence of genuine intentionality in such a case, I think we need to ask what would have to be added to the machine in order to give the symbols being processed more than mere 'as-if', observer-relative content.

The normal functionalist answer would be that we need to add some causal connections between those symbols and some non-symbolic features of the external world. In particular, we need to connect those symbols with real-world, non-symbolic inputs and outputs of the system that is processing them (Van Gulick 1980). On the input side we need to add some analogues of sensory organs, some transducers which take physical stimuli as inputs and give formal symbols as outputs. Ideally they should be 'smart' transducers, which are capable of reliably detecting not only patterns of physical stimuli but also external objects or states of affairs that would produce those patterns of stimulation to which the transducer is tuned. Such a transducer could be said to detect the presence of an external object, such as a moving red ball or a sharp cliff edge, if it produced a particular symbolic output only in response to the sorts of stimuli uniquely associated with such objects. Detailed accounts of such detector content have been offered by Dretske (1981) and Stalnaker (1984), among others, and are commonly assumed, if not always spelled out with rigour, in information-processing models of perception (Lindsay and Norman 1972).

In addition to such input connections, some connections will also need to be added between symbols and non-symbolic outputs or behaviours by the system, especially behaviours which produce regular systematic changes in the external world. There should be some causal relations between internal symbols and external behaviour in virtue of which the symbol moulds or shapes the behaviour to achieve a constant result as the external conditions change. The external symbols might, for example, control the system's motion so that it reliably tracked or followed an erratically moving target object. Of course to do so there will need to be systematic connections between the symbols produced by the system's detectors and those governing its behaviour. All these conditions in effect amount to transforming our original conversational computer into a robot.

Searle is quite right that you cannot get semantic properties from syntax alone. But the functionalist's point in transforming the original computer into a robot is to argue that what gives syntactic symbols their semantic properties is the causal relations that they bear to non-symbolic items in the world. And in particular, the functionalist claims that internal symbols acquire their intentional content on the basis of the causal functional roles they play in successfully mediating the organism's (or system's) adaptive interaction with its environment. What makes a pattern of activity occurring in my visual cortex the perception of a small red cylinder is that under normal conditions it is produced only by objects having those properties, and it can guide my behaviour relative to such an object in appropriate ways should I need to grasp it, roll it, fill it with water, or pick it out from among other objects differing from it in colour, shape, or size. In sum, the functionalist claim is that symbols gain semantic properties not only from their relation to other symbols but also from their causal relations to the non-symbolic world.

However, Searle insists, on allegedly intuitive grounds, that such relations alone will not suffice to produce any states with intrinsic, intentional content. He explicitly denies that transforming our symbol-processing computer into a robot will result in any states with intrinsic, as opposed to merely observer-relative, semantic properties. In his very much discussed article 'Minds, brains and programs', Searle (1980) offers a robot version of his famous (or infamous, depending on one's perspective) Chinese room example. We are to imagine that the robot's transducers produce formal symbols on an internal display screen viewed by a homunculus. In response to these symbols, the homunculus types in other symbols according to a complex set of instructions, which are specified entirely in terms of formal or syntactic relations among symbols. Unknown to the homunculus, the symbols that he types produce behavioural responses by the robot that are appropriate to the environmental features producing the symbols on the display screen, though the homunculus remains ignorant about these causes. Searle's contention is that in such a case there is only one agent with intrinsically intentional states, namely the homunculus. And since, by the very definition of the case, he has no understanding of the semantic properties of the symbols that he manipulates and which he treats as merely uninterpreted formal items, it allegedly follows that there is no one in the case described who has any understanding of their semantic content. At most, those symbols have observer-relative content for us in describing the case; they have no semantic content for the homunculus or for the robot of which he is a part.

The crucial claim here is that the homunculus is the only agent in the case with any intrinsically intentional states. Searle (1980, p. 421) puts it as follows:

The formal symbol manipulations go on, the input and output are correctly matched, but the only real locus of intentionality is the man (i.e., the homunculus), and he doesn't know any of the relevant intentional states.

Why is the homunculus supposed to be 'the only real locus of intentionality'? And why does Searle regard as absurd any suggestion that the whole organized system consisting of robot plus homunculus plus instruction books might have intrinsically intentional states? The answers to both questions are roughly the same. Intentionality requires a capacity for having conscious subjective experiences. The homunculus has such experiences, but a combination of robot, homunculus, and instruction books is not the sort of thing that would have any such experiences.

We may seem then to have come full circle arriving back at (7) and the claim that a capacity for subjective experience is a necessary condition for having intentional states. If so, Searle's analysis of the robot case might seem to beg the question. Or at least the functionalist will allege that he and Searle are simply left with conflicting and theoretically motivated intuitions. Where Searle sees no intentionality because there is no subjective experience, the functionalist does see intentionality because there are states fulfilling the relevant functional roles. We are left at an impasse.

I am a functionalist, and that is where I have been inclined in the past to leave the issue, as a basic clash of conflicting intuitions between Searle with his first-person introspective perspective and those of us functionalists who have a more third-person view of mentality and intentional content. However, I no longer find that a stable resting place or a satisfactory point at which to walk away from the dispute. For though there do seem to be good reasons for assigning at least some sorts of intrinsically intentional states to systems solely on the basis of their functional organizations and relations to the environment, there none the less seems to be a significant dimension along which intentional states involving conscious subjective experience differ from intentional states not involving consciousness. I remain a functionalist. I do not regard a capacity for subjective experience as a necessary condition for having intentional states, but I do believe that considering the nature and role of conscious subjective experience can add considerably to our understanding of intentionality.

Thus I would like to close by making a few tentative suggestions about the connections between consciousness and intentionality. They are less conclusive and more programmatic than I would wish, but I hope that they will at least open the way for fruitful future investigation.

One way to bring out the possible link between consciousness and intentionality is to consider Searle's distinction between a symbol's having content for the system which processes it and its having content merely for an outside observer. This is one of the ways in which he explains the difference between intrinsic (or literal) intentionality and mere 'as-if' or metaphoric intentionality. In a case of mere 'as-if' intentionality, the symbols being processed have no content for the system which processes them; they are not really symbols for that system.

What seems to be lacking in the mere 'as-if' cases is any understanding of the symbols' content. The system treats the symbols that it processes as purely uninterpreted formal or syntactic items. Any semantic content they may have in virtue of their relations to the world seems to be irrelevant to how the system reacts towards them or operates upon them. By contrast, in those cases that Searle would regard as clearly involving intrinsic intentionality, such as a human being's conscious visual experience or conscious deliberation about a philosophical question, the relevant system or agent appears to have a substantial, if not necessarily complete, understanding of the semantic or intentional content of any symbols (or representations) used in his mental processes. Human beings understand how their visual experiences represent the world as being, and they generally understand the semantic properties of any words or images they use in their deliberative thinking. Or, at least, so it seems from the first-person perspective, which Searle urges us to adopt.

I think the description just given is half right. It is true that conscious agents do to a large degree understand the semantic or intentional properties of the symbols or representations they use in thought. But it is false that there is no such understanding of content unless a system has conscious subjective experiences. A capacity for having subjective experiences may guarantee that one has at least some such understanding, but a system need not have such a capacity in order to understand facts about the semantic properties of the symbols that it processes or uses in mediating its interaction with its environment. For the functionalist, understanding is always a matter of a capacity for practical success. To understand that a certain state of affairs obtains is to have the capacity to modify one's behaviour in ways that are specifically appropriate to that state of affairs with regard to one's goals or objectives. For a system to understand that there is a barrier four feet high dividing the field in which it is located is, in part, having the capacity to adapt its movements in ways that would enable it to get around, go over, avoid, dismantle, hide behind, or increase the height of that barrier, should it have the need to do so. Similarly, to understand that a specific symbol which one processes or uses has a given semantic content is a matter of being able to behave toward that symbol in ways that are specifically appropriate to its having that content. If one accepts such a functionalist picture of understanding, it will turn out that a great many systems have at least some understanding of the semantic properties of the symbols that they process, including many systems that have no conscious subjective experiences. In so far as a system acts toward a given symbol in ways that are appropriate given its semantic content, but that would not be appropriate if it did not have that content, the system can be said to that extent to understand the semantic content of that symbol, and, in so far as the symbol is one used by the system itself, we attribute to the system at least some measure of semantic self-understanding. An appropriate behaviour might involve reacting to a specific transducer-generated symbol in a way which produces an external response specifically appropriate to the external object whose presence is reliably indicated by that transducer output. Or it might include producing by purely internal processes a given symbol X in response to transducer-generated symbols Y and Z, where the behaviour guided by X is appropriate to an external state of affairs which is the logical consequence of the joint occurrence of the two states of affairs indicated by Y and Z respectively. The examples are simplistic but should serve to illustrate the point.

Three consequences follow from such a functionalist account of semantic self-understanding:

I. Semantic self-understanding is not an all-or-nothing matter. There will be *many degrees* to which a system can have such understanding. One system may understand the semantic properties of its symbols to a greater degree than another, in so far as it has a more complete or more subtle capacity to respond differentially and appropriately to the symbols that it processes in ways sensitive to their semantic content. In allowing for such differences of degree, semantic self-understanding is just like the understanding that a system or organism may have of some external state of affairs, which also allows for one system to have a more complete or subtle understanding than another.

II. Semantic self-understanding need not be explicitly represented or localized in an isolable state within the self-understanding system. Such understanding can be implicitly embodied in the system's over-all organization and in the structure of its symbol-processing and behaviour-regulating mechanisms. That is, the system's meta-semantic understanding need not be represented in a declarative way; some (or all)

of what it understands about its own intentional and semantic properties may be represented procedurally. This could be the case, for
example, with respect to its learning capacities. If learning is to produce
long-term changes in a behaviour-regulating mechanism that are
adaptive rather than merely random, the learning mechanisms that
produce those changes will have to be sensitive to the functional and
intentional properties of the structures that they modify. If a feedback
loop is to adaptively alter a rat's feeding habits in response to the onset of
nausea after eating a particular food, the relevant learning mechanism
must 'know' or 'understand' which nervous system structures direct
eating of that food, or appropriate changes in future behaviour will not
be produced. However, the required understanding may be entirely
embedded as procedural know-how in the organization of the feedback
circuits that produce the appropriately matched changes.

   III. As almost a corollary to II, semantic self-understanding need not
involve any subjective experience of understanding. In systems that
lack any capacity for conscious subjective experience there will, of
course, be no subjective experience of semantic understanding. But such
understanding can still be present, as long as the relevant system has the
required capacities to react to the semantic properties of the symbols
that it uses or processes.

   When I noted above that I was dissatisfied with my earlier view of
Searle's position as involving just a clash of conflicting intuitions, I
asserted that there was a dimension along which intentional states
involving conscious subjective experiences differ from those not involving such experiences. The relevant dimension is the one we have just
been discussing, namely the degree to which a system understands the
semantic content of the symbols or representations it uses or processes.
My conjecture is that subjective experiences involve a very high degree
of such understanding. If we call the relevant dimension semantic
transparency, the conjecture is that conscious subjective experiences
involve representations or complex symbolic structures which have a
high degree of semantic transparency for the system or organism having
the experience. For example, having a conscious visual experience, as
when I look out into a crowded city street, seems to involve my using or
having present in my mind a complex representation which has the
form of a three-dimensional manifold, which is locally differentiated in
a variety of ways. When I have such an experience I also understand, on
the whole, how that representation represents the world as being.
Indeed, such visual representations are so transparent in content that
we normally 'look right through' them. Our experience is of the external

world as represented. A similarly high degree of semantic self-understanding is also present with respect to the linguistic representations which are involved in our conscious deliberative reasoning.

In looking for an explanation of this transparency associated with conscious experience, we should look primarily to the kinds of processing associated with conscious representations. For it is by better understanding the processes that allow us to move almost instantaneously from one representation to another semantically related one that we are likely to gain insight into how the representations or symbolic structures associated with conscious experience differ from other sorts of representations. If so, then most analytic philosophers who have concerned themselves with conscious experience in recent years have been wrong in focusing primarily on qualia and so-called 'raw feels' such as the redness of phenomenal red or the subjective taste of a pineapple (Shoemaker 1975, 1981; Block 1980). They would do better to consider the dynamical aspects of experience and the processes that underlie them. In that respect there is much, I suspect, that might be learned from the phenomenological tradition (Husserl 1931). What then can the functionalist say about this subjectively experienced transparency? Functionalist understanding is always practical: it always involves a capacity for appropriate behaviour. Understanding a representation's content is being able to relate that representation to other representations and to items in the world in ways appropriate to its content. It is internal behaviour relating representations to representations which holds the best promise of accounting for the subjective experience of understanding. As long as understanding is analysed as some form of behavioural capacity, even if the relevant behaviour is all internal to the nervous system, the functionalist can hope to accommodate it within his general account of mind.

Thus the functionalist should resist any equation of phenomenal transparency with some form of non-behavioural, self-luminous understanding. The intuitive appeal of the self-luminous view may be undercut if the functionalist can account for the subjective experience of understanding in terms of smooth and seemingly automatic transitions among representations. The internal component of understanding need be nothing more than an ability to interrelate a great many diverse mental representations very quickly. If we view the mind as in part an organized system for manipulating representations, its understanding a visual representation's content would be a matter of its being able to appropriately connect that representation with other visual, non-visual, and non-perceptual representations. I understand that what I see as I look out into the garden is a tomato plant. The internal component of my understanding need consist in nothing more than the fact that, as

soon as attention focuses on the relevant portion of the visual representation, a number of other representations are accessed or activated
including linguistic ones concerning conceptual knowledge about
tomatoes. *How* this is done is not something to which I have linguistic
or introspective access, but there must be powerful processing mechanisms which produce these seemingly instantaneous transitions.

It is here that meta-psychological understanding plays a role in the
process. The capacities of the processors that generate the flow of
experience are clearly such that they satisfy the functionalist's requirements for meta-psychological understanding. It is they that in the first
instance make the appropriate connections among representations on
the basis of their content.

The personal-level experience of understanding is none the less not an
illusion. I, the personal subject of experience, do understand. I can make
all the necessary connections within experience, calling up representations to immediately connect one with one another. The fact that my
ability is the result of my being composed of an organized system of
subpersonal components which produce my orderly flow of thoughts
does not impugn my ability. What is illusory or mistaken is only the
view that I am some distinct substantial self who produces these
connections in virtue of a totally non-behavioural form of understanding. Perhaps on this point Hume (1965) was close to the truth. There is
no understanding self which produces the flow of thought and experience, but it is rather the other way around. What makes me an understanding self is that I am so organized that my internal workings consist
in part of an orderly swift flow of connected representations.

However, even if the functionalist can in the long run provide some
dynamic processing account of the way in which we understand our
own thoughts, at least one major question will remain unanswered:
Must a system employ phenomenal representations (that is, internal
representations with the sorts of qualitative properties typical of conscious perceptual experience) in order to be capable of such processing?
If the answer to that question is 'yes', then it would turn out that only
systems with phenomenal representations could really (or fully) understand the semantic properties of the internal representations that they
process. And thus, for reasons given above, only in such systems could
the representations being processed be said to be symbols for the
processing system itself. Such a result would vindicate Searle's intuition that a capacity for subjective phenomenal experience is a necessary condition for having states with intrinsic intentionality. However, I
see at present no argument to support such a claim or any reason to
believe that the answer to the outstanding question would be 'yes'.

Moreover, having representations with phenomenal properties in itself would not seem to guarantee the processing capacities required for semantic transparency. It seems possible to have experiences with sensuous phenomenal qualities in cases in which those experiences lack the sort of organized cognitive structure needed for transparency. Newly born infants probably have such experiences, as do those who gain sight after a long period of congenital blindness. In such cases, a period of learning or stimulated development is necessary to establish the connections and processing mechanisms that underlie normal understanding, but which usually escape notice just because they are so swift and automatic.

At this point our theoretical understanding of the relevant processes is so slight that we cannot really make any informed judgements about the range of systems in which such processing might occur. On the basis of first-hand evidence it seems clear that a high degree of semantic transparency can be achieved in systems that employ phenomenal representations. But that may be just one of many ways of achieving such transparency, and there does not seem to be any a priori reason for ruling out the possibility of systems (perhaps man-made computational devices) that understand the semantics of the symbols that they use, even though none of those symbols have any phenomenal properties. The fact that we or Searle may find such systems intuitively unimaginable may not reflect any empirical or theoretical constraints on the range of self-understanding systems, but rather the limits of our particular human ability to empathetically comprehend systems that are radically unlike us in the means by which they achieve the processing needed for genuine semantic self-understanding. That possibility ought to make us cautious about relying too heavily on our intuitions in reaching conclusions about what is or is not necessary for self-understanding, intrinsic intentionality, or having a mind. Intuitions can be valuable, but we need much better theories before we can draw any boundaries with confidence.

Let me summarize where things stand with respect to the relation between intrinsic intentionality and conscious subjective experience by listing seven tentative claims and conclusions, which seem highly plausible if one holds a basically functionalist view of mental and intentional states.

1. What is required for a system to have intentional states is that it have states that play the appropriate causal roles in mediating the system's interactions with its environment.

2. For a symbol-processing system to have states with *intrinsic* intentionality, as opposed to merely *observer-relative* intentionality, the

symbols that it processes must have semantic content for it, and not just for an external observer.

3. A symbol's having semantic content for a system is a matter of the system understanding that the symbol has that content. There are many degrees to which a system can understand the semantic properties of the symbols that it uses or processes.

4. Thus there are degrees to which a symbol processed by a system can have content for that system.

5. Systems that have no capacity for conscious subjective experience can none the less to some degree understand the semantic properties of symbols that they process. Semantic self-understanding is then possible to at least some degree without any conscious subjective experience.

6. Conscious subjective experience involves a high degree of semantic self-understanding, and the sorts of symbols or representations involved in conscious experience have a high degree of semantic transparency.

7. It remains a theoretically open question whether or not a capacity for conscious subjective experience involving phenomenal representation is or is not a necessary condition for achieving a high degree of semantic self-understanding (or semantic transparency).

Given the intractable nature of many of the problems with which they deal, philosophers are probably more inclined than people in other professions to count the mere clarification of a question as genuine progress. Answers to interesting philosophical questions are not easily come by. Perhaps with respect to the problem of consciousness, we are as yet all philosophers. Thus I hope that by sorting out some of the questions that professional philosophers have worried about, as I have tried to do in this paper, readers from other disciplines may see the way to answers where they might not before have recognized a question, as we work together to gain a genuinely scientific understanding of consciousness.

Acknowledgements

The present work is a revision of a paper with the same title delivered at a conference on 'Aspects of consciousness and awareness' sponsored by and held at the Centre for Interdisciplinary Studies (Zentrum für interdisziplinäre Forschung) of the University of Bielefeld, Bielefeld, West Germany in December 1986.

I have benefited from the comments on that original paper, which I received from the participants at the conference and from the editors of this volume. DawnMaree Girndt, Lisa Mowins, and Alastair Norcross provided valuable assistance in the production of the typescript.

References

Block, N. (1978). Troubles with functionalism. In *Cognition: issues in the foundations of psychology, Minnesota studies in the philosophy of science,* Vol. 9, (ed., C. W. Savage), pp. 261–325. University of Minnesota Press.

Block, N. (1980). Are absent qualia impossible? *The Philosophical Review,* **89,** 257–74.

Block, N. (1981). Psychologism and behaviorism. *The Philosophical Review,* **90,** 5–34.

Block, N. and Fodor, J. (1972). What psychological states are not. *The Philosophical Review,* **81,** 159–81.

Burge, T. (1979). Individualism and the mental. *Midwest studies in philosophy,* **4,** 73–221.

Chisholm, R. and Sellars, W. (1958). Intentionality and the mental. In *Concepts, theories and the mind–body problem, Minnesota studies in the philosophy of science,* Vol 2, (ed. H. Fiegl, M. Scriven, and G. Maxwell), pp. 507–39. University of Minnesota Press.

Churchland, P. M. (1981). Eliminative materialism and the propositional attitudes. *Journal of Philosophy,* **78,** 67–90.

Dennett, D. C. (1969). *Content and consciousness.* Routledge and Kegan Paul, Andover, Hants.

Dennett, D. C. (1971). Intentional systems. *Journal of Philosophy,* **68,** 87–106.

Dennett, D. C. (1978). *Brainstorms.* MIT Press, Cambridge, MA.

Dennett, D. C. (1981). True believers. In *Scientific explanation,* (ed. A. F. Heath), pp. 53–75. Clarendon Press, Oxford.

Dennett, D. C. (1985). Can machines think? In *How we know,* (ed. M. Shafto), pp. 121–45. W. H. Freeman, San Francisco.

Dretske, F. (1981). *Knowledge and the flow of information.* MIT Press, Cambridge, MA.

Eccles, J. C. and Popper, K. R. (1977). *The self and its brain.* Springer International, Berlin.

Grice, H. P. (1957). Meaning. *The Philosophical Review,* **66,** 377–88.

Grice, H. P. (1969). Utterer's meaning and intentions. *The Philosophical Review,* **78,** 147–77.

Haugeland, J. (1981). *Mind design.* MIT Press, Cambridge, MA.

Hume, D. (1965). *A treatise of human nature,* (ed. L. A. Selby-Bigge). Oxford University Press.

Husserl, E. (1931). *Ideas: general introduction to pure phenomenology.* Macmillan, New York.

Lindsay, P. and Norman, D. A. (1972). *Human information processing.* Academic Press, New York.

Nagel, T. (1974). What is it like to be a bat? *The Philosophical Review*, **83**, 435–50.

Nagel, T. (1979). *Mortal questions.* Cambridge University Press, New York.

Nagel, T. (1986). *The view from nowhere.* Oxford University Press.

Putnam, H. (1960). Minds and machines. In *Dimensions of mind*, (ed. S. Hook), pp. 138–64. New York University Press.

Putnam, H. (1975). The Meaning of meaning. In *Language, mind and knowledge, Minnesota studies in the philosophy of science*, Vol. 7, (ed. K. Gunderson), pp. 131–93. University of Minnesota Press.

Rorty, R. (1970). In defense of eliminative materialism. *The Review of Metaphysics*, **24**, 112–21.

Ryle, G. (1949). *The concept of mind.* Barnes and Noble, New York.

Schiffer, S. (1972). *Meaning.* Clarendon Press, Oxford.

Searle, J. (1979). What is an intentional state? *Mind.* **88**, 72–94.

Searle, J. (1980). Minds, brains and programs. *The Behavioral and Brain Sciences*, **3**, 417–24.

Searle, J. (1984). *Minds, brains and science.* Harvard University Press.

Shoemaker, S. (1975). Functionalism and qualia. *Philosophical Studies*, **27**, 291–315.

Shoemaker, S. (1981). Absent qualia are impossible—a reply to Block. *The Philosophical Review*, **90**, 581–99.

Skinner, B. F. (1953). *Science and human behavior.* Macmillan, New York.

Stalnaker, R. (1984). *Inquiry.* MIT Press, Cambridge, MA.

Stich, S. (1983). *From folk psychology to cognitive science: the case against belief.* MIT Press, Cambridge, MA.

Turing, A. (1950). Computing machinery and intelligence. *Mind*, **59**, 433–60.

Van Gulick, R. (1980). Functionalism, information and content. *Nature and System*, **2**, 139–62.

5

The (haunted) brain and consciousness

Edoardo Bisiach

A thorny problem for the unity of science—one should perhaps say for the unity of knowledge—is the apparent meta-theoretical impasse afflicting psychology; namely, the identity of the subject and object of inquiry. Behaviourists tried to eradicate this anomaly by resorting to another, in the form of a veto on hypothetical constructs regarding the contents of the black box. The black box, however, did not remain untenanted for long: starting within the regime itself (Tolman 1932, 1948) the reaction was afterwards intensified, as Hilgard (1977) pointed out, in the wake of the post-war commotion. Adverse side-effects of the counter-revolution were widespread underestimation of the legacy of behaviourism and the danger of 'the new freedom as an opportunity for free-floating uncritical fantasies about mental life' (Hilgard 1980, p. 15), contained within a spectrum ranging from conceptual confusion to more or less surreptitious attempts to restore the ghost to the machine.

Not even listed in the subject index of Kling and Riggs' 1971 edition of *Woodworth and Schlosberg's experimental psychology*, consciousness has regained an important position in post-behaviourist psychology. The issue of consciousness is also explicitly addressed by clinical neuropsychologists with reference to syndromes such as split-brain, blindsight, unilateral neglect of space, anosognosia, etc. Although the term 'consciousness' often leads to confused thinking, ostracizing it to exorcize ghosts and restrain wild speculation has already proven over-restrictive and ineffectual. A better way to keep our concepts in order would be by continuous care in their usage.

In this chapter, I first propose three different definitions of consciousness, with the aim of diminishing the danger of ambiguity. The relevance of these three aspects of consciousness for empirical science are then briefly discussed. An example is next provided of how ambiguity in the concept of consciousness can impair the interpretation of neuroscience data. I will then argue for the modularity of consciousness and show how this modularity has to be taken into account with respect to the problem of the timing of conscious events. Some reflections on the problem of the causal role of consciousness in cognitive processes follow. The chapter ends with a brief conclusion on the limits set on the scientific investigation of consciousness.

Definitions of consciousness

When initiating a debate likely to involve fuzzy or ambiguous concepts, one may either insist on the prerequisite of precise definitions or contend that, once the arguments have been taken care of, the definitions will take care of themselves. As a matter of fact, however, the process is circular, and the emphasis may be placed either way according to circumstances and personal propensities. None the less, I think that in discussing consciousness at least some preliminary distinctions are necessary in order to avoid equivocal arguments and debates that are out of register. Perhaps I am overstating preoccupations which are largely due to the rather extreme and anomalous phenomenology with which, as a clinical neurologist, I am acquainted. This may be true to some extent, although it still seems that much of the current literature on consciousness suffers from a persisting category mistake which I will try to expose in the following sections. I shall therefore propose three definitions of consciousness, corresponding to reference frames which can easily be confounded with one another.

The first, let us call it C_1 for short, is the phenomenal experience of a subject capable of perceiving and representing. Inaccessible to the external observer (it must not be confounded with its report), it constitutes the inner aspect, as it were, of complex physical events whose outer aspect lends itself to public observation within the limits of available technologies.[1] A somewhat disquieting consequence of the subjectivity of C_1 is that whatever physical system happens to be endowed with it cannot help being hopelessly unsure, at least in principle, of the extent to which it may be ascribed to any other system, however sophisticated. By contrast, a pleasing consequence of positing C_1 as mere subjective experience, indivisible from the physical events with which it has a relation of *identity*, is that it makes no sense to ask whether it is true or false, so that the problem of its 'corrigibility'[2] does not even arise. It can also be observed that if C_1 is held to be identical

---

[1] I do not presume to offer an unproblematic definition of C_1. The search for the most adequate definition would be, I am afraid, paralyzing or lead to the evocative, but hardly satisfactory suggestion, that consciousness can only be 'defined in terms of itself' (Angell 1904, quoted by Hilgard 1980).

[2] The reader unacquainted with this problem may consult, for example, Wilkes (1978). See also Dennett (1988, this volume).

with the *global* state of a physical system (e.g. with the state of the whole neural net whose activity happens to be conscious in C_1 terms) *at any given instant* of its representational activity, then idioms such as 'self-consciousness' or 'consciousness of consciousness' turn out to be semantically empty, inasmuch as physiologically absurd, in discussions of C_1.

The second definition of consciousness, let us call it C_2, refers to the access of parts or processes of a system to other of its parts or processes, though not to all (see the penultimate section). This might be for the purpose of a class of outward activities constituting some sort of print-out. It is worth while emphasizing right away that C_2 is not just a reductive operational definition of C_1, since it can safely be ascribed to several physical systems, including artificial ones.

The third definition, which I will label C_3, refers to 'non-physical' entities which are currently introduced with variable degrees of clarity. It ranges from outright Cartesian notions of 'immaterial mind' or 'conscious self' (e.g. Eccles, in Popper and Eccles 1977)—the more archaic 'soul' being apparently confined to the lexicon of sceptically inclined writers—to hints at a more vague 'metaphysical state [affecting] the same physical system on which it depends' (Underwood and Stevens 1979, p. vii).

Consciousness as an object of scientific knowledge

When, after Tolman, it is claimed that 'raw feels' are ineffable and that consciousness is not a subject of scientific inquiry, reference is clearly being made to C_1. By my definition, indeed, C_1 does not conform to the scientific requirement of public observability. This is what makes 'what it is like to be a bat' radically unknowable (Nagel 1974), and what is countenanced by the argument from inverted *qualia* (see Dennett 1988, this volume). Suppose your phenomenal experience of red is raised by the same stimuli and through the same brain processes which give rise to my phenomenal experience of green: the inverted quality of our experiences would forever be unnoticed and unnoticeable. Here, the 'unbridgeable gulf between consciousness and brain-process' (Wittgenstein 1953, p. 124) is something much more definite than a 'feeling': it is a real and ineliminable impasse. This, I think, is what Dennett (1978, p. xii) maintains when he claims that there are questions about consciousness that scientific theories alone—that is, unsupplemented by concepts from a philosophical theory—are, and forever will be, utterly unable to answer. The uncollectability of C_1 data for scientific purposes seems in fact to be the compelling moral of his essay 'Are dreams experiences?' (1978). In order to prevent misunderstanding, it ought to be clear from the definition given in the previous section that C_1 data do not lend themselves to a putative first-person science either. Indeed, they cannot be frozen in time for further examination from a privileged standpoint. Failure to appreciate this point leads to such incoherencies as those exposed by Dennett (1988, this volume). If this is so, it might be concluded, so much the worse for C_1. There is however at least one practical reason why we cannot merely ignore C_1. As will be argued in the next section, inappropriate allusions to C_1 still have a strong tendency to vitiate interpretations of human behaviour.

C_2, on the contrary, is a thoroughly legitimate object of scientific inquiry, though there might be questions about the indispensibility and the usefulness of the problematic term 'consciousness' in a great number of the statements in which it appears. Indeed, these could easily —and perhaps sometimes more clearly—be rephrased in terms of such hypothetical constructs as attention, representation, working memory, control, programming, etc. Mandler (1975, p. 229) maintains that the use of the word 'consciousness . . . avoids circumlocutions' and 'seems to tie together many disparate but obviously related mental concepts, including attention, perceptual elaboration, and limited capacity notions'. An objection to this claim is that it encourages a semantic indeterminacy which is likely to obscure mental concepts which are already far from clear. Therefore, it seems better to adopt the opposite stand and to renounce this short-hand, first approximation use of the terms 'conscious' and 'consciousness'. The use of these terms will here be restricted to reference to the *monitoring of internal representations*. This usage of the term does not map C_1 on C_2; in fact, it does not presuppose it.

Before demonstrating how careless reference to consciousness may create theoretical problems, one cannot avoid taking a position about the pertinence of C_3. C_3 may be viewed as a hypostasis of C_I, more or less explicitly contrived to fit some version of dualism. It has been claimed that any metaphysical option is consistent with any scientific finding (Maxwell 1976, p. 318) since the various forms of dualism or monism are unfalsifiable. Though this may be true for parallelist or epiphenomenalist solutions of the mind–body problem, it is false for interactionism. Since interactionism is bound to specify falsifiable hypotheses about psycho–physical intercourse, it is conceptually entitled to fight in the very arena of empirical science. Thus C_3 cannot be dismissed from science a priori, but only whenever it proves unnecessary for the explanation of behaviour or whenever its use can be demystified as being the category mistake (Ryle 1963, pp. 17–25) of positing *interaction* between an entity under a given description (i.e.

mind, 'emergent' consciousness, etc.) and the same entity under another description (i.e. some activities of the nervous system).

Pitfalls in the explanation of brain activities in terms of consciousness

C_1, C_2, and C_3 might seem to be rather intractable concepts, for they show a marked proclivity to revert, almost inadvertently, into one another. This is, indeed, what makes some readings much like the impossible objects familiar to students of perception. There are many examples of this; among them, the controversy about the remarkable work on the subjective chronometry of certain brain events by Libet *et al.* (1979) is quite pertinent.

Libet *et al.* obtained their data from patients with electrodes cerebrally implanted for therapeutic purposes. Three kinds of stimuli were employed: single-pulse electrical skin (S) stimuli, and trains of pulses applied to sub-cortical structures (SC)—thalamus, lemniscus medialis —or to the cortical somato–sensory area (CS). The trains of SC and CS pulses were calibrated in intensity so that they needed equal duration —in the order of hundreds of milliseconds—to exceed the threshold for subsequent verbal report. By coupling skin stimuli to cerebral stimuli of either kind with identical onset, it was found that stimuli in S–SC pairs were judged to be simultaneous by the subjects, whereas stimuli in S–CS pairs were judged asynchronous, S being the leading one. It must be noted that S and SC stimuli give rise to an almost immediate 'primary evoked potential' in the cortex, an electrical response which does not occur after direct cortical stimulation. The fact that verbal reportability of S stimuli requires a certain amount of cortical processing, the duration of which cannot be directly measured, does not seem to be crucial for the explanation of the results obtained by Libet *et al.* Indeed, the hypothesis that S stimuli might require shorter neural processing times than CS stimuli to reach the threshold for verbal report cannot explain the subjective asynchrony of simultaneously delivered S and CS stimuli, since SC stimuli, requiring the *same* amount of time required for CS stimuli to reach neuronal adequacy for verbal reportability, were judged synchronous with S stimuli having identical onset time. By the same token, the results cannot be explained by appealing to the fact that sensations due to S stimuli were described by the subjects as being sharper in onset than sensations due to SC or CS stimuli. Since the clue to the understanding of the data must logically be sought in a feature common to the processing of S and SC stimuli, but absent from the processing of CS stimuli, the only candidate in sight to act as a time-tag seems to be the primary evoked response. As for the interpretation of this chronometric information, Libet *et al.* (1979, p. 217) claim that their experiments 'provide specific support . . . for the existence of a subjective temporal referral of sensory experience by which the subjective timing [of sensations evoked by S and SC stimuli] is *retroactively antedated* to the time of the primary evoked response' (my emphasis). The idea of such 'retroactive antedating' stems from the fact that the primary evoked response is *per se* insufficient to generate subsequent verbal report of any related experience, which requires further neural events to develop *after* the primary evoked response is obtained. The conclusion reached by Libet *et al.* might be read as suggesting that a sensation (in C_1 terms), though dependent upon *subsequent* neural events, arises in the precise moment in which the primary evoked response takes place. As Churchland (1981) rightly remarks, this would imply an assumption of backward causation.

Taken at face value, the statements of Libet *et al.* are capable of more than one interpretation. Therefore, presenting them as suggesting a backward step in time made by a non-physical mind—a thesis unhesitatingly espoused by Eccles (1977)—is tendentious. However, this is indeed the conclusion which the authors themselves seem to be willing to force upon the reader. First, in their Figs 1–3 (pp. 199–203), Libet *et al.* (1979) actually make the onset of the subjects' 'experience' *coincide* with the primary evoked response. Second, they dispute an alternative explanation, suggested by MacKay in a discussion with Libet (1979, p. 219), to the effect that 'the subjective referral backwards in time may be due to an illusory judgment made by the subject when he *reports* the timings'. Third, and more significant, Libet *et al.* (1979, p. 220) hint at 'serious though not insurmountable difficulties' for the identity theory, caused by their data. Neither in the 1979 paper, nor in Libet's (1981) reply to Churchland, is any further light thrown on *how* these putative difficulties could in fact be surmounted.

Churchland set out to demolish Libet's argument on its own grounds. She showed that the quaint hypothesis of a sensation occurring earlier than the brain state necessary for its production could more simply and more reasonably be substituted by the hypothesis of a postponement of sensation produced by cortical stimulation relative to sensations produced by skin stimulation. She did not dwell upon whether discussions on this ground have any neurophysiological plausibility at all, though her hint as to the output order of a computer programmed with a time-sharing algorithm suggests a radically different approach to the issue. The same is true for Hoy's (1982) intervention in the discussion. By asserting that 'one cannot infer much about the actual time of an experiencing from reports where that experiencing is the object of some other state of experiencing' (p. 259), Hoy is within a hair's breadth of transposing the whole matter onto the terra firma of information-processing psychology.

The main equivocation in Libet's (1981) discussion of Libet *et al.*'s (1979) results lies in the facts that (a) an (allegedly unitary) *subjective* event is mapped on to the *objective* time scale to which the primary evoked response and neuronal adequacy for verbal reportability of a stimulus are related, and (b) consciousness ('sensations', meant as C_1 data relative to the stimuli at issue) is judged on the basis of verbal reports which are tacitly assumed to be reliable messengers of an utterly implausible subject of *homogeneous* sentience (see the next two sections).

Warnings about accepting verbal reports as faithful witness of a subject's acquaintance-wise experiences have been repeatedly voiced (e.g. by Dennett 1978). Nevertheless, this assumption often seems to work underground, and it has even been suggested that C_1 is ultimately shaped by inner speech (Eccles, in Popper and Eccles 1977). It must therefore be reiterated (and it will be again in the section after the next) that this problem cannot be addressed in scientific terms, since no valid criterion of C_1 is available, nor can any diagnostics to this effect be envisaged as a promissory outcome of future technology.

For scientific explanations of *brain activity and behaviour*, there are no inner objects of experience that presuppose a separate, unanalysed subject who has intentional states. The 'subject' *is* information being processed along diverse paths, at various, not necessarily sequential, stages. There are unmonitorable aspects of this processing, as in the case of the Helmoltzian 'unconscious inferences' about depth perception, or in the case of visual-spatial processing in a blindsight patient (Weiskrantz *et al.* 1974). Others can be monitored and signalled (by various speech acts or their equivalent); for example, the internal representations involved in imagery and certain percepts. The latter aspects, however, are never a carbon copy of earlier processing stages, since they require further processing which implies change—and even loss—of information which varies according to the selected output. This idea is at least as old as Comte's argument against introspectionism (see Boring 1950, p. 634; Marcel 1983). Thus, the task of psychologists and brain scientists is not to look for 'experiences' but to ascertain from observable behaviour the quantity and the form of information which can be addressed directly (i.e. consciously, in the C_2 sense) within a complex structure by different components, at different times, and for different purposes.

The modularity of consciousness

Data from investigation of brain-damaged subjects show dramatically how C_2 is far from being a unitary process (and how vain it would be to try to map C_1 onto C_2). Geschwind (1965, p. 638) remarked that Goldstein was perhaps the first to lay stress upon the significance (for the issue of the unity of consciousness) of the dissection of mental activities resulting from callosal lesions. Since then, the investigation of split-brain symptomatology (e.g. Gazzaniga 1988, this volume) has afforded a basis for the notion that both hemispheres are equipped with systems subserving C_2 functions, which can operate independently of one another. It might be suggested, and in fact it has been by Eccles (1976, and in Popper and Eccles 1977), that this duality is only apparent and that consciousness is a function of the left hemisphere, the right behaving much like an 'automaton'. Once again, this is a swift departure from scientific discourse, since C_1 is being referred to by Eccles. Indeed, the left hemisphere has the capability of expressing C_2 through language. But the illusion of its dominant role in consciousness, grounded on its verbal competence, ought to be soon dispelled if one thinks of the profound disorder of awareness which is manifest in the syndrome of unilateral neglect of space due to *right*-hemisphere lesions. When this disorder is severe, awareness of the left side of space—the side contralateral to the lesion—is grossly impaired. The patient even fails to form an adequate mental representation of it. For example, if required to form and describe the visual image of a familiar place, he may fail to report details relative to the left side of that image (Bisiach and Luzzatti 1978; Bisiach et al. 1979, 1981; see Bisiach and Berti 1987 for a comprehensive review). Brain structures unaffected by the lesion prove to be totally unable to register that there is a disorder. Comparable lesions of the left hemisphere do not produce such dramatic effects. This shows that right-hemisphere activities indeed play a primary role in the generation and monitoring of conscious brain events.

A further fractionation of C_2 functions within the activity of a single hemisphere is suggested by studies on anosognosia. It has long since been known (von Monakov 1885) that the failure of a specific brain function may be completely ignored by the patient if the responsible lesion involves what Kinsbourne (1980) calls, after Pavlov, its 'cortical analyser', that is, the structure where that function—e.g. vision—finds its highest processing level. Thus a patient may be selectively unaware of his blindness, or deafness, or hemiplegia if the different brain lesions which are responsible for one or the other of these disorders impair that level. At the same time, the patient may be fully aware of a concurrent disorder (e.g. of dysphasia) if this is due to the disruption of more peripheral processing stages. This fractionation of anosognosia into function-specific forms shows that monitoring of inner activity is not accomplished in the nervous system by a unitary, superordinate entity watching the workings of its slave mechanisms and able to detect faults in their operations as soon as they occur. Monitoring of inner activity is rather to be viewed as a function distributed across the different analysers to which it refers (Bisiach *et al.* 1986). Indeed, from a neuropsychological perspective, C_2 is more properly viewed as a collective name for a bundle of dissociable processes, conforming to Gazzaniga's 'sociological' concept of consciousness (see Gazzaniga and LeDoux 1978).

The view of consciousness as an entity distributed across the mosaic of individual analysers has particular significance in the case of processes working according to analog principles, such as those advocated by Kosslyn (1980) as a functional property of mental representations. This property seems to have an actual correspondence in the physiology of brain structures. As previously mentioned, phenomena of unilateral neglect show that a spatially circumscribed brain lesion may not only involve a spatially circumscribed *sensory* loss, revealing the analog structure of relatively peripheral sensory apparata, but may also involve a circumscribed lacuna in *representational* space. This may be true even for the representation of verbal items, as demonstrated by the inability to adequately spell the left half of words, observed by Baxter and Warrington (1983) and by Barbut and Gazzaniga (1987) in patients with left hemineglect. From these data we may infer that mechanisms subserving conscious representational activity which themselves operate as analogs are present in the brain. This topological correspondence between representational and neural space seems to constitute the more striking illustration of an identity of mental attributes such as consciousness and extensional properties of the brain: space can fractionate something which has extension, but not something which is unitary.

There are other phenomena which support the notion of the composite character of C_2 and present problems for the idea of a unifying role of language in the monitoring of mental occurrences, inherent in Eccles' view. Hilgard (1973) obtained a dissociation in outer manifestations of awareness of painful sensations through hypnosis, such that verbal denial of pain was in utter contrast with the contemporaneous written report of it. This fact, which seems to constitute an insurmountable paradox in C_1 terms, shows that a division of C_2 is possible, not only relative to the mosaic of sensory-specific analysers, but also with reference to the stock of responses among which an observable criterion for consciousness is selected: in some instances, awareness—as conveyed by speech—may turn out to be only a shallow mask hiding a much more substantial obliviousness. Clinical neurologists, in fact, are acquainted with anosognosic patients who manifest what Anton (1899) had called a 'dunkle Kenntniss' (dim knowledge) of their disorder; these are patients who verbally acknowledge their hemiplegia though engaging, at the same time, in impossible acts like standing and walking. A bright, alert patient of mine, for example, just admitted to the hospital following the sudden onset of a left hemiplegia of vascular origin, immediately after having complained of her inability to make any movement with the affected limbs asked her husband to bring her her knitting-needles, so that she could kill time while constrained to bed.

The timing of consciousness

A further aspect of the problem is the 'when' of consciousness, the particular slipperiness of which may, as in the case of the controversy about the interpretation of Libet *et al.*'s (1979) data, foster faulty conclusions. In an investigation still in progress, my co-workers and I have found that in right brain-damaged patients the detection of visual stimuli in the field contralateral to the lesioned hemisphere varied markedly according to the kind of response (verbal vs. motor) required of the patient. Stimuli were spots of light flashed by L E Ds (light emitting diodes). Trials were always initiated by the patient himself by pressing a central key, which produced either a single lateralized stimulus or two simultaneous stimuli, one on each side of the visual fixation point. On other trials no visual stimuli appeared. Two different responses were required in different blocks of trials. In one condition, the patient had to report the number of stimuli ('one', 'two', or 'none') verbally; in the other, he had to press a second key when a single stimulus appeared in either visual hemifield, whereas he had to refrain from reacting in case of double simultaneous stimulation or in the absence of any stimulus. As expected, in both conditions the patient showed an impaired performance in the visual hemifield contralateral to the side of the brain lesion: the detection of visual stimuli, virtually error-free in the right visual field, was markedly defective in the left; in other words, on many left single-stimulus trials the patient responded 'none' in the verbal condition or did not press the key in the motor condition. Furthermore, on double stimulation, the patients showed the well-known phenomenon of 'extinction': that is, the rate of detection of left-field stimuli declined if a concurrent stimulus was given in the right visual hemifield (in this case, the failure to detect the left stimulus was reflected by 'one' responses to double stimulation in the verbal condition and by faulty usage of the response key in the motor condition). However, the detection of left stimuli, in cases of both single and double stimulation, was significantly inferior for motor responses. This shows that there are processing stages in which information relative to the visual input may be heavily influenced by the nature of the required response.

At this point two interrelated questions may arise: (1) does the type of required response affect accessibility to consciousness even from relatively early stages of perceptual processing? Or else, (2) is it possible that a conscious stage is reached, independent of the response, after which awareness of the stimulus persists, or fades, or is 'repressed' depending on further processing?

Let us first consider a similar but even more crucial example in which the kind of response *was not predetermined* (Bisiach *et al.* 1985). The subjects were right brain-damaged patients who showed neglect of the left side of space. On each trial a single 200 msec flash of light appeared, either red or green and either to the right or left of fixation. Simultaneously, two response keys, one on each side of space, were lit—one red and one green. The subject's task was to press the response key of the same colour as the stimulus, irrespective of the side. Therefore, four possible stimulus–response mappings, two crossed and two uncrossed, occurred. One aspect of the results is relevant to the present context. Patient F S, who had an extensive right frontal lesion, was able to report verbally the colour of the flashing diode in his right visual hemifield (projecting to his unlesioned left hemisphere) on all trials of a preliminary session in which no motor responses were required. Furthermore, he gave accurate reactions to all right-stimulus trials whenever a *right* motor response was required. However, when the response required to right stimuli was on the *left* side, he gave no reaction at all on half the trials, sometimes spontaneously declaring that no stimulus had occurred.

Let us now return to the two questions stated above. The first, in C_2 terms, is an empirical question for which it is possible, at least in principle, to find an adequate answer. The second has a greater conceptual involvement. One might suggest that, in experimental conditions like those presented to patient F S, consciousness of a stimulus is only reached when a response is initiated, being back-referred, however, to the time of, say, the primary cortical potential evoked by the stimulus (since, if questioned, the patient would obviously assert that he was conscious of the stimulus before responding to it). Whenever a reaction is pre-empted, as in the case of responses toward a target located in the neglected half of space, there would be no consciousness of the stimulus. This suggestion may look implausible, since we are intuitively willing to admit that there can be consciousness of a stimulus in the absence of any overt response whatsoever. Therefore—the objection might continue—denial of supra-threshold stimuli in circumstances like those under consideration would be better interpreted as an inhibitory effect of the action mode which suppresses or overrides any experience of the stimulus after its fleeting appearance in consciousness and which prevents recovery of episodic memory of it.

There is no need to complicate the matter further, since the moral to be drawn is simple enough: if we try to rephrase the foregoing paragraph by substituting C_2 for 'consciousness', it would turn out to be rather nonsensical. In fact, according to the definition given on p. 103, C_2 is not susceptible to being 'back-referred in time'. Moreover, we infer its presence or absence on the basis of definite behavioural criteria; different criteria relate to different mechanisms for self-monitoring which may be selectively impaired. Absence of C_2 relative to visual stimuli was found in patient F S on trials in which the monitoring of the relevant information involved a disordered mechanism. The concomitant verbal denial can be explained by assuming that a fault in the relevant action system prevents any further processing (e.g. for phenomenal report) of that representation which would have been the basis of such a phenomenal report. There are other phenomena susceptible of an analogous explanation. In the case of optic aphasia, for instance, a motor response may be contaminated by the verbal disorder, so that the patient draws a misnamed object with some features derived from the wrong name (Lhermitte and Beauvois 1973).

The conclusion is that in trying to explain neuropsychological disorders we cannot turn from the Spinozian 'inner perspective' of the mental events lived by our subjects to the functionalist level. The only thing we can do is to ascertain indications of the states undergone by particular information at different loci and at different times of processing, without losing sight of the very important fact that introspective reports 'tend to force intrinsically parallel notions into a serial straitjacket' (Johnson-Laird 1983, p. 470).

The alleged 'functional' (causal) role of consciousness.

The boldest claims made today about consciousness relate to the specific function (or functions) which consciousness is purported to fulfil within the system of mental activities, and to its evolutionary significance.

Sperry (1977, p. 117) conceives of consciousness as a 'dynamic emergent of brain activity, neither identical with, nor reducible to, the neural events of which it is mainly composed'. In his view, consciousness is

... not conceived as an epiphenomenon, inner aspect, or other passive correlate of brain processes but rather to be an active integral part of the cerebral process itself, exerting potent causal effects in the interplay of cerebral operations. In a position of top command at the highest levels in the hierarchy of brain organisation, the subjective properties [are] seen to exert control over the biophysical and chemical activities at subordinate levels.

Thus, Sperry concludes, consciousness, heretofore 'explicitly excluded from the domain of science on materialist principles ... is now put within the province of science and is something that cannot be ignored where science wants an explanation of higher brain activities'.

Some critics might contend that this claim is the most recent fully fledged edition of the ghost in the machine. Were it actually such, it would be quite harmless, if it were not lent some apparent credit even *in partibus infidelium*. Dewan (1976), indeed, prompted by Sperry's views, suggested the 'virtual governor' emerging from interconnected a.c. generators (Wiener 1961) as a simile for a strictly physicalist account of a consciousness emerging as a holistic property of brain activity. However attractive, the simile fails in at least two respects: first, as argued in the preceding two sections, consciousness seems far from being a holistic property, and, second, the 'virtual governor' is a very nice instance of an emergent self-regulating mechanism, but—as Dewan (1976) himself seems ready to admit—to attribute 'sentience' to it would be a trick of science fiction.

The issue of the functional properties of consciousness has also been openly faced within the current paradigm of cognitive psychology. That a ghost—a ghost in the box—could even creep into the tidy information-processing flow-chart, might arouse some concern. This apprehension is not entirely groundless since the danger of the category mistake of implying a mapping of C_1 on C_2 might apply to arguments concerning the causal role of consciousness.

Shallice (1978) distinguishes three lines of speculation about consciousness in cognitive psychology. The first of these focuses prevalently on the perceptual aspects of consciousness (e.g. Marcel 1983) and is basically concerned with the nature of the step which would make a percept conscious—such that it is explicitly reportable in some way. This approach does not seem to be *necessarily* committed to take any stand regarding the question of the causal role of consciousness. The second approach relates the concept of consciousness to limited capacity processing mechanisms. It can be linked with the third, which hinges on the question of control. This last approach is proper to Johnson-Laird (1983) and to Shallice himself (1972, 1978), who more than other theorists has addressed the issue closest to the output side of mental processes.

In the models of both Shallice and Johnson-Laird the concept of consciousness is so much interwoven with the software that it raises the particularly pressing question of whether consciousness has any causal role *qua* phenomenal experience. A positive answer to the question is neither explicitly suggested nor denied by either author. It is however possible to check whether such an answer is in some way supported by their models.

According to Shallice, potential actions are governed by high-level programs (action systems) in multi-reciprocal competition. Anarchy is prevented by the dominance of a single, strongly activated action system which becomes dominant through inhibition of the competitors. Inputs to action systems come from perceptual and motivational systems. That which is taken as input by the dominant action system is said to correspond to consciousness. Consciousness would therefore refer to 'the dual functions of (a) providing the activation that enables the action-system to become dominant and of (b) setting its goal' (Shallice 1972, p. 390). An important specification, anticipating possible objections, relates to the existence of action systems specific to perception and presiding over actions such as 'looking for' or 'listening for', which do not necessarily have overt behavioural manifestations (Shallice 1978, p. 140).

In Johnson-Laird's (1983) model a system of parallel processors carries out non-conscious mental processes which may end in deadlock or other pathological configurations. A superordinate operating system intervenes to prevent chaos by channelling parallel processes into the serially ordered stream required for the adequate patterning of goal-directed behaviour. The operating system corresponds to consciousness.

In the models outlined by Shallice and by Johnson-Laird there is thus a particular event (1) exerting a crucial causal role upon subsequent events and (2) characterized by the peculiar attribute of consciousness. But how are these two properties related? Is consciousness itself a *prerequisite* for the operation of selection, which in both models constitutes the critical episode, or a *consequence* of the selection, by which the critical property, distinguishing conscious from non-conscious processing, is merely created? In the first case it would have a causal status, whereas in the second it might be regarded as epiphenomenal. The first alternative seems to run into a web of difficulties. There are sudden, inconsequential changes in the 'stream of consciousness' whereby new representations and thoughts intrude, unwilled and unheralded by any related content (see Oatley 1988, this volume). Accepting the thesis of selection by consciousness, such an erratic functioning would imply an enigmatic consciousness placed like Janus Bifrons astride the boundary between two separate fields, acting as a filter of its own contents and then feigning innocence.

Another problem for ascribing causal properties related to the selection and serial ordering of cognitive processes to consciousness is constituted by some phenomena of sonnambulism and of epileptic 'automatism', as well as by instances of problem-solving activity occurring during sleep, such as the well-known case of the invention of the benzene ring by von Kekulè. Even more problematic are those circumstances in which the chain of events conducive to overt behaviour seems to run across different states of consciousness, as might be the case in the post-hypnotic accomplishment of tasks imposed during hypnosis or in the following anecdotal observation relative to an epileptic patient whom I was able to observe several years ago. The patient, who was afterwards found to show a left frontal epileptogenic focus on EEG, had frequent logoclonic seizures associated with a slight rightward turning of the head. The seizures (bursts of stereotyped repetition of words) occurred during normal speech and lasted for several seconds, after which the patient resumed her talk as if nothing had happened; she denied any abnormality in her behaviour and looked visibly disconcerted by the examiner's questioning. Since this contrasted with the fact that at the time of the seizures she did not show the vacant stare peculiar to epileptic absence, a pen was noticeably put inside the breast-pocket of her dressing-gown during one of the attacks. When it was over, the patient on request gave back the pen quite promptly and naturally. Much more decisive are the experiments carried out by Marcel and his co-workers (reviewed in Marcel 1983) and by Allport *et al.* (1985), which show that higher level semantic processing may occur, which remains totally opaque to subsequent conscious states upon which it has a definite priming effect. So, if we consider a chain of conscious episodes (thoughts, images) it is clear that any one of them is shaped not only and not always by preceding conscious episodes.

Therefore, it seems that neither Shallice's nor Johnson-Laird's models imply a definite causal role for consciousness. Moreover, the appendices to Shallice's 1972 and 1978 papers relative to the mutual entertainment of action systems by inhibitory loops, show that the dynamics which promote the emergence of a dominant action system may be conceived in all likelihood as non-conscious. This deduction seems strongly corroborated by Shallice's (1978, p. 147) admission of non-conscious highest-level control devices, whose explicit admittance in a comprehensive information-processing diagram of brain activity would be welcome. On the other hand, after careful examination of the supervisory mechanisms whose candidature for consciousness he contemplates, Johnson-Laird (1983, p. 474) even comes to question the actual ascribability of consciousness to the functions at issue. This is very far from postulating a causal role of consciousness in mental activity.

By asserting that 'Any scientific theory of the mind has to treat it as an automaton', Johnson-Laird (1983, p. 477) seems to take a position similar to that expressed at the beginning of the second section about the pertinence of C_1 to natural science. Although this position maintains that a causal role of C_1 is scientifically unthinkable, and dispels the ghost of a disembodied consciousness hanging in the IP flow-chart box like the Cheshire cat's smile, it leaves open the question of the causal role of C_2.

If the arguments put forward in the foregoing paragraphs are valid, much of the evolutionary advantage which is often predicated for 'consciousness' must be attributed to mechanisms which *entail* consciousness (C_2) as one of the properties of a state of information being processed by the brain; the property, for example, of being able to be spoken about, etc. Of course, however engendered, this property is not devoid of consequences. The most obvious refers to intersubjective communication. More problematic, but no less crucial, is the use of conscious representations for intrasubjective communication; that is, for those thought processes of which the recursive procedures described by Johnson-Laird under the rubric of self-awareness are an instance: the data base on which such procedures are executed, indeed, is in Johnson-Laird's model conscious (C_2) by definition.

The functional importance of the recruitment and inner utilization of a conscious data base is far from clear. A considerable amount of recursive computation similar to that considered by Johnson-Laird in relation to conscious processes might occur non-consciously, e.g., in the activity preceding the choice of a move in chess. It has not been clarified to what extent pathological loss of conscious access to an otherwise available data base leads to deterioration in cognitive activities (see, in this volume, Marcel 1988 and Weiskrantz 1988). In the case of loss of visual imagery, I had the opportunity of observing one of these patients (Basso *et al.* 1980). Following a left-hemisphere infarction involving the occipital and temporal lobes, he complained of having lost all forms of visual imagery, including hypnagogic images and dreams. His disability did not seem to affect his everyday life at all, with one puzzling exception. If he tried to represent explicitly to himself the setting in which he had to perform an activity, no matter how habitual, e.g. looking for what is needed to set the table, he would get lost and have to start again trying to accomplish the task more automatically.

This notwithstanding, it would be quite implausible to maintain that conscious (C_2) qualities remain mainly epiphenomenal. Therefore, their role and evolutionary advantage can safely be countenanced and investigated without extravagant resorts to phenomenal (C_1) principles.

For fiscal reasons, the staff of whisky distilleries have no access to the spirit throughout its processing. There are however 'windows' which allow monitoring of critical phases, so that decisions can be taken to modify further processing to obtain the desired results. Although calling it 'causal' might not be fully appropriate, the role of such inspection equipment is crucial. Some monitoring of brain events might likewise be crucial for adaptive behaviour, even in cases in which the choices taken on its basis happen to follow non-consciously.

I will not go deeper into the problem of a functional account of consciousness, since this would exceed the scope of this chapter.

Conclusions

The study of consciousness has considerably intensified during the last decade or so, but the state of the art is such that whoever ventures into the area for the first time is likely to feel much as Pierre Besuchov, the well-known character from *War and Peace*, on the Borodino battle field. Two main factors contribute to the initial dismay. First, the approach to the issue is basically multidisciplinary and the involved parties differ considerably in background, area of expertise, and personal as well as group bias. In the second place, the same problems are simultaneously presented as non-existent, or insoluble, or happily solved in a number of contrasting ways. This might contribute to perpetuating the halo of mystery surrounding consciousness.

Yet it is not perfectly clear whether (and, if so, why) consciousness should be any more mysterious than a stone. It has one aspect which is suitable for scientific treatment, while another stands totally and firmly outside of it, which is no less real nor natural. Knowledge of the former (C_2) has considerably advanced with the advent of cognitive psychology and neuropsychology; its further development will perhaps help us to better realize just how unatteinable is the latter (C_1).

We may have to learn how to live with the idea that some of the questions set by commissurotomy, blindsight, unilateral neglect of space, etc. will remain forever unanswerable: without direct experience we will never know what is it like to be a patient affected by unilateral neglect.

Acknowledgements

I wish to thank Dan Dennett and Tim Shallice who read an earlier version of this chapter and provided comments which aided in refining and rewriting many parts of it. I am also grateful to Frances Anderson and Tony Marcel who carefully reviewed the English. Above all I thank Tony Marcel for helping me to clarify some of my thoughts to myself.

References

Allport, D. A., Tipper, S. P., and Chmiel, N. R. J. (1985). Perceptual integration and postcategorical filtering. In *Attention and performance XI* (ed. M. I. Posner and O. S. M. Marin), pp. 107–32. Lawrence Erlbaum Associates, Hillsdale, NJ.

Angell, J. R. (1904). *Psychology: an introductory study of the structure and functions of human consciousness*, p. 1. Holt, New York.

Anton, G. (1899). Ueber die Selbstwahrnehmung der Herderkrankungen des Gehirns durch den Kranken bei Rindenblindheit und Rindentaubheit. *Archiv für Psychiatrie und Nervenkrankheiten*, 32, 86–127.

Barbut, D. and Gazzaniga, M. S. (1987). Disturbances in conceptual space involving language and speech. *Brain*, 110, 1487–96.

Basso, A., Bisiach, E., and Luzzatti, C. (1980). Loss of mental imagery: a case study. *Neuropsychologia*, 18, 435–42.

Baxter, D. M. and Warrington, E. K. (1983). Neglect dysgraphia. *Journal of Neurology, Neurosurgery and Psychiatry*, 46, 1073–8.

Bisiach, E. and Berti, A. (1987). Dyschiria. An attempt at its systemic explanation. In *Neurophysiological and neuropsychological aspects of spatial neglect* (ed. M. Jeannerod), pp. 183–201. North-Holland, Amsterdam.

Bisiach, E. and Luzzatti, C. (1978). Unilateral neglect of representational space. *Cortex*, 14, 129–33.

Bisiach, E. Luzzatti, C., and Perani, D. (1979). Unilateral neglect, representational schema and consciousness. *Brain*, 102, 609–18.

Bisiach, E., Capitani, E., Luzzatti, C., and Perani, D. (1981). Brain and conscious representation of outside reality. *Neuropsychologia*, 19, 543–51.

Bisiach, E., Berti, A., and Vallar, G. (1985). Analogical and logical disorders underlying unilateral neglect of space. In *Attention and performance XI* (ed. M. I. Posner and O. S. M. Marin), pp. 239–49. Lawrence Erlbaum Associates, Hillsdale, NJ.

Bisiach, E., Vallar, G., Perani, D., Papagno, C., and Berti, A. (1986). Unawareness of disease following lesions of the right hemisphere: anosognosia for hemiplegia and anosognosia for hemianopia. *Neuropsychologia*, 24, 471–82.

Boring, E. G. (1950). *A history of experimental psychology*. Appleton-Century-Crofts, New York.

Churchland, P. S. (1981). On the alleged backwards referral of experiences and its relevance to the mind–body problem. *Philosophy of Science*, **48**, 165–81.

Dennett, D. C. (1978). *Brainstorms*. MIT Press, Cambridge, MA.

Dennett, D. C. (1988). Quining qualia. In *Consciousness in contemporary science*, (ed. A. J. Marcel and E. Bisiach), p. 42. Oxford University Press.

Dewan, E. M. (1976). Consciousness as an emergent causal agent in the context of control system theory. In *Consciousness and the brain. A scientific and philosophical inquiry* (ed. G. G. Globus, G. Maxwell, and I. Savodnik), pp. 181–98. Plenum Press, New York.

Eccles, J. C. (1976). Brain and free will. In *Consciousness and the brain. A scientific and philosophical inquiry* (ed. G. G. Globus, G. Maxwell, and I. Savodnik), pp. 101–21. Plenum Press, New York.

Gazzaniga, M. S. (1988). Brain modularity: towards a philosophy of conscious experience. In *Consciousness in contemporary science*, (ed. A. J. Marcel and E. Bisiach), p. 218. Oxford University Press.

Gazzaniga, M. S. and LeDoux, J. E. (1978). *The integrated mind*. Plenum Press, New York.

Geschwind, N. (1965). Disconnexion syndromes in animals and man. *Brain*, **88**, 237–94, 585–644.

Hilgard, E. R. (1973). A neodissociation interpretation of pain reduction in hypnosis. *Psychological Review*, **80**, 396–411.

Hilgard, E. R. (1977). Controversies over consciousness and the rise of cognitive psychology. *Australian Psychologist*, **12**, 7–26.

Hilgard, E. R. (1980). Consciousness in contemporary psychology. *Annual Review of Psychology* **31**, 1–26.

Hoy, R. R. (1982). Ambiguities in the subjective timing of experience debate. *Philosophy of Science*, **49**, 254–62.

Johnson-Laird, P. N. (1983). *Mental models*. Cambridge University Press.

Kinsbourne, M. (1980). Brain-based limitations on mind. In *Body and mind. Past, present and future* (ed. R. W. Rieber), pp. 155–75. Academic Press, New York.

Kling, J. W. and Riggs, L. A. (1971) *Woodworth and Schlosberg's experimental psychology*, (3rd edn). Methuen, Andover, Hants.

Kosslyn, S. M. (1980). *Image and mind*. Harvard University Press, Cambridge, MA.

Lhermitte, F. and Beauvois M-F. (1973). A visual-speech disconnexion syndrome. Report of a case with optic aphasia, agnosic alexia and colour agnosia. *Brain*, **96**, 695–714.

Libet, B. (1981). The experimental evidence for subjective referral of a sensory experience backwards in time: reply to P. S. Churchland. *Philosophy of Science*, **48**, 182–97.

Libet, B., Wright, E. W., Feinstein, B., and Pearl, D. K. (1979). Subjective referral of the timing for a conscious sensory experience. *Brain*, **102**, 193–224.

Mandler, G. (1975). Consciousness: respectable, useful, and probably necessary. In *Information processing and cognition. The Loyola symposium* (ed. R. L. Solso), pp. 229–54. Lawrence Erlbaum Associates, Hillsdale, NJ.

Marcel, A. J. (1983). Conscious and unconscious perception: an approach to the relations between phenomenal experience and perceptual processes. *Cognitive Psychology*, **15**, 238–300.

Marcel, A. J. (1988). Phenomenal experience and functionalism. In *Consciousness in Contemporary Science*, (ed. A. J. Marcel and E. Bisiach), p. 121. Oxford University Press.

Maxwell, G. (1976). The role of scientific results in theories of mind and brain: a conversation among philosophers and scientists. In *Consciousness and the brain. A scientific and philosophical inquiry* (ed. G. G. Globus, G. Maxwell, and I. Savodnik), pp. 317–28. Plenum Press, New York.

Nagel, T. (1974). What is it like to be a bat? *The Philosophical Review*, **83**, 435–50.

Oatley, K. (1988). On changing one's mind: a possible function of consciousness. In *Consciousness in contemporary science*, (ed. A. J. Marcel and E. Bisiach), p. 369. Oxford University Press.

Popper, K. R. and Eccles, J. C. (1977). *The self and its brain. An argument for interactionism*. Springer, Berlin.

Ryle, G. (1963). *The concept of mind*. Penguin Books, Harmondsworth, Middx.

Shallice, T. (1972). Dual functions of consciousness. *Psychological Review*, **79**, 383–93.

Shallice, T. (1978). The dominant action system: an information-processing approach to consciousness. In *The stream of consciousness. Scientific investigations into the flow of human experience* (ed. K. S. Pope and J. L. Singer), pp. 117–57. Plenum Press, New York.

Sperry, R. W. (1977). Forebrain commissurotomy and conscious awareness. *The Journal of Medicine and Philosophy*, **2**, 101–26.

Tolman, E. C. (1932). *Purposive behavior in animals and men*. Appleton-Century-Crofs, New York.

Tolman, E. C. (1948). Cognitive maps in rats and men. *Psychological Review*, **55**, 189–208.

Underwood, G. and Stevens, R. (1979). Preface. In *Aspects of consciousness*, Vol. 1, (ed. G. Underwood and R. Stevens), pp. vii–ix. Academic Press, New York.

Von Monakow, C. (1885). Experimentelle und pathologisch–anatomische Untersuchungen ueber die Beziehungen der sogenannten Sehesphaere zu den infrakorticalen Opticuscentren und zum N. Opticus. *Archiv für Psychiatrie und Nervenkrankheiten*, **16**, 151–99.

Weiskrantz, L. (1988). Some contributions of neuropsychology of vision and memory to the problem of consciousness. In *Consciousness in Contemporary Science*, (ed. A. J. Marcel and E. Bisiach), p. 183. Oxford University Press.

Weiskrantz, L., Warrington, E. K., Sanders, M. D., and Marshall, J. (1974). Visual capacity in the hemianopic field following a restricted occipital ablation. *Brain*, **97**, 709–68.

Wiener, N. (1961). *Cybernetics*. MIT Press, Cambridge, MA.

Wilkes, K. V. (1978). *Physicalism*. Routledge and Kegan Paul, Andover, Hants.

Wittgenstein, L. (1953). *Philosophical investigations*, (trans. G. E. M. Anscombe). Blackwell Scientific Publications, Oxford.

6

Phenomenal experience and functionalism
Anthony J. Marcel

This essay is about consciousness as phenomenal experience. Its contention is that reference to consciousness in psychological science is demanded, legitimate, and necessary. It is demanded since consciousness is a central (if not *the* central) aspect of mental life. It is legitimate since there are as reasonable grounds for identifying consciousness as there are for identifying other psychological constructs. It is necessary since it has explanatory value, and since there are grounds for positing that it has causal status. However, the relationship of certain aspects of consciousness to the functionalist approach, which currently dominates and unites natural science, is problematic. Those aspects discussed here are phenomenal experience and content. Either functionalism will be able to deal with the problems posed, or a purely functionalist psychology will be inadequate. Psychology without consciousness, without phenomenal experience or the personal level, may be biology or cybernetics, but it is not psychology.

Why discuss consciousness?

Why discuss consciousness at all? To any ordinary person this may seem a bizarre question since any aspect of mind would seem to be within the domain of enquiry of psychologists and philosophers of mind, and consciousness is such an aspect. Yet, as Dennett (1981) has pointed out, a historical reversal has come about in our theoretical attitudes to consciousness. At the time of Locke mind was virtually synonymous with conscious mentation; the notion of non-conscious mentation was inconceivable, if not meaningless. However, currently most theories of cognition make no call at all upon consciousness. In artificial intelligence models, the processing of information and the execution of tasks are carried out by systems for which no one would want to claim any phenomenal experience or subjectivity. In most psychological modelling of human cognition no mention is made of consciousness or intention. The words may be used and the rest of what is modelled may depend on them, but no serious analysis of their roles is given. Information-processing theorists seem to regard them as unnecessary or not their concern. Several authors who claim to be discussing consciousness turn out to be discussing some function (e.g. control, storage) whose necessary connection with phenomenal experience or awareness is not at all apparent.

The reason for this state of affairs may well lie in the major approach that unites artificial intelligence and psychology, and indeed certain levels of biology. What unites the various fields that constitute what is currently called cognitive science is functionalism or at least one version of it. Of course there are people in psychology, biology, and linguistics who are not functionalists. Functionalism (or perhaps more properly *Turing machine functionalism*) maintains that every mental or physical assertion is expressable in an abstract and physically neutral language that designates functions and functional relations. The functionalist level of discourse is clearly useful and even necessary for psychology, biology, and computation since it does not commit one to particular instantiations, especially not to particular physical instantiations. The fact that functionalist discourse is also independent of what it deals with, computes on, or represents is seen as an advantage by some, but as a fallacy by others, particularly by those who see the mental as essentially intentional (see the last section of this chapter; and Van Gulick 1988, this volume).

Now functionalist models may provide some confusion. There are many who feel that a computationally realized model will not exhibit phenomenal experience. One answer to this is 'Why should it?' We do not expect a computational model of the weather to actually produce wetness when it produces the functional equivalent of rain. This is indeed Johnson-Laird's answer (personal communication). But it is not clear that this is either an appropriate answer or an appropriate analogy. It is not an appropriate answer because in most models there is no attempt to model consciousness or any equivalent, while in meteorological models there is an attempt to model precipitation or its equivalent. It is not an appropriate analogy, because while it would seem to imply that a (particular) instantiation, presumably physical, would exhibit the property, we have no idea how to verify this. Further, no answer is given to those who maintain that phenomenal experience is one of the properties that distinguish the mental from the physical, or who at least enquire what principle yields phenomenal experience from physical systems. Thus it may be that the reason for the absence of consciousness is due to the assumptions of functionalists. Of course, there are those (Dennett is a good example) who do not make such assumptions, but attack in a most thoughtful fashion the very requirement to address phenomenal experience and the concept itself. But this is not true of the vast majority who are actually engaged in research on and theorizing about human behaviour. In fact I suspect that most psychologists and computational theorists have nothing to say about such concepts or issues, probably because they have not thought about them or do not believe that they are of any importance, or believe that some aspect of them, privateness or subjectivity, prevents them from being studied or debars them from 'science'. Therefore it is not merely rhetorical to ask the question 'Why discuss consciousness?' Four responses to this question will be given here.

By 'mental life' we mean conscious *mental life*

The title of a book published by Miller in 1964 was 'Psychology: the science of mental life'. This seems a reasonable, if not exhaustive, indication of the domain of psychology. Ordinary people, and indeed psychologists, want accounts of and explanations of their mental life. Even if such accounts ultimately rest upon underlying non-conscious mechanisms, what one is being called upon to explain are aspects of consciousness: percepts, memories, emotions, dreams. Indeed one's *mental life* consists in the projects, aims, worries, pains, personal relationships, feelings about self *of which we are conscious*, or of which we can become conscious. It is upon this level of existence that we base our intended interactions with the world and our attempts to change it. This statement is not at all inconsistent with the notion that much, perhaps most, of the mind (or at least what contributes to our behaviour and its form) is non-conscious. As Lashley (1958) pointed out, the processes that realize percepts, problem-solving, memory are not themselves conscious. In addition, most behaviour, which is the legitimate concern of psychology, is of itself irrelevant to consciousness. However, these phenomena are not what we call 'mental life'. The relationship of the psychodynamic unconscious to consciousness is slightly different. Its description is in terms no different from our conscious mental life: we can become conscious of it, it has been conscious, and, were it not for mechanisms of denial, it would constitute our conscious experience. But to return to the centrality of consciousness, if there are no expectations of a person in coma on a life-support system regaining consciousness (i.e. a state in which there is evidence of phenomenal awareness), we question the point of their continued life. Clearly, at the very least, no account of mind which omits consciousness can be a complete account. The next response suggests that it would not even be an adequate account.

Consciousness has causal status

Consciousness would deserve attention even if it were an epiphenomenon, if it had no effect on behaviour, since it is such a preeminent
characteristic of our minds. Recently, however, psychologists have
suggested several ways in which consciousness plays a role in our
behaviour. Mandler and Kessen (1974) have suggested that it underlies
the solving of novel problems, and Shallice (1972) has related it to the
fact that our behaviour is coherent and organized. But these suggestions
have largely been expressed by identifying consciousness with functionalist concepts which do not seem to imply phenomenal awareness.
Such concepts would be computationally instantiable without subjective experience or direct knowledge of their operation by the system in
which they were embodied. It is my contention in this chapter that
phenomenal experience itself, even if it is epiphenomenal to some
functions, has effects. A later section will attempt to specify and discuss
several ways in which phenomenal experience both is essential to our
minds as they are and has consequences for our behaviour. There is no
suggestion that it is essential to human behaviour as a whole; we could
survive (in a simpler world) without it. But our behaviour would be quite
different from what it is if we did not have phenomenal experience; we
could not live in this socially and technologically arranged world. Even
if it plays its role because of our beliefs, that makes it no less causal.

Consciousness is what psychologists actually examine

Perceptual psychologists adhering to entirely functionalist approaches
concern themselves either with the role or nature of sensory information (Gibsonians) or with the nature and structure of the mechanisms of registration and description of sensory information (information
processing and artificial intelligence). Such approaches need not
(according to their paradigm assumptions) and often do not concern
themselves with phenomenal experience. However, almost all other
psychologists interested in perception, even if they do not admit to it,
are intimately concerned with consciousness. Phenomenal experiences
(or strictly speaking, introspective reports of phenomenal experiences)[1]
are their primary data. Their dependent variables are aspects of what one
sees, hears, feels consciously. Alternatively they use response measures
based on and reflecting the conscious percept. Button presses represent,
or can be used to perform the same function as, introspective speech
acts, conveying that 'this one appears to be *x*' or 'this is when I think that

---

[1] Throughout this essay, when phenomenal experience is discussed *as data*, what is
being referred to is introspective report, conveyed verbally or otherwise.

y'. As I have pointed out elsewhere (Marcel 1983*b*), subjects in experiments where they have to respond to stimuli will not do so *willingly* if such stimuli are not consciously apprehendable to them. Thus even though most theoretical interest has focused on non-conscious aspects of processing, most of the measures used require the subject to be conscious and they rely on and reflect conscious percepts. When we discuss perceptual illusions in normal people, and disordered perception in clinical populations, we are discussing conscious perception. So, whether we like it or not, much of the data we have from perceptual psychologists for the last 100 years is about consciousness. This applies to much research on memory as well. Much of our data concerns either what people consciously remember or what they are conscious of *as memories*. Given the recent work on amnesia, it is not clear to exactly what extent the problem is one of experiencing memories *as memories* for the amnesic. Both Tulving (1985) and Weiskrantz (1988, this volume) acknowledge this. To treat the effects of past events on organisms, their storage of information or their ability to reproduce what has been put into them under the term 'memory', while legitimate, applies to many inanimate objects as well. But certainly episodic memory, intentional recall, and reminiscence are all phenomena that are essentially conscious. As regards language comprehension, Tyler (1988) argues that what many of our current techniques measure is conscious comprehension. Again, emotions are only emotions by dint of their phenomenal qualities. Without such categorical experience (*which* emotion it is) there is only bodily disturbance or excitement (though probably a non-conscious cognitive component is often what produces that disturbance or excitement; e.g. the recognition of a situation as belonging to a particular class such as threat or caring). Now it is clear that not all of what psychologists examine is really some aspect of consciousness or dependent on awareness. For example, we can exempt the execution (as opposed to initiation) of a motor skill, or habitual procedures. However, since even some motor skills require consciousness for their scheduling, successful non-conscious processes may depend on consciousness (see, in this volume, Umiltà 1988 and Johnson-Laird 1988). In sum, what is being said here is that in many cases we should acknowledge that what we are studying either is or involves consciousness. Such an acknowledgement would not only legitimize but require us to discuss consciousness.

The concept has ideological implications

Scientific concepts exist in a social context. On the one hand they reflect the conceptual structures and attitudes of the culture, as concepts of hierarchical structure in so much of nineteenth-century science reflected the ideas of both Romanticism and the Enlightenment. On the other hand they contribute to and are used in social doctrines. Just as some reasons for *not* discussing consciousness are ideological (consider the relationship of industrial Taylorism to behaviourism, or Stalinism's treatment of Luria as opposed to Pavlov), so other reasons *for* discussing consciousness are also ideological. It is not yet clear exactly what ideological status is conferred on us by the ascription of consciousness, because we do not have anything like an adequate view of consciousness. But I do suspect that views of beings without consciousness have certain consequences. (And such a view is implied by a psychology that aspires to a complete account of its topic but makes no mention of consciousness.) Non-conscious beings by definition do not have feelings, do not have pain. And how we are willing to behave to other beings usually depends on this ascription. Hoffman (1982) has argued that empathy (experiencing the feelings of another) is the basis of our social morality and our justice systems. Certainly legitimation of adverse treatment of others has often been achieved by conceiving of them as categorically different, and in the case of animals by denying that they have phenomenal experience. Empathy is of its essence conscious. Underlying it is knowing or imagining 'what it is like' to be a human being or the particular other in question (Nagel 1974). In addition, non-conscious beings are not agents, at least in the sense that they are not held responsible for their behaviour. Whether or not we become like the images we are given of ourselves (given by, for example, 'scientists') is debatable, but images of ourselves as essentially machine-like hardly seem likely to lead to behaviour and attitudes consonant with liberal, humanist ideals. The natural science tradition of treating its subject matter as objects licences the treatment of experiencing agents as objects.[2]

The natural science approach has not so far dealt adequately with aspects of mind such as content, subjectivity, meaning. (This problem is returned to in a later section). If one wishes to have an adequate approach to mind these will have to be tackled and one cannot avoid

---

[2] Note that I am not suggesting that the assumption of consciousness would alone and of itself prevent undesirable manipulation or treatment of others and submission to such behaviour. Indeed, regimes and ideologies that elevate consciousness have been responsible for terrible atrocities (e.g. Fascism, the Inquisition). If one actually wants to make someone else suffer, one can only do so if one assumes that the other person is sentient. Nor will the assumption of a lack of consciousness necessarily lead to undesirable social behaviour. However, while the assumption of consciousness may not be a *sufficient* condition to ensure desirable social arrangements, it is surely a *necessary* condition. It is not only the treatment of animals and patients under general anaesthetic that devolves on the issue of their phenomenal consciousness, but that of almost any marginalized group in society, such as those classified as insane or racially distinct or 'them' as opposed to 'us' (Thomas 1984).

them if one focuses on consciousness. A completely natural science type of psychology implicitly denies any meaning to what it studies. Most of the important personal decisions we make are based upon views of our personal and social worlds as meaningful. An attempt to deal with consciousness, even if it explains meaning in other terms, need not be inconsistent with the assumption (of meaning) with which our lives are led, as long as it does not ignore meaning. Such an attempt would at least have the chance of being intelligible to ordinary people, since it would relate to their experience of themselves. An entirely natural science psychology which deals only with public, contentless phenomena has little such chance. Further, it is doubtful whether people could *use* such a science as effectively as the former. Quite seriously, the 'inner-game' type of instruction in perceptual motor skills, based on inner phenomenology rather than outer observation of oneself, is infinitely more effective than conventional instruction (e.g. in skiing). This is also true of biofeedback. Even if augmented external feedback aids in identifying or becoming conscious of the inner sensation, it is on the inner sensation itself that one comes to base one's response. It is much easier for people to base personal actions on concepts that relate directly to their own experience than on those which do not. In one case we might have an applied psychology that people can use, in the other case we have an applied psychology that is used on people.

The topic: what it is and how we can know it

Some reasons have been given why psychology needs to treat consciousness as a topic proper to its domain of theory and investigation. However, this immediately raises three issues. First, we need to indicate what is and is not being referred to by the term. While Weiskrantz's point (1988, this volume) is well taken that a definition should be the end point rather than the starting point, and while this requires an adequate theoretical framework which does not yet exist, none the less it will help to at least clarify what is under discussion in this chapter. If nothing else this ought to avoid unnecessary confusion. The second issue concerns the criteria by which we are to recognize the occurrence of consciousness. Since we are referring to something which is internal and not itself an overt piece of behaviour, this is of some importance. Third, if something is to be a topic of psychology, those psychologists who conceive of themselves as natural scientists will ask what can be known about it in an objective sense. Now there are many hypothetical constructs in science which themselves are not directly observable or measurable. And there are also plenty of concepts (especially in physics) which are demanded purely for their explanatory value. But it would certainly help if phenomenal experience were not quite as ineffable or unmeasurable as some suppose it to be. This section attempts to address these three issues. What is offered is not complete and is inevitably contentious.

Clarifying the referent

Natsoulas (1978) has distinguished seven concepts of consciousness. However, what is being referred to here by the term consciousness is primarily phenomenal experience. This includes sensation, but covers rather more than what Tolman (1932) called 'raw feels', and what in the philosophical literature are called 'qualia' (Dennett 1988, this volume). It is that which, over and above information or internal representation, we refer to as known directly or non-inferentially when we report our states or feelings. Whether it *is* directly or non-inferentially known is another question; but it *seems* so—and this seeming is the most important point. It is our experience and this is essentially subjective. When someone says they have a headache or are thirsty, they are referring to something, and what they are referring to is not, say, the expansion of cerebral blood vessels or a state of the stomach or hypothalamus, but rather a certain type of sensation. There are two reasons for focusing on phenomenal experience as the main referent of consciousness. First, if we did not have phenomenal experience we would not have a concept of consciousness at all. This is because we know it in a first-person rather than a third-person manner, by acquaintance rather than by description, whereas Bisiach's 'C$_2$' (1988, this volume) or equivalent functionalist concepts such as 'working memory' are known only in a third-person manner. Phenomenal experience is thus the *raison d'être* of the concept of consciousness. [But see Wilkes (1988, this volume) for the historical and linguistic relativity of this relation.] Second, phenomenal experience is that aspect of consciousness which is the main stumbling block to a totally functionalist or natural science approach to mind (Nagel 1974).

Now it is difficult to give a satisfactory objective referential definition of awareness or phenomenal experience. This invites the type of critical approaches voiced in Allport's (1988) and Dennett's (1988) papers (both in this volume), which are most useful in sharpening our usage, but are also much easier than the converse, the positive definition. However, we can sharpen what we are referring to by contrastive distinctions. Indeed such distinctions are necessary to avoid confusions, many of which abound in the recent literature.

The first distinction is between non-reflexive and reflexive consciousness. Traditionally, non-reflexive consciousness refers to phenomenal experience or sensation, while reflexive consciousness refers to (at least) two slightly different things: (a) our awareness of our phenomenal experience, our considerations of our sensations, and (b) our awareness of self. Is reflexive consciousness really another type or sense of consciousness? This question can be considered to cover also the concept of self-consciousness, which Oatley (1988, this volume) treats as separate. In the first case, where we are referring to knowledge of or consideration of one's self or one's sensations or actions (past, present, or future), we are really discussing knowledge or thought, irrespective of how the knower or thinker comes to know about the topics. This is important, since, as is discussed elsewhere, that which is known in the form of phenomenal experience is known in a direct first-person way, whereas we can also know things in a third-person way, by learning of them indirectly. In the second case, where we are referring to awareness of self, it is just that the object or focus of phenomenal experience is the self or one's own sensations. This is not a different *type* of experience, it is only a matter of the object of its focus. Thus in many perceptual-motor skills one's performance alters if one focuses attention outside of oneself 'spatially', rather than on one's own feelings and movements (see Duval and Wicklund 1972; Csikszentmihalyi 1978). It would seem, then, that there is no special category or type of consciousness that is reflexive, but rather that one is either referring to knowledge or to a difference in content. However, it will be proposed in a later section, as in an earlier paper (Marcel 1983*b*), that the nature of phenomenal experience depends on the conceptual knowledge that we entertain about ourselves.

Several further distinctions need clarifying. The distinctions are between dichotomies which are often loosely and mistakenly taken to be interchangeable or coextensive. The first and most obvious distinction is between (a) consciousness in the current sense of phenomenal awareness, whose conterpart is that of which at any moment we are not aware and does not have phenomenal status, and (b) the psychodynamic or psychoanalytic use of the term, whose counterpart is the dynamic or systemic unconscious. The first difference between these dichotomies is that while what is phenomenally non-conscious may be only perceptual information, which is momentarily being processed, an individual's dynamic unconscious refers not merely to his knowledge of the world, but includes longstanding tacit beliefs, interpretations of experiences in terms of their 'significance', and a set of relationships and dispositions (desires, fears, beliefs) to prototypical characters, entities, and situations which give them their significance. Such concepts and entities are supposedly represented unconsciously in symbolic terms which underlie the language and semiology of our psychodynamic lives. By contrast, such characterizations cannot necessarily be made of all that which, for cognitive scientists, is not phenomenally conscious. The second difference between these two dichotomies is that the psychodynamic unconscious is itself often conceived of in the same representational terms as consciousness, with its own intentionality and its own phenomenology; i.e. as another consciousness but hidden from our reflexive consciousness in the sense of our knowledge. This can be illustrated by the readiness of psychodynamicists to talk of unconscious pain, which is a phenomenological term. To talk of non-conscious pain in our current usage is a contradiction in terms. The third difference is that, for the psychoanalyst, to be conscious of something of which we were previously unconscious can amount to explicit vs. tacit knowledge (i.e. acknowledgement that something is the case; see below), whereas the phenomenological sense requires a direct phenomenal experience.

Another dichotomy which is often taken as coextensive with conscious–non-conscious is that of explicit vs. tacit knowledge (Polanyi 1966). Tacit knowledge is that knowledge which one does not know that one has. It underlies most of our activities. Examples in language are use of presuppositions, Gricean rules of conversation (Grice 1967), and responses to indirect speech acts; an example from motor skill is our ability to keep a bicycle upright while turning at different speeds. Polanyi (1966) makes the point that the representational form of explicit knowledge is often qualitatively different from its tacit counterpart. Thus, in the case of balancing while cycling, the explicit knowledge takes the form of a mathematical equation (or its verbal equivalent) representing biophysical relationships. Tacit knowledge can be made explicit, but this is not at all the same as it being made conscious. I may be able to describe how I ride a bicycle, how I read orally, or how I hold a coherent conversation. But such a description is a *third-person* description; it is not internally or directly known. By a third-person description is meant the description one can give of another object, essentially the same as an objective scientific description or theory. By contrast, knowing by consciousness is first-person, knowledge by acquaintance. Indeed, those aspects of phenomenal experience which tempt people to describe it as ineffable often render it quite inexplicit.

This leads on to a further distinction. What is verbalizable is sometimes treated as equivalent to what is conscious, and vice versa. This is seen in recent work on protocol analysis and introspection (Ericsson and Simon 1984) and on knowledge elicitation. It is at best an oversimplification. First, the nature of a sensation is often not adequately conveyable by words. Second, answering a question often does not require (prior) consciousness of the answer. Giving well-known information such as one's telephone number as a response to a request is an illustration of this. This is actually a quite critical point since many memory researchers and psychodynamicists consider answering a question and what is said as equivalent to what is conscious. Often what is said, although it is intentional, is not a reflection of consciousness at all. In psychoanalysis what a person can or cannot say about themselves may reflect what they know about themselves explicitly or tacitly. But this does not necessarily reflect what they know directly in a first-person way as phenomenal experience. This is a quite separate point from whether there are intentional states such as desires or fears which are unconscious as opposed to being consciously experienced as such.

Criteria

There is really only one criterion for phenomenal experience. This is a person's report, direct or indirect, that they have a sensation of one or another kind, that they are or were conscious in one or another way (though this means being conscious *of* something). Direct reports include statements such as 'I have a headache, or an itch', or 'I feel hungry'. Indirect reports include statements such as 'I see a light' or 'I hear a car', where the person is not directly reporting the sensation, but means by the statement that they *consciously* see or hear something. They do not mean that they conclude from observation of their behaviour that they perceive something. In the latter case they might say 'I must have seen/heard x'. Of course, people can be uncertain as to whether they are having an experience or not, or as to its precise nature. But even the expression of uncertainty conveys something of the nature of their experience. There is no claim here as to whether people are either corrigible or incorrigible as to their experience (see Dennett 1988, this volume), since it makes little sense to talk of the corrigibility of experience. However, provided that the person is not lying, there is little reason to doubt the validity of a report *that* there is phenomenal experience. This of course relies on the notion that reports are descriptions, that they validly refer rather than being just information-bearing 'verbal behaviour', and that what they refer to is phenomenal experience rather than some other informationally isomorphic representation (see Dennett 1982).

Such reports do not have to be verbal. In many experiments as well as in real-life communicative situations, people carry out actions such as button-pushing as a way of performing some conventionally arranged speech act, such as *asserting that* two figures appear equiluminant right now, or *answering that* 'yes, my snap judgement is that the word just presented was in the previously presented list'. [I owe this point and these examples to Dennett (1982).] Indeed, awareness itself can be indicated by a subject's ability to make a discriminative response as to whether or not they are aware of the presence of a stimulus and which corresponds to objective presence or absence of a stimulus, *and when the subject says they are not guessing.* When the subject says that they are guessing, we cannot take such discriminative responses as reliable indicators of consciousness. In blindsight or in masking or other sensory threshold experiments, people can discriminate with greater than chance accuracy when they claim to be guessing. So it is crucial that the convention of the speech act be agreed between experimenter/ interpreter and subject. For in the latter case the response is not a description, it does not mean 'I do or do not perceive *x*'. We appear to be entitled to conclude that in such cases the subject is influenced by perceptual processing of which they are not aware. However, in cases such as these the subject can be said to be conscious of the act (e.g. of guessing), though not of the cause for the particular category of the act.

It has to be realized, of course, that the criterion of report is asymmetric. The inability to report some stimulus or sensation does not imply lack of its consciousness. There are several reasons why someone may forget what they were conscious of even for those things in the very immediate past of one second ago, especially if the particular phenomenology is brief. This applies also to denials of awareness.

Interestingly, in recent attempts to achieve reliable and valid psychophysical indices of conscious awareness, it has been concluded that a subjective criterion, such as a confidence rating or the assertion of guessing or not, as mentioned above, is the appropriate criterion. Cheesman and Merikle (1985) distinguished the objective threshold from the subjective threshold. The former is where forced discrimination performance reaches chance; the latter is where the subject's confidence in his performance reaches chance, and is higher than the former. What is compelling is that it is the subjective threshold which indicates the point in stimulus conditions where qualitative differences are obtained in the effects of the stimuli in question. Below the subjective threshold polysemous stimuli activate multiple meanings, above it only one meaning (Marcel 1980); below this threshold the proportional occurrence of stimuli has little effect, above it strategic effects are induced (Cheesman and Merikle 1986). These kinds of data seem to support the criterion that what is conscious is what a person is prepared to acknowledge with a criterial level of confidence as the referrent of a true description.

Can subjective experience be objectively known?

Can we know what a particular phenomenal experience of (another) particular individual is? This is one version of the 'other minds' problem, and the usual answer is that, since phenomenal experience is subjective, qualitative, and largely ineffable, we cannot know another's experience. All that was suggested in the preceding parts of this section is that we can (sometimes) know *when* another individual has a phenomenal experience, without reference to the particular nature of that experience except in rather broad categorical terms (visual or auditory, pleasant or unpleasant). Well, for the purposes of everyday communication and within the limits of natural language descriptions, we do not seem to do too badly. When someone says that they have a headache or names an emotion they are experiencing, it appears that we understand them accurately enough. 'Understand them accurately enough' does not imply that the sensation *we* know under such a referential description is the same as theirs. It only implies that the sensations *we* call by those names are associated with the same behaviours and are subject to the same remedies as those associated with those sensations similarly named by the other person.[3] However, medical practitioners successfully rely on descriptions of phenomenal experience (quite apart from objective behaviour such as vomiting or stiffness of a limb) in order to carry out differential diagnoses. What is persuasive about this is that often the patient has never had the particular phenomenological symptoms before, and so cannot have learnt to correlate the language terms with the sensation. Thus we have here an instance of people spontaneously describing phenomenology in such a way that the description is consistent across individuals. Even if the physician supplies the terms to be chosen between ('Is the pain sharp or dull?'), consistent choice in naive individuals indicates that the descriptors can be reliably applied.[4]

[3](a) This can be shown by the 'inverted spectrum' argument (Locke 1690; see 1959 edn.).

(b) There seems to be no way that we can ever know for certain that another person is having a *sensation* as opposed to, say, a *brain state*. In this chapter the assumption is made that the other person is *referring to* something (by a term like headache) and that what they are referring to is a sensation. To converse with terms referring to phenomenal experience and to do so with mutual consistency suggests that the semantics of such terms are the same for both conversants.

[4] Medical students in the United Kingdom are taught to encourage the patient to give as much spontaneous history of his case as possible, lest questions become effectively 'leading questions' which bias the patient's report (Loftus 1979). Of course, initial diagnoses can be confirmed later by medical tests. However, in some cases the patient's phenomenology is a *more reliable* indicator than physical measures: angina is more reliably revealed by the description of the pain than by ECG measures.

Good examples of the phenomenon are provided by abdominal disorders. Ulcers

When it comes to being precise, however, we do not do well at all, as Dennett (1988, this volume) points out. Certainly, as regards science, there would seem to be no way at all of publicly knowing or expressing what a fried egg tastes like to particular individuals, or what it is like to be in a psychological state such that we could identify that phenomenal experience or compare it between individuals, except by relying on natural language terms, which lack the necessary precision. However, I am not convinced that we can say nothing about the nature of particular phenomenal experiences. It is true that we can do very little about characterizing a sensation in a purely and absolute referential sense. But we are probably able to do quite a lot about characterizing it in a relational sense. Suppose that we carried out second-order psychophysical scaling (Shepard 1975) on a set (or domain) of sensations. One such technique involves individuals making comparative judgements or relational ratings on n dimensions over a set of sensations (e.g. by paired comparisons). The relative positions between sensations on each dimension can be converted from an ordinal to an interval scale by Thurstonian scaling techniques (Thurstone 1927). These dimensional positions can then be combined so that the set of sensations can be plotted for each individual in an n-dimensional space. (While the origin and scaling of each dimension is arbitrary and relative rather than absolute, they can easily be standardized.) We could then assess the intra-individual consistency and reliability of such maps, and we could compare between individuals. There is nothing less objective about such relational maps than there is about first-order psychophysics. Indeed all psychophysics is relational. For the case of the coffee-tasters discussed by Dennett (1988, this volume), we would be able to tell whether and how the taste of a certain coffee compared to other tastes had changed or not, and, separately, in the same relational sense, whether preferences had changed. As it so happens, this approximates to what has been carried out for pitch of musical notes and note sequences (Krumhansl and Kessler 1982; Shepard 1982). Relationships between the experience of different pitches do seem to be remarkably consistent over individuals, such that the auditory (non-emotional) experience of

produce sharp, localized, stabbing pains; gall-bladder complaints produce pain that is intermittent, comes in waves, builds up to a crescendo and dies away followed by a painless period (generally true of instances of intermittent smooth-muscle contractions: labour contractions are the same in character but different in location); peritonitis produces a generalized ache all over the abdomen, as severe as ulcer pain but more continuous. Cardiac disorders may produce pain much like ulcer pain (a constricting sensation), differentiated mainly by location and extent (extending into a limb). The phenomenal description is what differentiates a true cardiac case from a cardiac hysteric, whose phenomenal description conforms to the naive expectation (e.g. 'sharp pain in left side of chest'). I am grateful for discussion of this to Bob Bingham, Consultant Anaesthetist, Hospital for Sick Children, Great Ormond Street, London.

music, whatever it is like, is the same for different people within a culture. The procedure has also been conducted to examine whether people's experience of an image of something is the same as their experience of its perception, for shapes, colours, and odours (Shepard 1975). Thus we *could* tell what a fried egg tastes like to an individual, at least in relation to other tastes, and whether it tastes the same (in relation to other tastes) to two different people. In such procedures, whether people use descriptive terms equivalently is assessed by whether their use of rating scales is consistent.

From the foregoing it may appear that we still cannot say exactly *what* a particular sensation is like. However, it is not clear that our abilities here are much more constrained than they are in the ability of language to refer to external, 'objective' objects. As Olson (1970) has pointed out, the semantics by which natural language refers to the world are largely relational. To this extent we can publicly know what an individual's experience is like. This seems inconsistent with some of Dennett's (1988, this volume) conclusions in his attack on qualia. Whatever their status, we appear to be able to measure them to some extent. We may not be able to find out what it is like to be a bat (Nagel 1974), but we can find out *something* of what it is like to be another human.

Separating phenomenal experience from processing: two distinctions

Focusing on phenomenal experience as consciousness implies a contrast with that which is non-conscious, but this and the foregoing discussion confound two kinds of distinction. The first kind of distinction is that of levels or kinds of description. Thus one may choose to describe the same behaviour *either* in terms of its phenomenological aspects *or* in terms of its information-processing or even neurophysiological aspects. One chooses one's type of description and account according to one's purposes. But note that many theorists would assume that the different terms proper to each level or domain of discourse refer to the same thing. This reducibility assumption may not be valid. The second kind of distinction in no way seeks to deny that consciousness is describable at a processing level, but it distinguishes processing without consciousness from processing which is also describable as conscious. In both of these cases, there may be constructs at the phenomenological level which either are causal in themselves or which cannot be reduced to other domains of discourse without loss of explanatory power.

A major problem is that cognitive scientists, psychologists, neurologists often fail to distinguish the two kinds of distinction. When one asks if it is necessary and legitimate for scientific explanations of behaviour to refer to consciousness, it seems one is referring to the first distinction. But to answer the question we have to refer to both distinctions. It was argued in previous papers (Marcel 1983a, 1983b) that psychologists may have thought they were only choosing a level of description according to the first distinction, but that in fact in so doing they have either ignored the second distinction or assumed it to be invalid; that is, they have used phenomenal experience and behaviour that is consciously generated and organized, assuming that it told them directly about processing that is non-conscious. The assumption of the identicality of conscious experience and processing, which underlies this inferential procedure, was termed the 'identity assumption'. The status of this assumption has great relevance in our response to questions about the necessity and legitimacy of reference to consciousness. If the assumption of identity were valid, our debate would be mostly centred on when it is appropriate to use which type of description. But if it is invalid, then we can ask about the causal status and role of consciousness, as distinct from processing without any subjective experience.

The falsification of the identity assumption has been discussed fairly fully elsewhere (Marcel 1983a, 1983b). But, to recapitulate briefly, there are two criteria: dissociations and qualitative differences. Two examples of dissociation can be mentioned here. The first is in visual masking. In essence, under certain conditions where a person has no awareness of a stimulus and cannot distinguish its presence from its absence, that stimulus can none the less be shown to have affected the processing of other stimuli at both structural and referential levels of description. Thus its representation at these levels is achieved independently of the person's consciousness. The second example is 'blindsight' in cases of cortical blindness. Following Weiskrantz's (1986) work, two people have been extensively investigated (Marcel and Wilkins 1982) in whom neurological damage to the cortical projections from the eyes to the left occipital lobe had resulted in a right homonymous hemianopia; that is, they cannot detect, are unaware of, bright, stationary light sources to the right of the centre of their field of view. With one or two exceptions (Humphrey 1974; Weiskrantz 1977), most previous views of cortical blindness have emphasized neuroanatomical function, assuming that the intact tectal projections in such people mediate visual location and movement, while the damaged cortical projections would normally mediate shape perception. However, we have found that, when forced to reach for objects in the blind field, preparatory adjustments of the wrist, fingers, and arm are suited much better than by chance to the shape, orientation, size, location, and distance of the objects. We have also been able to bias the interpretation of an auditorily presented, polysemous word (BANK) by a preceding visual presentation of an upper-case word related to one of its meanings (RIVER/MONEY) in the blind field. These findings imply that extremely sophisticated descriptions of three-dimensional shape and of the structure and order of strokes composing letters have been achieved. We can also produce phenomenal vision of shapes in the blind field, in terms of after-images and illusory contours in Kanisza figures, when they are related to complementary stimuli simultaneously presented to the sighted field. We thus believe that the main problem in cortical blindness concerns the phenomenal experience of qualities which have been adequately analysed but remain non-conscious. This provides a second example of the dissociation of processing from phenomenal experience.

Several putative qualitative differences between conscious and non-conscious representation of the same event have also been discussed (Marcel 1983*b*). In speech perception the effective non-conscious descriptions of the signal appear to be acoustic, whereas consciously they are phonemic. In vision, we consciously experience a world of rigid bodies characterized by Euclidean geometry, while there is evidence that the same world may well be described non-consciously in terms of non-metric projective geometry. Another example from vision is that phenomenological measures of visual persistence after stimulus offset are inversely related to stimulus duration and to stimulus intensity (the longer and brighter a stimulus the less time does it appear to persist), while information-processing measures show no such relationship (Coltheart 1980). A further qualitative difference comes from visual masking (Marcel 1980). When we are aware of a polysemous word (PALM) we are aware of only one meaning at a time and, indeed, such a word will only facilitate processing of subsequent words related to that meaning. However, when such a word is masked so that we are unaware of it, it will facilitate subsequent words related to *either* of its meanings, irrespective of any other biasing context. It appears that non-consciously both meanings are effectively represented simultaneously, while consciously only one meaning at a time is represented.

Thus by dissociation and by qualitative differences it seems that we can find empirical criteria to distinguish processing with and without phenomenal experience.

To return to the theme of this section, if we can separate conscious from non-conscious processing, two questions arise. First, what is the role of phenomenal experience (or at least processing with that quality), if any, in behaviour? What causal status or consequences does it have? Second, what sort of theoretical approach can we have to the relation between the conscious and the non-conscious? This latter question of course has two senses: (a) the relation between conscious experience and non-conscious processing, and (b) the relation between aspects of mind picked out by the phenomenological level of discourse and those picked out by the functionalist level of discourse.

Causal status: possible roles for consciousness

If we really wanted to study perception as functionalists we could so do without reference to phenomenal experience by looking at what different types of descriptions and representations of different aspects of the environment are achieved, what functions they have, and how they are achieved. This would be a combination of the Gibsonian and artificial intelligence approaches. And this type of study would use entirely indirect techniques; i.e. it would not rely on phenomenal experience and might even try to prevent it. If we wished to study memory in the same way, we would restrict ourselves to examining the effects of past experience on future performance, much as is done with studies of memory and learning in animals.

On p. 124, one reason given for discussing consciousness was its potential causal status. Indeed for most functionalists this would be the main reason for giving any attention to the topic at all. Certainly people taking a biological approach have often felt that phenomenal experience is an epiphenomenon. However, it must be emphasized that I am not suggesting merely that the functionalist equivalents of consciousness (short-term memory, focal attention) have causal status. This is the strategy adopted by Shallice (1988), Umiltà (1988), and Johnson-Laird (1988) (all in this volume). I am suggesting that phenomenal experience as such has causal status.

It is important to distinguish two senses of causal status. The narrower and usual sense of causation is that in which A can be said to cause B only if A, apart from any correlate of A, always and necessarily leads to B (given other necessary conditions). However, we can also attribute causal status in a broader sense if A *permits* or *enables* B, without necessarily leading to it, whatever other conditions obtain. A does not have to be a necessary condition for B. This approximates the historical sense of cause. A tool like the wheel or writing has causal status in its very existence, not because it causes something directly, but because it can be used and it *permits* certain things. The present contention is that human behaviour is different given that we are conscious than it would be were we not conscious: consciousness makes a difference. Several ways can be suggested in which phenomenal experience has consequences for our behaviour, makes it what it is. If any of them are valid, reference to consciousness is necessary for an adequate explanation of human behaviour.

One can further distinguish the uses to which something can be put from its *function*, in the sense of what it is *for*. There seems no obligation to discuss the 'purpose' of consciousness, since [in contrast to Oatley's (1988, this volume) position] in evolutionary terms, this seems a senseless exercise if one wants to avoid teleology. In *The blind watchmaker* Dawkins (1986) takes this position with regard to biological evolution. He suggests by way of example that the word-processor is essentially the same machine as was used 30 years ago for calculation. It is just that a new use has been found for it. But we would not have the type of word-processors we have today if the older generation of computers for calculation had not existed. This is the paradigm for the present sense of causal status.

If a person does something for a certain reason, that reason can be said to be one of the causes of the action. Beliefs can be causal in this sense. Indeed one of the ways proposed here in which consciousness has causal status is via beliefs about phenomenal experience. Such beliefs do not have to be true or well-founded, nor do they have to refer to coherent concepts for them to be causal. Consider beliefs about demons: their ability to inhabit humans, that they should be destroyed, and can be, by destroying their human host, and that their presence is detectable by whether a tied-up human floats in water. Irrespective of the truth or even meaningfulness of such beliefs, they were causal in the seventeenth century: many people were killed as witches as a consequence. Some beliefs can be tacitly held. Even if phenomenal experience (what it feels like to perceive, think, etc.) is not directly causal, it could be causal through tacit beliefs about it.

Let us examine how this might work. There are many behaviours which, as a matter of fact, we will only perform if we (believe that we) know particular things. For example we are normally only willing to reach for an object if we believe that the object is there; we are only willing to answer questions about spatial arrangement if we believe we know about the relevant spatial arrangement. (There are, of course, many types of action for which this is not the case.) For us, the most reliable indicator that we know certain types of things is that such knowledge takes the form of first-person phenomenal experience (percept, image, or sensation). However it is plausible that the reason for phenomenal experience having this status is a belief. Suppose that having certain knowledge in non-conscious form is in fact quite sufficient to cause a certain behaviour. Suppose, however, that a belief is held that the particular knowledge must be in phenomenal form (has to be known under a particular, phenomenal, description) for the be-

haviour to be executed. If such a belief can modulate the actual causal
effect of the non-conscious knowledge on behaviour, then it makes the
relevant phenomenal experience a necessary condition for the knowledge to have its causal effect. Thus if such a belief were not held (e.g. a
different culture) the knowledge in non-phenomenal form would of
itself cause the behaviour. It can be supposed that both of the above
types of belief can be held tacitly. None the less they can give phenomenal experience causal status. Note that in this case it is the
phenomenal experience itself which is causal, not some functionalist
construct that does not have phenomenal quality. Of course all of the
above, including the beliefs, can in principle be translated into discourse
about brain states and processes. Some brain states would correspond to
non-conscious knowledge and others to the same knowledge in phenomenal form. However since the relevant beliefs refer to knowledge
under phenomenal description, the brain would not only have to be
sensitive to which of its states were relevant, but might have to 'know'
that they referred to phenomenal experience as such. That is, I am not
sure whether one can escape some form of dualism. As far as I know,
there has been almost no discussion of how to treat beliefs which are
about phenomenal experience.

It should be clarified that the four proposals which follow remain
suggestions. No strong arguments are presented that they are necessarily the case. They express what are hopefully satisfactory characterizations of certain human behaviours. However, it should also be made
clear that the suggestion is that it is phenomenal experience on which
these behaviours depend.

Self-monitoring

The first thing that phenomenal experience enables is a certain kind of
self-monitoring. Clearly one does not need consciousness for all kinds of
self-monitoring. Non-conscious machines do it and we do it non-
consciously in editing our speech production and in executing many
motor intentions. But there are some kinds of self-monitoring that seem
to rely on consciousness.

Let us start with four possible examples. First, as Weiskrantz (1988,
this volume) points out, amnesic patients who have no conscious
memories are severely impaired. They simply cannot lead an organized
life. Baddeley and Wilson (1986) have suggested that lack of conscious
(episodic) autobiographical memory in amnesics produces an impaired
sense of self and personal identity. The amnesic patient without conscious memories does not know where he is, when he is, 'who' he is.
Second, patients with cortical blindness but with blindsight may make

appropriate visually based actions with respect to the part of space corresponding to the blind field, if they can be induced to move, but otherwise simply do not allow themselves to behave as if they had any such visual capacities. Lack of visual phenomenal experience prevents one from knowing that one can see, and this severely impairs one. Third, it would appear from the arguments of Shallice (1988), Umiltà (1988), and Johnson-Laird (1988) (all this volume) that if one cannot consider courses of action consciously, then one would be almost unable to carry out planning of action. Fourth, and most speculative, consider pain. Pain is by definition conscious, a phenomenal experience; tissue damage is not pain. If we did not have painful or pleasant sensations we would not be able either to gauge our state, our abilities, or to learn about harmful situations. It is true that simple organisms, of whose consciousness we have no evidence, can be conditioned to avoid certain situations. But the evidence of conditioning without awareness in humans is quite discredited (Brewer 1974). Also, the football player who is so involved in the game that he feels no pain from a broken leg, continues to play when the adaptive response would be to stop. This paradigm can be generalized to perception in general.

We happen to live the kind of lives we do lead, of a relatively organized kind, by reference to a representation of our environment, of our relation to it, of our past, and of our present moment-to-moment self-state. Without access to such representations we would be much more dependent on our immediate circumstances and more rigid. Indeed some of the representational systems we use are mental inventions, cultural artefacts, both internal (counting systems, conceptual segmentations of our activities) and external (contrast the mental life of cultures with and without writing). These representations are mental models whose roles in individual cognition, social interaction, and emotion have been emphasized respectively by Johnson-Laird (1983), Humphrey (1983), and Oatley and Bolton (1985). In modelling our self-state and relation to our environment it is necessary fairly continuously to monitor and update what information is provided by our interoceptors and exteroceptors, and to relate it to our records of where we are in time and space and to our shorter and longer range plans. It is such relating operations that permit ordered existence and behaviour and it is in such relating operations that our mental lives consist.

Metacognition and learning

Metacognition is the term that has come to be employed to describe a range of activities that require knowing that you know or perceive, knowing what it is that you know, accessing your knowledge, judgements based on your knowledge or percepts, reflecting on or paying attention to your phenomenal experience, and putting into operation one set of processes or another. Thus judging that two words do or do not mean the same thing or being able to discuss their meaning is separated from their practical comprehension and use for everyday purposes. Clearly that which is accessed, considered, or commented on in meta-cognitive tasks is either phenomenal experience or consciously enter-tained explicit knowledge. The concept of metacognition has been particularly evolved in developmental psychology, where many Piagetian tasks require decontextualized reflection and report rather than contextualized behaviour (Donaldson 1978). Interestingly, many clinical neuropsychological tests rely on metacognition. Indeed it has been suggested that a number of neurological disturbances are primarily metacognitive. Milberg and Blumstein (1981) showed that aphasic patients who apparently could not comprehend certain words or judge their semantic similarity to other words nevertheless showed semantic priming effects from these same words in a lexical decision task (see also Tyler 1988). Therefore such patients' apparent inability to compre-hend could potentially be redescribed as inability to consciously com-prehend and as a deficit in intentional conscious access to the meaning [see Marcel (1983*b*) for a review of neurological syndromes interpretable in this fashion].

I should like to suggest (along with many others) that metacognition plays a vital role in learning of certain types. Take the example of learning to read with an alphabetic script. Such scripts represent speech at the grain of phonemes. Many children who speak and perceive speech adequately have difficulties in paying attention to speech at that level of analysis (they cannot understand what is going on in language games like Pig Latin or Backslang where the order of phonemes in words is changed), and have great difficulty in learning to read. Morais *et al.* (1987) have shown that difficulty in paying attention analytically to speech sounds (termed 'phonological awareness') is highly related to difficulties in reading acquisition. Unless one can pay attention at an appropriate level to one's own sensations or actions, one will have great difficulty learning a new, essentially arbitrary, skill which is based on that level of representation.

It has been suggested by others that metacognition also plays a more general role in development. Via metacognitive conceptualization, the conscious apprehension of one's behaviour, new cognitive structures can be formed. The term 'horizontal décalage' refers to the phenomenon whereby in terms of a certain principle a child behaves inconsistently in different situations (at the same stage of development) with regard to a supposed cognitive structure embodying that principle. Karmiloff-

Smith (1979) has suggested that learning a new skill or cognitive procedure will proceed in specific bits (and cannot be generalized) until a theory of the subject matter or activity is formulated. Such a theory may be overgeneralized and lead to apparent horizontal décalage (e.g. inappropriate use of the past tense or plural morpheme in English), but can be reformulated. But crucially it is the formation of theories which leads in many cases to new structure-driven behaviour. Such theories start from consideration of one's own behaviour; that is, they start from conscious reflection (even if they are not verbally expressable).

Another, more dramatic view of the role of metacognition has been proposed by Leslie (1987). In this view, the infant starts with an ability to represent the world as described by its sensory apparatus, a first-order representation. It then acquires the ability to have a second-order representation which affords knowing that it knows or perceives. This ability frees the child from taking what it perceives or imagines as necessarily representing reality, and underlies the capacity for symbolic play where possible worlds can be entertained and one thing can be used to represent another. It is also supposed to permit the child a theory of mind. That is, by being aware of its own perception, states of mind, behaviours the child can now conceive of others' minds and of the possibility of another's beliefs *as beliefs* (i.e. as being possibly wrong). The capacity to entertain imaginary worlds also allows the decontextualization of knowledge that underlies much of what formal education is supposed to contribute (Donaldson 1978). For example, in the context of the research of Carraher *et al.* (1985), such conscious reflection allows the child to learn about mathematics as opposed to merely how much specific quantities of fruit (which the child may sell in the market) cost. Thus, to the extent that such second-order representations are equivalent to a consideration of phenomenal experience, consciousness may well play a crucial role in our normal development, at least for certain cultures.

Tasks

In current cognitive science there has been something of a return to faculty psychology. In the analysis of human abilities, by the use of componential analyses and neuropsychological dissociation, the current approach is to analyse our behaviour into individual mechanisms and their relation to each other. This can be seen in several areas: artificial intelligence, neuropsychology, information-processing models. But in spite of the apparent success of this enterprise, there is both an inconsistency and a fundamental failure of explanation.

Artificial intelligence systems are designed to do specific jobs. Any one system will carry out different tasks only if the differential flow of control is built in. The modular systems discussed by Fodor (1983), and implicit in much information-processing modelling, are supposedly automatic and the only behaviours they will carry out are reflex and rigid. Most importantly and pertinently, what are presented as information-processing models in the cognitive psychology literature, which are supposed to account for various behaviours, are not really models, since they appear to do everything but actually do nothing. Examples of such models are found applied to perception (Marr 1982; Palmer 1983; Treisman 1986), attention (Broadbent 1958), memory (Sperling 1963; Atkinson and Shiffrin 1971), language processing (Morton and Patterson 1980; Coltheart 1981; Shallice 1981). To illustrate the failure of such models, we can take as paradigmatic the models of language processing. Figure 6.1 exemplifies their typical form, where boxes and arrows represent the flow of information through representations and processes which recode those representations or store and retrieve them.

Suppose we present such a system with a stimulus such as the word 'dog', written or auditory. There is nothing in such models which will tell us whether the output will be: 'Yes, it's a word', 'No, it's not a piece of furniture', the response 'dog', 'it has three letters'; or whether the response will be written or verbal; etc., etc., etc. Of course all the

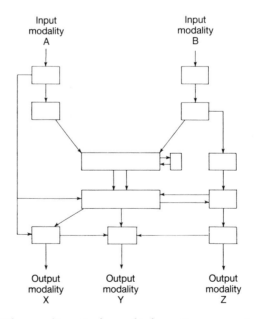

Fig. 6.1. The paradigmatic form of information-processing models.

processors will go to work and recode and pass on information. But the way in which such models are formulated ensures that they cannot, selectively, perform *tasks*. A 'task' is the use of resources in such a way that the end result (the goal) has a specific logical relationship to the starting point, a relationship different from that in other tasks. One reason for this inability in such models is that usually what their authors want them to do is to account for *many* tasks. But in attempting this they confound models of 'resources' (specialist processors and representational systems) with models of 'information flow' (command and control designations which use resources in a specific contingent structure and order). There is nothing in such models which represents a person's intention (self-generated or corresponding to compliance with instructions) nor how that intention is realized. Thus a real performance model would not confound, but combine and explicate the interaction of resources and control flow and add to them the source of such selective organization.

If we are to have accounts of cognition in terms of modules which are automatic, then we must have a 'theory of tasks' in addition. How is it that when a person is asked to name the colour of the ink in which a word (another colour name) is printed, they can actually do the task, even though we know (from the Stroop effect) that they automatically compute the phonology of the name of the word?

The anwer is that the organization and realization of effective information processing (including the choice and execution of appropriate action and its being based on the appropriate data structure or representation) depends at least in part on consciousness: an originally conscious organization, a conscious intention temporally and situationally appropriate, and a certain degree of conscious monitoring. It is not being suggested that the control processes are themselves conscious, nor that such organization necessarily relies on consciousness once we have initially learnt to do the task. However, consciousness is necessary at several points:

1. In many cases, when tasks that are not biologically given are first acquired, the organization (in a high-level description) needs to be conscious.
2. The intention to perform a non-habitual or non-stimulus-driven task needs to be conscious. If we are not conscious of instructions to perform a task, such instructions will be ineffective [even in post-hypnotic suggestion it turns out that the actor was conscious of the instructions even if he cannot remember them later (see Hilgard 1979)]. An exception to this is probably our personal long-range life goals.

3. We need to be conscious of the task itself (i.e. the intended result and its relation to the initial situation) at the time of performance; otherwise automatisms will result in the form of slips of action (Reason 1979).

A similar position to this is taken by Umiltà (1988, this volume). In addition, the problems voiced here, and theoretical solutions to them, have been discussed by Norman and Shallice (1986). However, what is suggested here is not only the necessity for a particular functionalist programme, but that, for humans, phenomenal awareness plays a role *per se* in the learning and execution of tasks.

Voluntary actions on the environment

No claim is being made here about actions in general, with regard to consciousness. But a claim is being made about so-called voluntary actions that are performed with respect to the environment. A typical example would be picking up a pen when I wish to write and I see the pen on the desk. Consider the following situation, which has to be treated as a thought experiment since we have not carried it out in any rigorous form. If a person with cortical blindness and blindsight in one hemifield is very thirsty and a glass of water is placed such that it falls within that person's sighted field, there is little doubt that she will either reach for it and drink it or will ask if she can take it. Now suppose the glass of water is placed so that it falls within the blind field. Remember that, from our own work, the object is apparently sufficiently well described visually to be identified and to permit adequate grasping (but non-consciously so). What will the person do? Will they do the same as when the stimulus is in their sighted field? Or will they reach out but not know why (until they contact the glass manually)? Or will they do nothing? The contention here is that they will do nothing—partly on the basis of anecdotes communicated by such people, partly on the basis of observation. In fact an example which fits this prediction is provided by patients with left hemineglect who neglect to dress the left side of their body. Presumably they are unaware of it.

The point is that people will not themselves initiate voluntary actions which involve some segment of the environment unless they are phenomenally aware of that segment of the environment. There are two kinds of reason for this proposal. One is that in such actions the critical segment of the environment is a logically necessary part of the intention. To the extent that the intention is conscious, it can not be well-formed if a necessary part of it cannot be in the necessary state (i.e. conscious). The other reason is that, *to the extent that someone is paying attention to their behaviour*, they do not normally allow then selves to perform actions without reason. This is not to deny that ther can be non-conscious motivators of action (Nisbett and Wilson 1977). But normally people do not allow the initiation of an attended action without being able to acknowledge the motivator or at least a plausible one. Further, there is no claim here that people *could not* do so nor that they *could not learn* to do so, nor that people in a radically different culture would not do so; only that people in our culture would not do so. The reason for this may be located in our beliefs about actions and their rationality. Some support for this may be taken from cases of neurological patients who are letter-by-letter readers. In such cases the time taken to read a word increases linearly with the number of letters. However, as Shallice and Saffran (1986) have shown, such people can be induced to make lexical decisions or semantic categorizations in a shorter space of time than it takes to read the word aloud. If the problem of some of these patients is in attaining a *conscious* percept of the whole word, then it seems that they *can* base actions on their non-conscious representation but are normally unwilling to do so. If a word is visually presented to a normal person, but masked such that the subject is unaware of it, yet it provides semantic priming (Marcel 1983*a*), it makes little sense to ask the subject to 'read' the word ('What word?'). If we are not prepared to base a voluntary action on things we are not aware of, then, even if it depends on our beliefs, consciousness has an effect, is causal.

Relations between conscious and non-conscious

Conscious experience and non-conscious processing

The first approach to the relationship between conscious and non-conscious is directed toward the difference between that processing which at any one time has phenomenal qualities and that which does not. What one is conscious of is very limited, while the amount of internal and external sensory processing going on is large, as is the number of stored memories and ongoing goals. The issue here is equivalent to the relation between the states of what is and is not attended.

Looked at in one way this relationship is unproblematic (except for how voluntary attention is achieved and realized). The conception of the conscious status of a representation either as one part of a system having access to some portion of its data base or processes [Bisiach's 'C$_2$' (1988, this volume)], or as one representation being a basis for action or memory (see Allport 1988, this volume), presents, in principle, no computational trouble. Two problems do arise though. One is why such a representation should have phenomenal and first-person qualities. The other is that, as mentioned above, there are numerous reasons for suspecting that the conscious and non-conscious representations of the same thing are qualitatively different in terms of representational primitives. If it were not for these issues one could conceive of the relationship in purely functionalist terms: how one process takes as input the output of another process. But these two points suggest that we are dealing with two domains, if not two discourses. However, we cannot leave it like that because, for one thing, the two domains are dependent on one another: that which has phenomenal properties is derived from that which (apparently) does not; and, if the suggestions made above about causal status are valid, then behaviour, or whatever corresponds to behavioural intentions, is often dependent on phenomenal experience. Leaving aside for the moment the question of phenomenal experience, if the second problem is valid, then the relation between the two states (e.g. when one becomes conscious of what was hitherto some non-conscious representation) requires a cognitive mediation. Our conscious experience is to a large extent in terms of conceptual categories which do not seem to apply to our non-conscious representations. It also seems to be mediated by tacit beliefs and attitudes. The application of such beliefs and attitudes in the processing that transforms a non-conscious to a conscious representation (Helmholtz's 'unconscious inference'), e.g. in attending, implies a model both of the world and of the self. Some of these concepts and the model they constitute are acquired or are socially or culturally transmitted. Consistent with this notion are data that suggest that consciousness as we know it emerges developmentally (Izard 1980) and that phenomenal experience of the same stimulus or internal state differs as a function of culture (e.g. whether or not certain physiological states are experienced as pain, or the form in which certain physiological states are experienced as emotion).

The main aspects of this view of the relationship, developed more fully elsewhere (Marcel 1983b), can be sketched briefly. Monitoring is carried out continuously on the processing that the system is carrying out. This monitoring is (attentionally) selective. What is monitored is largely determined by the currently dominant goal.[5] The monitoring

---

[5] Several people, including Shallice himself, have seen Shallice's (1978) and my own (1983b) views as quite different. This is an exaggeration (as Shallice has agreed, personal communication). The main difference is in what aspect of consciousness we have chosen to focus on for explanation. Shallice focused on the role of goal-directed actions in determining what we are conscious of. I focused on the mechanism by which what would

is conducted via a model of the self and the environment (i.e. what is monitored depends on the categories constituting the model). This model is largely acquired and is therefore cultural in two senses. It contains concepts available to a particular society, and, since it is also derived from social interaction, it is a social model ('I' am examining or influencing 'my sensations' or 'self') and embodies cultural conceptions of self. For such reasons the model may include the very belief in phenomenal experience. Such monitoring is equivalent to reflexive consciousness. States of parts of the system become conscious only when they are monitored; but clearly a call can be made for monitoring. The phenomenal quality of such data depends on the constructs, categories, and beliefs constituting the model.

The view outlined above is related to the ideas of Mead (1934), Humphrey (1983), and Johnson-Laird (1988, this volume), but differs from each. It shares with Mead's view the notion of social derivation and the role of a self-concept. While Oatley (1988, this volume) sees these factors as leading to another type or aspect of consciousness, my own proposal is that socially derived constructs and a self-concept are basic to phenomenal experience itself and its quality. Incidentally, lest there be any confusion, there is no conflict between something being of social origin but instantiated in the brain. The brain is a necessary but not sufficient condition for consciousness.

Along with Humphrey's (1983) view, this approach also places emphasis on reflexive consciousness. We usually assume that non-reflexive consciousness is logically prior to reflexive consciousness. Indeed it seems reasonable to say that you cannot consider or be aware of your phenomenal experience without there being phenomenal experience. But this is because the form of words implies or reflects a conception of the 'I' being in a perceptual relation to its experience. Humphrey emphasizes this relationship because he feels that it enables certain types of behaviour, largely social. In this chapter there is no

---

otherwise be non-conscious data becomes phenomenal experience. Shallice has emphasized the role of the temporally most dominant action system in determining what we are conscious of (the input to that action system). This relationship is similar to the suggestion in my own paper that what we are conscious of is under the functionally most useful description (i.e. for the current purpose). My own emphasis has been on the process that turns information into phenomenal experience, and its constructive nature. The main apparent substantive difference between the two views is in the nature of that process. Shallice favours activation whereas I favour matching. But yet again this is a difference in focus. I also suggested that *what* is matched is partly determined by activation. What I was attempting to capture by the notions of construction, matching, and beliefs is the nature of our phenomenal experience and its relation to the nature of non-conscious sensory and motor information, whereas what concerned Shallice was what becomes conscious rather than its form. The other difference is in the role of inhibition, a more peripheral issue for me than for Shallice.

attempt to deny this type of relationship. Indeed it is the basis of the proposals about metacognition and voluntary action put forward in the preceding section. However, according to the present view it is the operation of reflexive consciousness which creates phenomenal experience from otherwise non-conscious data. To the extent that reflexive monitoring can only be directed on the basis of a model of what is to be monitored, which is surely partly socially constructed, our phenomenal experience is the realization of a social construct. The self-model of course is tacit rather than explicit.

While structurally this view is very similar to Johnson-Laird's proposal in the functional concepts and their relations, it is radically different in terms of the status and source of consciousness. Johnson-Laird proposes that consciousness is a characteristic of a certain type of system; and, if human beings are correctly characterized by his description of that system, then they are necessarily conscious. Instead the view of this essay is that consciousness depends on the beliefs embodied in the model of self. Different beliefs about the self produce different selves.

Content and functionalism: social and natural science

The previous section attempted to offer a view of how non-conscious information can be turned into phenomenal experience when it is the object of attention. This is essentially a functionalist approach to the relation between conscious and non-conscious. The second problem of relationship raised by their separation is the relationship between aspects of mind picked out by phenomenal experience and those picked out by functionalist levels of analysis. The problem is that there are variables in the former which as yet have no (consensually accepted) realization in the latter. Pre-eminent among these is 'content', intentionality, meaning. Since Brentano (1874) proposed that intentionality was what distinguished mental from physical phenomena, philosophers have argued about how it can be realized in functionalist systems or whether it exists. What concerns me here is the importance of this for psychologists.

It was pointed out at the beginning of this chapter that functionalist analyses are indifferent to what the function operates on, its 'content'. Indeed as functionalists, psychologists concern themselves with the processes of perception, memory, problem-solving, action, without concerning themselves with *what* is being perceived, remembered, solved, enacted. Clearly their assumption must be that the content makes no difference. But if it does make a difference, then are the theories of any generality, and what of functionalism? The present proposal is that content and meaning are central aspects of our mental lives and are not as yet adequately dealt with in functionalist discourse. Further, the content and significance of our percepts and actions are not only relevant to our choice of actions and how the world is experienced, but determine and influence the processes that realize them.

Before discussing this at a general level let us consider two specific examples. They concern neurological patients, but essentially the same type of effects have been observed in children learning at school (Laboratory of Comparative Human Cognition 1983) and in several cognitive tasks in children and adults (Carraher *et al.* 1985; Lave *et al.* 1984). The first example comes from the work of Poizner *et al.* (1984). Congenitally deaf people who after left-hemisphere lesions become apraxic were asked to perform the movements for certain actions (such as drinking) and the sign-language gesture which represented those actions, where the movements for both mime and sign are exactly the same. They were unable to perform the movements as mime but quite able to perform them as sign. The status of the intentional description of the action has a reliable effect.

The second example concerns research, in which I am currently involved, whose primary aim is to examine to what extent, in patients with problems of action and movement, motor control is influenced by the status of their intended actions (as decontextualized behaviour or as part of an action with a naturally motivated purpose). Three levels of performance have been distinguished, the third serendipitously. The patients involved, when asked to perform an action such as picking up a vertical cylinder or copying letter-like shapes, have problems of sequencing, selective finger control, smooth movement, lifting a weight, or may be quite unable to perform at all. (None of their inabilities are due to problems of perception or comprehension.) This is level-1 performance. Some of these patients, by contrast, will perform the same behaviour with marked superiority or greater ease when it takes the form of a natural action, such as lifting a glass of water to drink while having a meal, or writing to dictation (level-2 performance). However, the most surprising findings are that in the patients' homes and in certain non-test situations, particular cases of the actions in question are performed even better (level-3 performance) than their previous superior level of performance as natural contextualized actions. Indeed, when filmed, independent judges consistently differentiated the levels of performance of the same action. Analysis and further experimentation indicates that very little of the differences can be attributed to 'context' effects or to attention. In our attempts to find what was common to level-3 performance, the single, reliable, differentiating characteristic (quite unsought) turns out to be situations which render the action

socially or personally significant. Two examples will give the flavour
of this. When, on our arrival at her home, one of the patients offered
us tea, this produced better (level-3) performance in grasping and
lifting crockery than when she was clearing up (level 2). Another patient
showed this difference between writing her diary or noting her next
appointment (level 3) and writing to dictation (level 2). Note that the
significance of offering tea to a visitor is only definable in terms of how
the culture categorizes activities (what they represent) and how it
represents such activities as signifying states of the self in relation to
others. This is equally true of the basis on which the personal significance of particular types of writing (for example) depend. It is also most
probable that the intention of the action differs in a commensurate
manner for the three levels of performance:

a. Extend my arm so that my fingers extend around the cylinder, grasp
    it, raise it, etc. (level 1); satisfy my thirst (level 2); serve tea to guests
    (level 3).

b. Perform a downward stroke of such an extent, perform a horizontal
    stroke on top, etc. (level 1); write the word x, followed by the word y,
    etc. (level 2); convey what I managed to do today, or that my
    intention is to do a on Monday (level 3).

The point of this illustration is twofold. First, it appears that, just as in
the studies on other psychological functions mentioned above, the
content of an intention, *what* is being realized by motor processing,
makes a difference to whether the processing can be carried out and how
it is carried out. Second, one of the differences made by the content is
according to the significance of that content. There is little doubt that
such significance *can* be realized or captured in functionalist terms,
though we need to know rather more than we do about how brains and
information can represent. But meaning in this sense can only be fully
defined at the personal level of discourse (as opposed to the subpersonal
level of information processing and most functionalism), and its source
is social or cultural. [This last point has been discussed more fully
elsewhere (see Marcel, 1988).]

Why does this create a problem? The problem of content was stated
above, and the issue of intentionality for functionalism and its relation
to phenomenal experience are discussed by Van Gulick (1988, this
volume). The problem of significance, though overlapping with that of
intentionality, poses some other problems concerned with levels and
domains of discourse. Meaningfulness, significance, and symbolization
seem to depend on two kinds of thing (see Marcel, 1988): (a) a model of

the self and others in which goals and evaluations are represented (Oatley and Bolton 1985), and (b) sets of categories relating to one another in an ordered way within and between sets. But the way that the categories relate, which is what determines significance (e.g. the employer–employee relationship may signify the father–child or mother–child relationship in terms of power or caring), depends on the way that the social world or culture arranges or defines them (Lévi-Strauss, 1962). The problem, then, is this. It appears that characterizations which permit an adequate description of such phenomena are picked out at the personal level or derived from the social sciences, and the variables exist at the personal level and originate in the social world. Yet the consequences of these characterizations, their realization, are found at the subpersonal level in natural science measures, here concerned with biomechanics, kinetics, and accuracy of movement.

There is no claim here that content and meaning are not neurally instantiated (though there is a claim that not all aspects of mind originate from biology). However, at the moment there is an important difference between two kinds of approaches to mind. On the one hand, natural science seeks to account for phenomena as objects, including functional objects, in terms of characterizing the nature of the phenomenon (physical or functional), in terms of determinist rules relating cause and effect, in terms of other levels of description, usually more analytic, i.e. reductionist. On the other hand, what can loosely be called social science attempts to treat content and its aspects such as meaning and structure. Hermeneutics is not part of natural science. Consider two examples. Freud (1909; see Strachey 1976), in the case of 'the Rat Man', attempted to make sense of the patient's bizarre slimming programme by suggesting that it *represented* an attempt to get rid of a rival called Richard (the German for 'fat' is 'dick'). In Dickens' novel *Great Expectations*, Miss Havisham's strange behaviour (she wears a wedding dress and one white shoe and lives in a room decked out for a feast) is made sense of when we learn that she learnt of her fiancé's desertion while dressing on her wedding day. In both cases what is crucial is the meaning of the behaviour. The natural scientist seeks a causal account. In the examples above, the significance of the behaviour, its meaning, is not itself the cause of the behaviour. At the very least, sometimes in human behaviour we need to start by understanding meaning, which may mediate causal connections.

On the whole, natural science is antipathetic to this enterprise, and not just because controlled experiments cannot be carried out to adjudicate the validity and veridicality of accounts. It is also because in general (a) the paradigm tends to assume that its objects have no meaning, are merely caused, and (b) its objects, as objects, cannot have content: other objects cannot have meaning or significance for them. The apparent exceptions to the first point, which are to be found in functional approaches or ethological biology, are not really exceptions. The biologist does not assume that for the herring-gull chick the red spot on the parent's beak signifies or stands for food; it only has this status for the biologist. Indeed Dennett (1986) has suggested that the only intentionality is this derived intentionality. But then how does the biologist come to have the concept? The answer surely is that he derives it from his own intentionality, just as we could never understand what it would be for another person to have emotional experience unless we had it. As for the second point, Dennett (1978, Chapter 12) argues for the functionalist position, suggesting that, although the intentional stance has advantages, it is nothing but an 'as-if' stance.

If we want to account for *what* people do and experience, as opposed to merely how they do it and come to experience it, then we shall have to have some recourse to what is so far not comfortably dealt with in functionalist science. The problem is to find a language or paradigm common to both process and content, and to the personal and subpersonal levels, which does not do violence to either. Whether functionalism can achieve this is uncertain, for, if it means dealing with particular instantions and particular intentions, then that would seem to violate the functionalist enterprise. Whether there need be no distinction between social and natural science is also problematic. However, we should not, as we have done until recently, continue to ignore things just because our paradigms do not adequately deal with them. That sort of ideological hegemony in science is merely reactionary. We should not be keeping consciousness out or pretending it is something else. We should be welcoming it.

Acknowledgements

I am grateful for comments and criticisms on an earlier version of this paper by Edoardo Bisiach, Patricia Smith Churchland, Tony Dickinson, Carlo Umiltà, and Paul Whittle. I am also grateful for the useful conversations I have had with Alan Allport and Phil Johnson-Laird. None of these people is responsible for the views or arguments that appear in this paper.

References

Allport, D. A. (1988). What concept of consciousness? In *Consciousness in contemporary science*, (ed. A. J. Marcel and E. Bisiach), p. 159. Oxford University Press.

Atkinson, R. C. and Shiffrin, R. H. (1971). The control of short-term memory. *Scientific American*, **224**, 82–90.

Baddeley, A. D. and Wilson, B. (1986). Amnesia, autobiographical memory and confabulation. In *Autobiographical memory*, (ed. D. C. Rubin), pp. 225–52. Cambridge University Press.

Bisiach, E. (1988). The (haunted) brain and consciousness. In *Consciousness in contemporary science*, (ed. A. J. Marcel and E. Bisiach), p. 101. Oxford University Press.

Brentano, F. C. (1874). *Psychologie von dem Empirischen Standpunkt*, (trans. A. C. Rancurrello, D. B. Terrell, and L. L. McAlister). (ed. L. L. McAlister, 1973), Routledge and Kegan Paul, Andover, Hants.

Brewer, W. F. (1974). There is no convincing evidence for operant or classical conditioning in adult humans. In *Cognition and the symbolic processes* (ed. W. B. Weimer and D. S. Palermo), pp. 1–42. Erlbaum, Hillsdale, NJ.

Broadbent, D. E. (1958). *Perception and communication*. Pergamon Press, Oxford.

Carraher, T. N., Carraher, D. W., and Schliemann, A. D. (1985). Mathematics in the street and in schools. *British Journal of Developmental Psychology*, **3**, 21–9.

Cheesman, J. and Merikle, P. M. (1985). Word recognition and consciousness. In *Reading research: advances in theory and practice*, No. 5, (ed. D. Besner, T. G. Waller, and G. E. MacKinnon), Academic Press, New York.

Coltheart, M. (1980). Iconic memory and visible persistence. *Perception and Psychophysics*, **27**, 183–228.

Coltheart, M. (1981). Disorders of reading and their implications for models of normal reading. *Visible Language*, **5**, 245–86.

Csikszentmihalyi, M. (1978). Attention and the holistic approach to behavior. In *The stream of consciousness*, (ed. K. S. Pope and J. L. Singer), Plenum Press, New York.

Dawkins, R. (1986). *The blind watchmaker*. Longman, Harlow, Essex.

Dennett, D. C. (1978). *Brainstorms*. MIT Press, Cambridge, MA.

Dennett, D. C. (1981). Introduction. In *The mind's I*, (ed. D. R. Hofstadter and D. C. Dennett) The Harvester Press, Brighton.

Dennett, D. C. (1982). How to study human consciousness empirically. *Synthese*, **53**, 159–80.

Dennett, D. C. (1986). *The myth of original intentionality*. Report No. 142/1986, Research Group on Perception and Action, Centre for Interdisciplinary Research (ZIF), Bielefeld University, West Germany.

Dennett, D. C. (1988). Quining qualia. In *Consciousness in contemporary science*, (ed. A. J. Marcel and E. Bisiach), p. 42. Oxford University Press.

Donaldson, M. (1978). *Children's minds*. Fontana, Glasgow.

Duval, S. and Wicklund, R. A. (1972). *A theory of objective self-awareness*. Academic Press, New York.

Ericsson, K. A. and Simon, H. A. (1984). *Protocol analysis: verbal reports as data*. MIT, Press, Cambridge, MA.

Fodor, J. (1983). *The modularity of mind*. MIT Press, Cambridge, MA.

Grice, H. P. (1967). Logic and conversation. In *The logic of grammar*, (ed. D. Davidson and G. Harman). Dickenson, Encino, CA.

Hilgard, E. R. (1979). Divided consciousness in hypnosis: the implications of the hidden observer. In *Hypnosis: developments in research and new perspectives* (2nd edn), (ed. E. Fromm and R. E. Shor). Aldine, New York.

Hoffman, M. L. (1982). Affect and moral development. *New Directions for Child Development*, No. 16, 83–103.

Humphrey, N. K. (1974). Vision in a monkey without striate cortex: a case study. *Perception*, **3**, 241–55.

Humphrey, H. K. (1983). *Consciousness regained*. Oxford University Press.

Izard, C. E. (1980). The emergence of emotions and the development of consciousness in infancy. In *The psychology of consciousness*, (ed. J. M. Davidson and R. J. Davidson). Plenum Press, New York.

Johnson-Laird, P. N. (1983). A computational analysis of consciousness. *Cognition and Brain Theory*, **6**, 499–508.

Johnson-Laird, P. N. (1988). A computational analysis of consciousness. In *Consciousness in contemporary science*, (ed. A. J. Marcel and E. Bisiach). p. 357. Oxford University Press.

Karmiloff-Smith, A. (1979). Micro- and macrodevelopmental changes in language acquisition and other representational systems. *Cognitive Science*, **3**, 91–118.

Krumhansl, C. L. and Kessler, E. J. (1982). Tracing the dynamic changes in perceived tonal organization in a spatial representation of musical keys. *Psychological Review*, **89**, 334–368.

Laboratory of Comparative Human Cognition (1983). Culture and cognitive development. In *Handbook of child psychology: Vol. I. History, theory, and method* (4th edn), (ed. P. H. Mussen), pp. 295–356. John Wiley, New York.

Lashley, K. S. (1958). Cerebral organization and behavior. In *The brain and human behavior*. Proceedings of the Association for Research on Nervous and Mental Disease. Williams and Wilkins, Baltimore, MD.

Lave, J., Murtagh, M., and de la Rocha, O. (1984). The dialectic of arithmetic in grocery shopping. In *Everyday cognition: its development in social context*, (ed. B. Rogoff and J. Lave), pp. 67–94. Harvard University Press.

Leslie, A. D. (1987). Pretense and representation: the origins of 'theory of mind'. *Psychological Review*, 94, 412–26.

Lévi-Strauss, C. (1962). *La pensée sauvage*. Plon, Paris.

Locke, J. (1959). *An essay concerning human understanding*. (ed. A. C. Fraser), Dover, New York.

Loftus, E. F. (1979). *Eyewitness testimony*. Harvard University Press.

Mandler, G. and Kessen, W. (1974). The appearance of free will. In *Philosophy of psychology*, (ed. S. C. Brown), Macmillan, London.

Marcel, A. J. (1980). Conscious and preconscious recognition of polysemous words: locating the selective effects of prior verbal context. In *Attention and performance*, Vol. VIII, (ed. R S Nickerson). Erlbaum, Hillsdale, NJ.

Marcel, A. J. (1983a). Conscious and unconscious perception: experiments on visual masking and word recognition. *Cognitive Psychology*, **15**, 197–237.

Marcel, A. J. (1983b). Conscious and unconscious perception: an approach to the relations between phenomenal experience and perceptual processes. *Cognitive Psychology*, **15**, 238–300.

Marcel, A. J. (1988). Electrophysiology and meaning in cognitive science and dynamic psychology. In *Psychodynamics and cognition*, (ed. M. Horowitz), pp. 169–89, University of Chicago Press.

Marcel, A. J. and Wilkins, A. J. (1982). Cortical blindness: a problem of visual consciousness or of visual function? Paper presented at Fifth International Neuropsychology Society European Conference, Deauville, France.

Marr, D. (1982). *Vision*. Freeman, San Francisco.

Mead, G. H. (1934). *Mind, self, and society*. University of Chicago Press.

Milberg, W. and Blumstein, S. E. (1981). Lexical decision and aphasia: evidence for semantic processing. *Brain and Language*, **14**, 371–85.

Miller, G. A. (1964). *Psychology: the science of mental life*. Hutchinson, London.

Morais, J., Alegria, J., and Content, A. (1987). The relationships between segmental analysis and alphabetic literacy: an interactive review. *Cahiers de Psychologie Cognitive*, **7**, 415–38.

Morton, J. and Patterson, K. E. (1980). A new attempt at an interpretation, or, an attempt at a new interpretation. In *Deep dyslexia*, (ed. M. Coltheart, K. Patterson, and J. Marshall), Routledge and Kegan Paul, Andover, Hants.

Nagel, T. (1974). What is it like to be a bat? *Philosophical Review*, **83**, 435–51.

Natsoulas, T. (1978). Consciousness. *American Psychologist*, **33**, 906–14.

Nisbett, R. E. and Wilson, T. DeC. (1977). Telling more than we know: verbal reports on mental processes. *Psychological Review*, **84**, 231–59.

Norman, D. A. and Shallice, T. (1986). Attention to action: willed and automatic control of behaviour. In *Consciousness and self-regulation*, Vol. 4, (ed. R. J. Davidson, G. E. Schwartz, and D. Shapiro), Plenum Press, New York.

Oatley, K. (1988). On changing one's mind: a possible function of consciousness. In *Consciousness in contemporary science*, (ed. A. J. Marcel and E. Bisiach), p. 369. Oxford University Press.

Oatley, K. and Bolton, W. (1985). A social–cognitive theory of depression in reaction to life events. *Psychological Review* **92**, 372–88.

Olson, D. R. (1970). Language and thought: aspects of a cognitive theory of semantics. *Psychological Review*, **77**, 257–73.

Palmer, S. E. (1983). The psychology of perceptual organization: a transformational approach. In *Human and machine vision*, (ed. J. Beck, B. Hope, and A. Rosenfeld). Academic Press, New York.

Poizner, H., Bellugi, U., and Iragui, V. (1984). Apraxia and aphasia in a visual–gestural language. *American Journal of Physiology*, **246**.

Polanyi, M. (1966). *The tacit dimension*. Doubleday, New York.

Reason, J. T. (1979). Actions not as planned. In *Aspects of consciousness*, Vol. 1, (ed. G. Underwood and R. Stevens). Academic Press, London.

Shallice, T. (1972). Dual functions of consciousness. *Psychological Review*, **79**, 383–93.

Shallice, T., (1978). The dominant action system: an information-processing approach to consciousness. In *The stream of consciousness*, (ed. K. S. Pope and J. L. Singer). Plenum Press, New York.

Shallice, T. (1981). Neurological impairment of cognitive processes. *British Medical Bulletin*, **37**, 187–92.

Shallice, T. (1988). Information-processing models of consciousness: possibilities and problems. In *Consciousness in contemporary science*, (ed. A. J. Marcel and E. Bisiach), p. 305. Oxford University Press.

Shallice, T. and Saffran, E. (1986). Lexical processing in the absence of explicit word identification: evidence from a letter-by-letter reader. *Cognitive Neuropsychology*, **3**, 429–58.

Shepard, R. N. (1975). Form, formation and transformation of internal representations. In *Information processing and cognition: the Loyola symposium* (ed. R. Solso). Erlbaum, New Jersey.

Shepard, R. N. (1982). Geometrical approximations to the structure of pitch. *Psychological Review*, **89**, 305–33.

Sperling, G. (1963). A model for visual memory tasks. *Human Factors*, **5**, 19–31.

Strachey, J. (ed. and trans.) (1976), *The standard edition of the complete psychological works of Sigmund Freud*, vol. 10. Norton, New York.

Thomas, K. (1984). *Man and the natural world*. Penguin Books, Harmondsworth, Middx.

Thurstone, L. L. (1927). Psychophysical analysis. *American Journal of Psychology*, **38**, 368–89.

Tolman, E. C. (1932). Concerning the sensation quality: a behavioristic account. *Psychological Review*, **29**, 140–45.

Treisman, A. M. (1986). Features and objects in visual processing. *Scientific American*, **255**, 106–15.

Tulving, E. (1985). Memory and consciousness. *Canadian Psychologist*, **26**, 1–12.

Tyler, L. K. (1988). Spoken language comprehension in a fluent aphasic patient. *Cognitive Neuropsychology*, **5**, 375–400.

Umiltà, C. (1988). The control operations of consciousness. In *Consciousness in contemporary science*, (ed. A. J. Marcel and E. Bisiach), p. 334. Oxford University Press.

Van Gulick, R. (1988). Consciousness, intrinsic intentionality, and self-understanding machines. In *Consciousness in contemporary science*, (ed. A. J. Marcel and E. Bisiach), p. 78. Oxford University Press.

Weiskrantz, L. (1977). Trying to bridge some neuropsychological gaps between monkey and man. *British Journal of Psychology*, **68**, 431–45.

Weiskrantz, L. (1986). *Blindsight: a case study and implications*. Oxford Psychology Series, 12. Oxford University Press.

Weiskrantz, L. (1988). Some contributions of neuropsychology of vision and memory to the problem of consciousness. In *Consciousness in contemporary science*, (ed. A. J. Marcel and E. Bisiach), p. 183. Oxford University Press.

Wilkes, K. V. (1988), ——, yìshì, duh, um, and consciousness. In *Consciousness in contemporary science*, (ed. A. J. Marcel and E. Bisiach), p. 16. Oxford University Press.

7

What concept of consciousness?

Alan Allport

Some people—including more than one contributor to this book—refer to *the* 'phenomenon of consciousness'; they appeal to its 'unitary nature'. 'Whatever else you may doubt, you cannot doubt *this*', they are inclined to say, accompanying their words with some sort of vigorous gesture, as though pointing inwards towards their own mental life, their own 'phenomenon of consciousness'[1].

What are they pointing to?

People in the grip of this intuition appear to be quite sure that they know what they are referring to when they refer to consciousness. For them the problem of consciousness just is: how to explain it? Is it the product of a certain computational architecture; a question of particular spatio–temporal configurations of neural activity; a product uniquely accessible to some unitary 'central executive' or 'operating system'? To others, myself among them, this reference to 'it', to a unitary, underlying phenomenon or state of affairs is bewildering. I find that I have no clear conception what people are talking about when they talk about 'consciousness' or 'phenomenal awareness'; nor, for that matter, when they talk about its linguistic–conceptual Siamese twin, 'the conscious self' ('the subject of experience'; that which itself has 'access' to 'conscious information'; to which a privileged subset of the information represented in the brain is 'accessible', etc., etc.).

In all this terminology—certainly in all of it that refers to a unitary 'phenomenal awareness', given to (or had by) a unitary 'subject'—there appears a persistent temptation to invoke a being, *within* the mind/brain but not coextensive with it, which itself perceives (has 'access' to) the data of consciousness. Who or what is this being? If one is to credit this entity with the causal powers apparently required of it, even if implicitly, by some of its protagonists—not only *all* forms of 'conscious

---

[1] I recognize the gesture, and the intuition that it expresses, well enough, as I have frequently been enticed by the same intuition myself. Nevertheless the intuition is a misleading one. I believe that to accept it is to commit oneself to a serious conceptual mistake. From there on down, matters will only get worse, or at least more confused.

awareness', but also all forms of 'conscious decision-making', intention, emotion, etc.—there can not be much mistaking the identity of such a creature: it is the familiar theoretical bogey of psychological theory (*bad* psychological theory), the *homunculus*—the little man in the head. Moreover, the causal powers that must be attributed to it—to the 'subject of conscious experience'—appear no less than those properly (and uniquely) attributed to *persons*.

Criteria

The purpose of this chapter, however, is not to warn about the dangers of unintentionally incorporating homunculi in our psychological theories. My enterprise is more restricted and, I hope, more constructive. I am looking for criteria.

Suppose someone comes forward with a putative 'theory of consciousness'. How might we set about evaluating such a theory? Better, suppose we have at least two rival theories. How can we begin to adjudicate between them?

Clearly, first of all, we need to know *what it is that is to be accounted for* by such a theory: what is the scope of the theory? We must at least be able to tell when we have a definite example of the phenomenon that the theory is intended to explain. How do we recognize an instance of the phenonenon when we have one? (How do we recognize a *non*-instance?) It will hardly do for the protagonists of such a theory to respond: 'Well, it's a theory that explains *this*', simultaneously gesturing somehow towards their own inner experience, their own 'phenomenal awareness'. I take it that no putative 'theory of consciousness'—no scientific theory of consciousness—is intended by its author, A, as a theory solely of A's *own* consciousness[2], but as generalizable to all other cases of the 'same' phenomenon—whatever that phenomenon might be supposed to be. Consider, for example, a theory of consciousness expressed in terms of a specific computational architecture, such as the theory proposed by Johnson-Laird (1983, and 1988, this volume). A strong implication of such a theory is that *any* system embodying the proposed computational architecture should thereby be endowed with consciousness. Indeed, this appears to be much the strongest and clearest prediction that the theory proposes. So *how should we test it?* What would we expect to find in the behaviour of the robot—the computational system—which we

---

[2] Such a theory would arguably be of limited interest to others not acquainted with A. It would also have the unsatisfactory property that *A* was, logically, the only person in a position to confirm or disconfirm the theory, a situation (to put it mildly) unpleasantly conducive to insider dealing.

had built according to these theoretical specifications, that would tell us whether the robot now enjoyed 'phenomenal awareness' or not?

Or suppose that we discover exactly the specified computational architecture, neuronally embodied, not only in human beings (we should *certainly* expect this, according to the theory) but also, say, in chimpanzees; but that we do *not* find it in dolphins. (Or in chimpanzees *and* dolphins, but not in rats? Or in all vertebrates, but not in flatworms? Etc., etc.) How should we establish whether those species fortunate enough to be endowed with the right sort of brain organization, embodying the theoretically required computational architecture, were indeed also blessed with consciousness (or 'phenomenal awareness'), while the other species, lacking some aspect of the computational architecture specified in the theory, were sadly denied any phenomenal awareness?

Unless we have criteria for identifying instances of the phenomenon that our theory supposedly explains, then, bluntly, *we do not know what we are talking about.*

Exactly the same need for criteria arises if the theory is applied only to humans. Is the experimental subject, engaged in a heavily demanding dual task, 'phenomenally aware' of the events occurring in *both* the concurrent tasks? *Equally* aware of both? How do we decide? Indeed what does it mean to assert one thing or the other? Does the human infant have 'phenomenal awareness'? Immediately at birth? Or only after some later developmental (linguistic?) stage is reached? Is it acquired gradually, or all at once? *How could we tell?* Does the right hemisphere of the callosally sectioned ('split-brain') patient enjoy 'phenomenal awareness'? What does such a claim specifically entail, and what (if anything) does it deny? These are all questions (there are, of course, many, many more) to which we require a definite answer if we are to have any usable idea of what our putative 'theory of phenomenal awareness' (or 'theory of conciousness') is supposed to be about. Most emphatically, if the theory is to be in any way *testable*, if it is to have consequences which could be empirically falsfied, then we must have independent criteria for identifying the presence or absence of the phenomenon that our theory is intended to explain.

Which phenomenon?

My own view is that there *are* no such (general) criteria, because *there is no such phenomenon.* That is, there is no unitary entity of 'phenomenal awareness'—no unique process or state, no *one*, coherently conceptualizable phenomenon for which there could be a single, conceptually coherent theory.

I do not deny the reality of phenomenal awareness, just as I do not deny the reality of *life*, or of *understanding*. Of course there is life; of course there is understanding. But 'life' is not a unitary phenomenon, susceptible of explanation in terms of a single explanatory principle. Neither is understanding. As Sloman (1985, 1987) has persuasively argued, our ordinary concept of 'understanding' denotes a complex set of prototypical capabilities or conditions, different subsets of which can be exemplified in different people, different animals or machines, yielding a complex space of possible systems. Our ordinary concepts, like 'understanding', are not suited to drawing global boundaries in such a space. To hope to do so, to ask 'Which capacities or which conditions are necessary for *real* understanding?' is to attribute a spurious precision to a concept in ordinary language. So too, I am convinced, with 'conciousness' and 'phenomenal awareness'. Wilkes (1988) and Smith Churchland (1988) (both in this volume) make this same point, in different ways, with great force and clarity. Also in this volume, Dennett (1988, 'Quining qualia') challenges head-on what many have wistfully (wishfully?) supposed to be the irreducible core of 'phenomenal awareness' —the so-called *qualia* or suchnesses of experience. I find his analysis, or rather his demolition of this incoherent notion, refreshing, and indeed liberating. *What* qualia, indeed?

Breakdowns of consciousness

How should we go forward? Like 'understanding', like 'life', our everyday concept of 'phenomenal awareness' denotes (with the wonderful imprecision characteristic of ordinary language, particularly when it comes to 'mental' concepts) a great range of different states of affairs. The important and exacting task that confronts us, therefore, if we are to make any scientific headway with the concept of 'consciousness', is to describe, to characterize in functional detail (including, of course, to provide behavioural and/or physiological *criteria* for the identification of), these different phenomena. Only then may we hope to ascertain whether there is, in fact, any identifiable property or properties in common between them. Certainly, and most emphatically, we should not simply *assume* that there is.

One heuristic for making salient to oneself this multiplicity or diversity of phenomena, clumped together under the notion of consciousness (or 'phenomenal awareness'), is to consider the multiplicity of ways in which *ordinary* or *normal* 'phenomenal awareness' is found to break down, or to become dissociated. Several of the contributors to this book have described fascinating neuropsychological cases—people in whom different domains of behaviour appear to have become pathologically dissociated from one another. In certain callosally sectioned patients, for instance, sensory information delivered to the right hemisphere can evidently be 'reported' or acted on appropriately by means of the patient's left hand, but not by his right hand and not by means of speech. The question is then sometimes posed: 'Is *the patient* phenomenally aware of the information presented to the right hemisphere?' Well, the patient's *left hand* appears to be aware of it—'aware' in the sense that it can *act* on it, *'report'* it. But, can a *hand* be aware? How we proceed to answer the question as put ('Is the *patient* aware . . . ?') depends not only on who (or what) we take 'the patient' to be, but also—of course—on what we take 'being aware' to be. Whatever answer we propose will need some further definition of terms.

'At least to a first approximation . . . that of which we are conscious is that of which we can tell, introspectively or retrospectively' (Dennett, 1978, p. 152). Suppose we adopted this proposal of Dennett's as a starting point—a first approximation—we should still need to decide what was to count as 'able to tell'. In the case of the split-brain patient is 'telling' something that may legitimately be done with the left hand, and only with the left hand . . . ?[3] Can a global aphasic 'tell'? To whom? To himself? If so, who is the agent of such an act of communication—of 'telling'—and who is its recipient? Are they different from one another? Two different persons, or only two different parts of *one* person? If the latter, does it follow that a *part* of a person can legitimately 'tell'? Alternatively, to the extent that the aphasic cannot 'tell', is he also, to that extent, deprived of consciousness?

As regards the split-brain patient, the question is also sometimes put: 'Is the patient's *right hemisphere* aware . . . ?' Again, in this case, we are left with a subsidiary agenda of questions, requiring further linguistic legislation. For instance, what set of entities can be denoted by 'we' —one of us—in Dennett's rubric? ('In so far as the actions of the left hand are concerned, then the right hemisphere is evidently aware . . .'. Would this do?) Is this a *different* awareness from that predicated of (enjoyed by?) the *patient*?

Certain patients with lesions in primary visual cortex are able to point

---

[3]In *Brainstorms* Dennett (1978) puts forward a sketch for a cognitive 'theory of consciousness', which includes among its components a crucial 'public relations unit', PR. The way it is described, PR is specifically a sentence- (or language-) production module, but the theory makes no provision for any other form of motor output, except through PR. It is not clear how far this peculiar restriction is actually important to the proposal. Are *all* kinds of motor actions acts of 'public relations', and (therefore?) also potentially phenomenal 'reports'—testimony to 'phenomenal awareness'? If not, then *which* classes of action are? The 'theory' seems to leave all the real theoretical spade-work still to be done.

or reach quite accurately to stationary objects or light flashes in areas of the visual field in which they nevertheless deny 'seeing' anything (Weiskrantz 1988, this volume). They are apparently unable to report these stimuli verbally—or at least to report them as *visual* objects or events, although some patients at least are prepared to acknowledge that they experienced *something*, even though they would not call it 'seeing'. Are such patients 'phenomenally aware' of these stimuli? Once again the question requires special linguistic (semantic) legislation to be enacted before it can be answered. ('Aware, but not *visually* aware, of a visual object?' Is that right?). Notice that, whatever semantic rearrangements we decide to adopt in order to arrive at a definite answer to the question, they are a *different* set of rulings from the by-laws we needed to enact to deal with the callosal ('split-brain') patient.

Yet other patients, with right parietal lesions, apparently neglect objects appearing to their left; and they neglect the left half of objects in front of them (Bisiach 1988, this volume). Requested to bisect a horizontal line by putting a mark where he judges the mid-point of the line to be, a typical 'neglect' patient may place his mark as much as three-quarters of the way over to the right of the line. Is this because he is not 'phenomenally aware' of the left half, or left portion, of the line? Before attempting an answer to that question, consider the result of a further test with the same patient. Suppose that the patient is requested to treat another horizontal line, identical to the previous one, as the base of an imaginary square, and to draw the right-hand-side of that square, i.e. with a line going upwards from the right end of the base line. In this case, the same patient now draws a vertical line of *exactly* the right length[4]. Is the patient phenomenally *un*aware of the left half of the horizontal line, so far as bisecting lines is concerned, but phenomenally *aware* of it when it comes to drawing squares [or, presumably as a closer approximation to the relevant conditional, when it comes to *actions* in the right (upper) quadrant of space—i.e. in the patient's 'good' space]? Once again there is manifestly a whole agenda of linguistic–semantic legislation to be gone through, before any such question can be intelligibly answered. The point to recognize, however, is that the semantic rejigging that would be needed is, in considerable measure, specific and different for each of these different neuropsychological conditions. And there are, of course, many other such conditions, each of which bespeaks a somewhat different dissociation of 'normal consciousness'.

If consciousness, or phenomenal awareness, can become fragmented,

---

[4] I am indebted for this observation to my colleagues Peter Halligan and John Marshall in Oxford. See also the observations by Bisiach (1988, this volume), which pose similar semantic conundra.

or dissociated, in these very diverse ways, in each case requiring the semantic furniture to be rearranged over again before we can begin to decide what is or what is not to count as an instance of awareness (or its absence), should this not prompt us to consider the possibility that, even in the 'normal case', the intuition that we have to do with a *unitary* phenomenon is, in fact, a conceptual mistake; that what we referred to as 'phenomenal awareness' is in fact a congeries of many and diverse kinds?

Before conceding this (to some, at least, possibly discouraging) con-clusion, I should return to my expressed objective: the search for positive criteria. Depending on the success, or otherwise, of such a search, we might then attempt to re-evaluate whether we can seriously pretend to *know what we are talking about*, when we propose—or indeed when we dispute—a scientific 'theory of consciousness'. To anticipate, I shall consider three broad types of possible criterion for the presence (or absence) of 'awareness of X'. All three, it turns out, encounter serious difficulties; furthermore, any one of these criteria can be found dissociated from (i.e. in the absence of) the other two. Ordinary language attributions of perceptual awareness ('O was conscious of X'), however, sit comfortably only when all of these criteria agree: in effect, that is, when a person's behaviour is fully *integrated* and *coherent*. Understanding of the neuropsychological processes responsible for the coherence and integration of behaviour, I am led to infer, may be as close as we are likely to get to a scientific understanding of 'consciousness'.

I now turn to this objective.

A criterion of potential action?

In all of the examples, above, of phenomenal (perceptual) awareness, the patients were able to act on or respond to information that they had picked up from the sensory environment. In common usage, it seems, to be aware of something or conscious of something carries at least the implication that 'something' can guide or control my choice of *action*. One of the complicating features of these neuropsychological cases was that the information, so encoded, might be able to control actions of *one* kind, or in one domain of action, but not in another. Granted this very important qualification, however (to which we must certainly return), we might wish to consider one possible, very broad *criterion* for the application of the everyday notion of perceptual awareness: that is, a criterion for deciding whether person or organism O was aware of event or situation X, under some description of X. This broad criterion is that

O could, in principle, act directly on (that description of) X, do some-thing about X.[5]

The sense of this (so far, very loosely phrased) criterion is that the intuitive or everyday notion of 'perceptual awareness' is indexed to a *behavioural disposition*, a conditional readiness to act on the object of awareness. (The same applies when the object of awareness is non-sensory; e.g. an imagined or remembered object. When such an entity is 'brought into consciousness', so this analysis runs, it can be acted on or commented on. When *not* 'in consciousness', then that entity can *not* be directly acted on or commented on.) Of course, there is no requirement that the person or organism actually carry out such an action, or actually make such a commentary. The criterion requires only that, in principle, they *can* do so.

The rationale behind this broad criterion is simply that, if O could *not* (was *in principle unable* to) act on or respond to X in any way whatever, either at that time or later, then there seems *no plausible other or further reason* for asserting that O had been aware of X (under that description). Certainly, we could have no means of *knowing* whether she was.[6]

The motivation for considering such a criterion of potential *action* is that it might permit us to go beyond, and in so doing perhaps to clarify, Dennett's (1978) notion of awareness as 'that of which we can *tell*'. (*Telling*, we surely wish to say, is something that can be done other than, or not only, through *language*?) However, the attempt to sharpen up any such would-be criterion quickly runs into difficulties.

Our ordinary language statements to the effect that 'O was (or was not) aware of X' are always, necessarily qualified by the rider 'under some description of X', even if this rider is usually implicit. ('I was aware of something AS an obstruction ahead of me/AS a startling sound/AS a gesture of sympathy/AS a question/AS something in Italian. . . .')[7] Awareness under one description in no way implies awareness in terms of *other* descriptions, or other aspects of X. Moreover, every such description, however vague, appears to constrain the range of appropriate actions which the state of affairs under that description would afford—i.e. would render appropriate—given any particular goals on the

---

[5] Note an important distinction (to which we return in a later section) between actions made *directly* in response to X, in contrast to *indirect* modulating or biasing effects of X on responses to objects and events *other* that X. The criterion, just broadly stated, specifies direct response to X.

[6] Any putative *physiological* indicator of 'phenomenal awareness' (should such an entity be discovered, or proposed), according to this criterion, could logically acquire such a status only by virtue of its observed correlation with other, *behavioural* indices.

[7] Of course there is no implication that the subject necessarily represents such descriptions *verbally*.

part of *O*. ('If I wish to go forward, I shall have to go around the obstruction. . . .')

So, what actions or what behaviours may we (or may we not) use, by this broad criterion, as indicators of 'awareness'? Rather, the important question is, what principled grounds can we put forward for *excluding* any particular class of behaviour as an indicator of awareness? If *O* neatly avoids the obstruction, is this not an indication, *in some sense*, of his awareness of the obstruction? If he replies in Italian, does this not indicate that he *recognized* ('was aware of') the language? If *O*'s pupils dilate, following the startling sound, is this not an indication of *some* kind of awareness of the sound?

If not, then where, exactly, did we draw the line, and why?

The obvious point is that *different* behavioural indicators tell us *different* things about someone's state of awareness. His *walking* (say, stumbling over tussocks, colliding with lampposts, or, instead, deftly stepping over or around such obstacles) can tell us about his awareness of the ambient spatial environment; his *talking*, at the same time, may tell us something quite different about a different aspect, or domain, or 'level' of his awareness. Similarly, his *words* may tell one story, his *tone of voice* another. Which channel reflects his *true* (unitary?) 'phenomenal awareness'?

In practice we manifestly do use, constantly, all kinds of non-verbal behavioural indicators—including autonomic or 'involuntary' ones—as cues to or evidence of another person's state of mind, or state of *awareness*. What a person *says* about himself by no means dominates all these other channels as regards what we are prepared to believe about his 'state of awareness'. Even aside from any possible intention to deceive, in everyday situations we are only too familiar with the conflicting indications that different aspects of a person's behaviour can give. It would be absurd to suppose that we do (let alone that we *should*) always and only believe what someone 'tells', in well-formed sentences of a language, about his mental life.

One proposal is that we should draw the line between 'voluntary' and 'involuntary' actions; that is, that we rephrase the proposed criterion of potential action as: '*O* is *aware* of that on which he can act *voluntarily*'. The trouble with this proposal is that it simply shifts the problem onto that of distinguishing 'voluntary' from 'involuntary' acts. Moreover, the proposal risks circularity since, in ordinary usage at least, excepting cases of external duress, there seems little distinction to be made between a 'voluntary' action and one 'consciously directed'. Thus 'consciousness', in terms of the criterion of potential action, is indexed in terms of *voluntary* acts, but what is, or is not, 'voluntary' is then to be defined in terms of *consciousness*.

If a 'voluntary' action is one that the agent *could have withheld*, we have the apparently intractable problem of establishing, in practice, of any particular, unique action whether this was the case or not. How to do so? In so far as the antecedent conditions of a given act can be precisely *repeated*, we may be able to establish a statistical statement about the probability of that class of action in those conditions. Even then, to establish that the agent could have withheld the action we shall surely need to alter at least some aspect of these antecedent conditions: at least, that is, we shall need to motivate the agent *not* to perform that action, in which case the antecedent conditions are no longer the same as they were. It is far from obvious, therefore, how such a distinction between voluntary and non-voluntary acts could provide any practicable basis for distinguishing actions that were or were not based on—and therefore potential indicators of—'conscious awareness'.

We also have the complementary problem, in that the criterion of potential action requires only that O be able, *in principle*, to act on any information of which he is consciously aware. Presumably we should not want to say that O was aware of X only if he did in fact act on X. In that case, how are we to establish that O could have acted, when he actually did not? The problem is evidently similar to the one that we have just encountered. To establish, of a given, unique occasion, that O *could have* acted when he actually did not appears no more tractable than to establish that O could have withheld that action, when he did in fact perform it.

Unless some other, principled grounds can be put forward for excluding any particular categories of behaviour, as indicators of some kind of underlying 'awareness', there would appear to be only two possible, consistent courses of action. The first is that we accept all categories of behaviour—i.e. all forms of 'direct response' to X—as indicators of awareness ('consciousness') of X, *of some kind*, in which case, as already suggested, we will presumably need to qualify those different 'kinds' of awareness. If we follow this course, that is, we shall need to recognize that 'awareness of X' (in so far as this is something we can say anything about) can be 'awareness' in respect of *one* category of behaviour, but *not* of another. In this case the implicit conception of the 'contents of consciousness' as some kind of representational *object* or *data structure*, passively available for contemplation by a unitary inner eye (or inner 'I') which is itself quite independent of, yet somehow able to command, all categories of overt behaviour, is evidently in need of serious revision. Alternatively, *no* possible behaviours can count as indicators of awareness, in *any* sense, in which case we are clearly back where we started, and we have no way of knowing—i.e. of *telling*—what we are talking about.

A memory criterion?

In ordinary usage, however, it would appear that the occurrence of overt action—even highly organized action—in response to or guided by complex environmental cues is not necessarily sufficient for the attribution of 'conscious awareness'. Thus of those periods of activity for which we appear afterwards to have no episodic recall, such as periods of driving (dressing, toothbrushing, etc., etc.) while thinking about something else (or thinking about nothing?), ordinary usage suggests that the activity was performed 'unconsciously', hence (presumably) without 'phenomenal awareness' of the actions or of the circumstances that guided them.

The basis for this self-ascription, indeed its only basis, appears to be the absence of subsequent (self-reported) memory for the events or actions concerned. This usage, in turn, might suggest the makings of a second possible criterion of 'conscious' activity or of consciously per-ceived events—a criterion, at least, that seems to be commonly applied in everyday use both as regards oneself as well as other people. The criterion is that the person be able to *remember* those events or that activity later. There are, of course, degrees of remembrance; likewise, according to this criterion, as also in everyday usage, there are degrees of awareness: 'I was only half aware (or dimly aware) of my surroundings, or of my actions . . .'. The basis for such a claim, a claim which is necessarily retrospective, is also necessarily dependent on memory.

Clearly not all information available to the senses is encoded in memory. To the contrary, memory encoding is highly selective. Conse-quently, if we are to provide anything like a scientific explanation of 'awareness', as loosely indexed by this (retrospective) *memory* criterion, what we shall need, among other things, is an account of the mental/computational/physiological processes (including, most importantly, the *selection* processes) necessary and sufficient for the encoding of new episodic memories. At least this would appear to be an enterprise clearly within the mainstream of cognitive science, or cognitive neuroscience. We should be able to make some progress *here*.

The memory criterion, however, could be, at best, reliable only as a *positive* indicator. There are many possible reasons why I may be *un*able to remember past actions or events. To keep the issues separate, suppose we grant that the actions in question were efficiently carried out, that they served to achieve a valid long-term goal (e.g. having clean teeth, preventing dental caries, or whatever), and that their occurrence at this place and at this time was fully appropriate and would, ordinarily, be 'intended'. That is, let us assume that the action is carried out entirely normally. I thus exclude cases of absent-mindedness in which, *besides* the absence of subsequent episodic memory, the performance is also in some way malformed or inappropriate (see Norman 1981). I suspect that it is quite commonplace for this to be the case. Although I have no quantitative observations to support it, I would not be at all surprised if the great majority of 'absent-minded' (in the sense of unrecallable) actions were performed, from the perspective of an outside observer, indistinguishably from equivalent actions that *could* subsequently be recalled. For the purpose of the present argument, all I require is that this condition can, at least, occur. Yet (let us suppose), the subject has no recollection whatever of these actions or events.

Let us now ask, during such a performance, what was the subject's *moment-by-moment state* of perceptual awareness? (Was there any perceptual awareness at all, and, if so, of what, and under what description?) In terms of self-report, it seems, the question is simply unanswerable. Given that the subject has no memory of performing the actions at all (that he simply finds himself, as it were, driving through Chalfont-St-Giles/fully dressed/with gleaming teeth, smelling of dentifrice, etc.), then *a fortiori* he cannot recall the *presence or absence* of incidental perceptual states that may or may not have occurred in the course of those actions. That is, on the basis of introspection, or self-report, there are no more grounds for *denying* moment-by-moment perceptual awareness during the absent-minded action than for affirming it.

Such a criterion evidently cuts, if at all, only along one edge. If the person *is* able to recall ('tell' about) past actions or events, then, to that extent, ordinary language seems to require us to acknowledge that the person must have been 'aware' of those events; the complete absence of recall, or failure of recall, on the other hand, leaves us uncertain. Was he 'aware' or not? Our intuitions falter. Our doubts about this case then prompt yet further doubts about the previously mooted criterion of immediate ('*voluntary*'?) action. Were these absent-minded actions 'voluntary'? How do we decide? Again the criteria seem puzzlingly elusive. To be 'voluntary', must an action also be recallable by the person who made it? That would hardly seem to be an essential feature; yet, in the absence of any such recall (or potential recall) we may look in vain for other, more decisive properties. And, if this is so, the threatened circularity of the suggested partitioning of voluntary and involuntary actions, as indicators of underlying 'awareness', looks even more problematic.

Once again, as regards the putative memory criterion, ordinary language does not serve us with sufficient precision. To begin with, as we have already argued, to be of any use to us the memory criterion, like any other criterion, must be an *explicit* (essentially, a behavioural) criterion. The memory recall, or recognition, must be overt recall, and overt recognition. As with the neuropsychological cases, we are confronted with troublesome problems of semantic legislation over what is to count as a sufficient indicator of recall. Problematic boundary conditions arise in the behaviour of normal subjects, no less than in the behaviour of neuropsychological patients.

To illustrate, consider once more the widely studied laboratory task in which an array of letters, numerals, or words is briefly flashed on a screen and the subject is requested to report (or recall) a designated subset. When subjects are asked to comment on whether, and under what description, they were aware of the identities of the *un*reported items, the typical response is simply puzzlement. One cannot tell. 'Did you consciously identify those items, but then instantly forget them? Or were you never actually aware of what they were?' There seems no way for the subject to answer such a question 'from the inside'. Indeed, and even more to the point, a similar uncertainty is commonly expressed about the *correctly* reported items: one is unsure whether one actually 'saw' the identity of those items or merely guessed them. In other words, in such circumstances there is no private, 'phenomenal' court of appeal, to which the subject can refer, that carries greater intuitive authority, even for the subject himself, than the record of his own actions, as evidence of what he 'saw'. Sampled over many similar trials, the experimenter may be able to build up a statistical picture of what information was available, on any one representative trial, to determine the subject's behavioural choices. There does not appear to be any other, better, or more direct index of what information was, or was not, 'consciously perceived'.

A confidence criterion?

Some authors (e.g. Cheesman and Merikle 1985) have proposed a criterion of self-evaluation by the subject: the experimental subject must not only be able to indicate correctly (e.g. among a set of alternatives) the stimulus identity that was presented, he must also indicate *confidence* in his own report, if it is to count as an indication of 'phenomenal awareness'. Cheesman and Merikle call the conditions under which the subject begins to assign greater-than-chance confidence in the accuracy of his own reports the 'subjective threshold'. They demonstrate that this 'subjective threshold' by no means coincides with the 'objective' threshold of correct reports, but is far above it. Their proposal is an interesting one. Cheesman and Merikle are surely correct in that, in ordinary usage, the application of a criterion of being able to 'tell' (to report, or recall, or discriminate) also requires some acknowledgement, on the part of the subject, of a degree of belief or confidence in what he tells, if the telling is to count as an indicator of 'awareness'.[8] If this is correct, it should prompt us to consider what antecedent psychological, or neurobiological, conditions must occur for a subject to acknowledge confidence in his own memory recall (cf. Squire and Cohen 1984; Gabriel *et al.* 1986). Here, too, we would seem to be in a position to make further progress.

For a 'confidence' criterion of this kind to have any meaning, it would seem a logical requirement, from the outside, that its contrast term also be applicable; that is, it must be possible for the same individual, on another occasion, to signal that he does not feel such confidence. Without language, such indicators may be harder to read, but they are surely not thereby excluded. (For example, what risks is an individual prepared to take, on the basis of what he 'believes'/'perceives'/'is conscious of'?) Is it perhaps—among other factors—the limited degree to which contrasting indicators of such confidence or lack of confidence ('belief' or lack of 'belief'?), are available in other species that provokes our hesitation, or unwillingness, to attribute awareness (or lack of awareness) to a range of creatures other than our own kind? (To human infants? To certain kinds of computational device?) Indeed, if intimate acquaintance with individuals of other (non-linguistic) species enhances our ability to read such signals could this be an important contributory factor to the (surely very common) enhanced readiness to attribute 'consciousness' to these particular individuals?

Dissociations between criteria

Manifestly there are many different ways and means of 'telling', and many more kinds of (non-linguistic) behavioural indicators, regarding what information has or has not been effectively perceived or encoded by a given individual. When a neurological patient with hemi-neglect of the left side of space collides with the left door-post, it seems not unnatural, in ordinary language, to describe him or her as 'unaware' of the door-post—at least 'unaware' in respect of his own movements directed toward that part of space. As already noted, he need not, for that reason, appear unaware in every *other* respect.

In normal subjects also, as in the neuropsychological patients, there can be striking dissociations between different behavioural indicators.

---

[8] Clearly there are problems of boundary conditions. *How much* confidence is required for the subject's perceptual report (or forced-choice discrimination) to count as an indicator of 'phenomenal awareness'? The proper response here is, presumably, that phenomenal awareness is *always* a question of degree. Certainly by this criterion it is.

In multi-channel selective listening, for example, there are numerous demonstrations of autonomic (GSR) responses to semantically-defined target words occurring in the ignored or unattended channel, despite the experimental subject's failure to report their occurrence by means of a (pre-instructed) manual or vocal gesture (e.g. Corteen and Dunn 1974; Forster and Govier 1978).

With visually superimposed outline forms, such as line-drawings of objects or letters of the alphabet, the subject can be instructed to attend (say) to the red figure and to ignore the superimposed green one (or vice versa). Under suitable conditions, his immediately subsequent attempts to recall or to recognize the identity of the to-be-ignored, green figure may be no better than chance (e.g. Rock and Guttman 1981). Despite this, the identity, and at least some semantic attributes of the successfully ignored object, systematically affect his responses to a subsequent, semantically or categorically related target object. That is, his response to the related object is consistently delayed: there is a 'negative' priming effect (Allport *et al.* 1985; Tipper 1985). We find this negative priming effect, even when the figural properties of the (ignored) priming stimulus and of the subsequent probe stimulus, the response to which is now inhibited, are radically different; for example, when the (ignored) priming stimulus is a line-drawing and the probe stimulus is a (semantically related) printed word (Tipper and Driver 1988). The negative priming appears greatest if the subsequent probe stimulus has the same abstract identity as the prime (equivalently, for example, with upper- and lower-case versions of the same letter); it is also found, though reduced in amplitude, when the prime and probe stimuli are only conceptually related (e.g. 'hammer' and 'wrench', or 'trumpet' and 'guitar'). Negative priming by an actively ignored stimulus is also found when the behavioural indicators (the 'responses'), respectively, to priming and probe stimuli are in different domains, using different effectors—e.g. a vocal naming response in one case, a manual classification response in the other (S. P. Tipper, personal communication). Finally, as one of the most important antecedent conditions of this phenomenon, negative priming is obtained from an actively ignored stimulus if and only if that stimulus accompanies another stimulus which is itself successfully and explicitly identified (Allport *et al.* 1985).

In this experimental paradigm all the direct behavioural indicators that we have thus far been able to tap are in agreement: the subject is unable to *tell*—to report or to make forced-choice guesses about—the identity of the actively ignored object with better-than-chance accuracy, or indeed to acknowledge having seen but then forgotten its identity. In contrast, his subsequent behaviour is manifestly *affected by* the identity of that same, ignored object. It follows that the object's identity, and certain of its semantic properties, must have been encoded in some form in the subject's nervous system. On the whole it accords rather comfortably with conventional usage (and with the notional criterion of confident ability to recall) to say that the subject, in this case, did not 'consciously' perceive the identity of the ignored object, but that, nevertheless, its identity was 'unconsciously' encoded—or possibly 'unconsciously perceived'. Examples of this kind suggest another possible basis for classifying behavioural measures, namely in terms of a contrast between so-called 'direct' vs. 'indirect' behavioural effects. Direct effects are those in which the behaviour that is recorded is elicited ('directly') in response to a given stimulus, X; indirect effects are those in which the simultaneous, or prior, occurrence of stimulus X *influences* the direct response to stimulus Y, for example by facilitating or inhibiting the latency or the probability of successful response to Y. In terms of this, admittedly not very well-defined, classification, all three sorts of criteria discussed above (potential action, recall, self-evaluation) are clearly concerned only with 'direct' measures. *Indirect* behavioural effects, in terms of the intuitions underlying conventional usage which we are here attempting to explore, are *not* sufficient, in themselves, to act as indicators of 'phenomenal awareness' of stimulus X; i.e. aware-ness of the stimulus exerting such indirect effects (cf. the parallel example of memory facilitation ('savings') without direct, or explicit, acknowledgement of memory).

Unfortunately it is *not* always the case that different direct indicators (different modes of 'telling') are in agreement with one another. A provocative example is provided in a series of experiments by Cumming (unpublished thesis, 1972). Cumming's subjects were shown a horizon-tal row of five letters flashed in rapid succession, one after another. The spatio–temporal arrangement of the letters was designed to produce, at rapid rates of presentation, a form of metacontrast masking known as 'sequential blanking'.

(In this arrangement, the first letter to be displayed is in the *second* position from the left in the horizontal row; the offset of this letter is followed im-mediately by the onset of the next letter, in the position *fourth* from the left; there follow, in turn, each of the remaining three letters, namely in the centre of the row and in the leftmost and rightmost positions. At a critical, rapid rate of succession between the five letters, subjects who are requested, at leisure, to describe what they see (or rather, what they *saw*) in the display typically describe only three letters as present, namely in the centre and in the two outer positions. The first two letters in the temporal sequence are not reported. They are 'phenomenally absent'.)

Under these conditions, adjusted for each individual subject to pro-duce optimal 'sequential blanking', the subject was given a visual search task. The instructions were to press one key if the letter *J* (for example) was present in the display, and to press another key if no *J* was present. In each display sequence, five letters were physically presented, all different from each other. Under only modest pressure from the experimenter to respond rapidly, subjects consistently pushed the 'target-absent' key when the designated target letter in fact occurred in one or other of the two 'blanked' or masked positions. However, when urged to respond as fast as possible, even at the cost of making a good many errors, subjects now tended to respond to the occurrence of a target letter in the 'blanked' positions with a fast (and *correct*) press of the 'target-present' key, and then, a moment later, to apologize for having made an error.

What are we to say about these subjects' 'phenomenal awareness'? Are we to treat *both* their rapid, first response as well as their subsequent, and slower, acknowledgement of an error as equally valid acts of 'telling' what they saw? On those terms, we are presumably bound to say that the subject was indeed, momentarily, 'phenomenally aware' of the target letter, but that this awareness was short-lived and was not accompanied by any subsequent, explicit memory for the event. That is, the transitory phenomenal awareness was succeeded by instant oblivion, or (perhaps better) by instant replacement of one phenomenal state by a different and contradictory one, leaving no memory record of the first. Or do we decide, somewhat arbitrarily, to downgrade one form of 'telling' relative to the other? But which? If we propose to disqualify the *first*, rapid gesture on the grounds that, if it is to be made at all, it can only be made within a rather narrow time-window, we seem to violate some of our strongest intuitions about 'phenomenal awareness'—that awareness is forever fleeting, transitory, time-dependent. On the contrary, the *later* act of telling is inevitably more retrospective, more memory-dependent; and we already noted that the memory criterion was necessarily unreliable as a *negative* indicator of awareness. Neither solution appears obviously correct.

Many examples could be provided of the complementary dissociation: the inability to respond to discrete perceptual events 'on-line' (i.e. one for one, as they occur), although the subject *is* able to recall or report the events subsequently, 'off-line'. A classic instance occurs in the task of judging the number of successive, discrete events in a rapid train of stimuli—so-called 'temporal numerosity' judgements (White 1963). People are able to report (retrospectively) the numerosity of sequences of events occurring at rates far above those at which they could respond directly to each event—for example by vocal (or subvocal) counting or finger tapping. This example may stand, here, for a wide range of situations in which the flux of perceptual events exceeds the rate-limiting characteristics of 'on-line' response, but in which 'off-line' recall of those events is in some measure possible.

It is clear, therefore, that each of the three broad classes of criterion that we have examined may be dissociated pairwise from each of the others. We can have the possibility of immediate, 'direct' response in the absence of either subsequent recall or self-evaluation, as illustrated in Cumming's experiments, and in many cases of 'absent-mindedness'. We can have recall, including even confident self-evaluation of that recall, in the absence of potential, immediate response, as illustrated in judgements of temporal numerosity (among many possible examples). We can also have correct retrospective report, or recall, but lack any confident self-evaluation, as, typically, in many different tachistoscopic report tasks (already briefly discussed). And we can also observe the converse—the confident assertion by someone that he had been temporarily aware of a given stimulus identity (dream content, or whatever) but that he had forgotten it before he could get around to any direct response to, or recall of, that identity.

These three broad categories of behavioural criterion were deliberately chosen to be as comprehensive as possible. However, we have already noted that dissociations *within* these broad categories, as well as between them, are also commonplace. Furthermore, as with the neuro-psychological dissociations, when these everyday behavioural criteria are in conflict with one another we are typically inclined to hedge our everyday attributions of consciousness, or awareness of 'X'. Maybe there was awareness 'in *some* sense', but then maybe not, 'in *another* sense. . .'. Unless, or until, these different senses of the everyday concept of consciousness, or awareness, can be clearly identified and separated from one another, *and* clear criteria established for deciding when we have (or do not have) an instance of one kind or another, or of any kind at all, there seems rather little prospect of scientific utility in the 'concept of consciousness' that we actually have. *Which* concept? Or rather, *what* concept?

Ordinary language and the paradigm case

My search for criteria has proved disappointing. (I would be the first to acknowledge that the search has, so far, barely scratched the surface. Nevertheless it has been a genuine search. It is at least less than obvious to me what other, radically different, broad categories of criterion to try.) For some people, I find, this was evidently a foregone conclusion. 'Of course you will never find objective criteria for consciousness (or for "awareness of X")', they say, 'for the obvious reason that consciousness (awareness) is a purely and quintessentially *subjective* phenomenon'. The odd thing about such a claim is that those who make it evidently wish to apply this claim to 'consciousness' in others, as well as in themselves. In practice they use expressions of the form '*O* was conscious/was not conscious of *X*' (etc., etc.) just as regularly of other people as of themselves. How do they know when to apply such an expression if it denotes a 'purely subjective phenomenon'? (What *is* a 'purely subjective phenomenon'? Could we ever agree about the identification of one?)

The explanation, I suspect, is that it *is* possible to characterize the outlines of an ideal or paradigm case, in terms of behavioural dispositions, to which ordinary language expressions of this form ('*O* was conscious of *X*') clearly and properly apply. Such an expression applies in conditions in which the speaker believes that person *O* (either the speaker or someone else) *could, in principle* (and at the appropriate time), *have acted directly on X* (in any way that *O* chose compatible with those conditions); that, furthermore, *O could subsequently have recalled X; and* that, if asked, *O could* (in suitable circumstances, at the appropriate time, etc.) *have confidently acknowledged* or testified to his awareness of *X*. This is, of course, precisely the case in which ALL the different behavioural dispositions that we have considered are deemed to apply concurrently.[9] [To put this claim another way, *O*'s 'consciousness of *X*', according to this account of ordinary usage, denotes that entire state of affairs (internal to *O*) which constitutes the necessary and sufficient condition for *O* to have these behavioural dispositions towards *X*.]

As soon as we find grounds for believing that any one of these behavioural dispositions does not apply—for example, we find some category of *O*'s behaviour that is inconsistent with any one of these generalized dispositions towards *X*—then our everyday attribution to *O* of 'consciousness of *X*' falls into question, or has to be hedged with qualifications. (*O* could not have been *'fully'* conscious' of *X*; he must have been conscious of *X* only 'at one level', but not at 'another level'; etc., etc.) This is why the dissociations between different behavioural indicators, which we have discussed, appear so troublesome and problematic. They are problematic precisely because our everyday concept of consciousness is *defined on*, or *modelled on*, the *paradigm* case, in which *all* the criteria agree. However, the paradigm or defining case offers no principled means of deciding, when these different indicators

---

[9] What is 'subjective' here (if anything) are the speaker's grounds for such belief. The difficulty of establishing criteria for any such 'paradigm case' has more to do, I suspect, with the problems of establishing *behavioural dispositions* than with the ineffability of qualia.

are in conflict, which among them offers a valid indication of consciousness or lack of consciousness; attributions of consciousness, in so far as they are made, appear correspondingly strained.

The paradigm case, in which all these behavioural dispositions are in agreement, is arguably also the case in which a person's behaviour (or potential behaviour) is fully *coherent*, *appropriate*, and *integrated* with respect to *X*. According to this analysis, the expression 'conscious of *X*' is useful as a shorthand or portmanteau term to denote (the entirety of) those underlying conditions sufficient and necessary to integrated behavioural dispositions towards *X*. The expression is serviceable in much the same way that terms like 'integrated', 'coherent', and 'appropriate' are serviceable. It would presumably be a vain enterprise to look for *criteria*, in general, say, for 'appropriateness'. But this does not mean that one cannot meaningfully characterize a given piece of behaviour as 'appropriate' or 'inappropriate' to a given set of circumstances. Far from it. It does mean, however, that we are unlikely to come up with some universal, explanatory principle of the 'phenomenon of appropriateness' in human (or other) behaviour.

Concluding discussion—behavioural integration

In the paradigm case, just discussed, *O*'s behaviour is integrated and coherent, in that *all* of *O*'s actions can be guided by the *same* (or congruent) information, *X*. In contrast, throughout this chapter I have considered a variety of cases in which *O*'s behaviour is *not* fully coherent or integrated, but is in various ways *dissociated*. Some categories of behaviour can be guided by *X*, or directed towards *X*, whereas others cannot. In these cases it appears that information about *X* is not uniformly available to control different action systems; *and*, in these cases, ordinary language apparently refers to a dissociation of '*consciousness*'. It may be worthwhile, therefore, to explore the relationship between the ordinary language concept of consciousness and issues of behavioural integration.

Clearly, behavioural integration is essential for survival. An individual in whom the left hand disrupts what the right hand is doing, or embarks on some unrelated and incompatible activity, is in serious practical difficulty. Someone who says one thing and does another (that is, someone who cannot help so doing, and who cannot tell that he is doing so) is in serious social difficulty too. Moreover, in both cases we are inclined to refer to an impairment, or a dissociation, of 'consciousness'.

Many of the computational problems of a behavioural system, problems of perception and action, can be characterized as a search for the optimal solution (sometimes, for *any* solution) *consistent* with a potentially very large number of internal and external constraints (e.g. Waltz 1975; Winston 1984). How the massively parallel computational processes in the brain achieve overall coherence, and integration, is largely unknown. In current 'connectionist', neural-network models (e.g. Ballard *et al.* 1983; Rumelhart, McClelland *et al.* 1986), the fundamental process that is captured or realized by such networks is a kind of parallel 'search'—a process of gradual *settling* towards a state of the network in which all the constraints operating on, and within, the network are reciprocally satisfied. (For this reason, such models are sometimes described as 'constraint-satisfaction networks'. The network is incapable of remaining, stably, in any activation state in which the states of different subcomponents, or subparts, of the network are mutually *in*consistent.) Here, then, in barest outline, is one very general class of mechanism for the achievement of computational consistency.

Interestingly, the paradigm case of 'phenomenal awareness' is seldom, if ever, satisfied under conditions of time pressure. Paradigmatic 'phenomenal awareness', as well as global self-consistency within neural networks, appears to require a greatly reduced rate of flux of external (and internal) events.

Clearly, *local* consistency in such networks can be achieved faster than global or overall consistency. The brain is very highly modular in organization: where neural computations can be done locally, there are evidently advantages (particularly of speed) in keeping them local; thus there are evolutionary pay-offs not only for globally integrated processing but also (particularly under time pressure) for relatively local processing. Different subsystems in the brain are close-coupled only if they *need* to be.

However, the functional interconnectedness of different cognitive subsystems in the brain must be subject also to very rapidly acting, *dynamic* modulation for the demands of action. One mechanism that appears to be important in creating the relative coherence of 'normal' (or optimal) sensory–motor organization is a process which I have called 'selection for action' (e.g. Allport 1980*a*, 1980*b*, 1987). Many kinds of action can, potentially, be controlled (or guided, or triggered) by a very large range of different environmental stimuli. When many such, potentially conflicting stimuli are present—when picking apples from a tree, to take a familiar example—information about all but a particular subset (typically, all but *one*) of these objects must be, in some way, selectively decoupled from the *direct* control of the action (reaching out and grasping one apple), although such information may still be used,

180 *Alan Allport*

indirectly, to modulate that action—e.g. so as not to knock other apples off the branch on the way, etc. The behavioural evidence suggests that, in these circumstances, the information that is *not* directly selected for action—i.e., in this example, for the action of picking an apple—is also selectively decoupled from potential, direct control of *other* categories of action, and also from formation into a subsequently recoverable, explicit memory. In other words, the pre-conditions of 'selection for action' and of 'selection for memory' are very closely related. In terms of the ordinary language criteria that I have tried to explore here, O is 'not conscious'—at most, O is conscious only 'in *some* sense' (not the *paradigm* sense)—of those objects that are not directly selected-for-action[10].

Clearly, many different mechanisms in the brain contribute to the coherence or integration of its end-product, behaviour. The important point is that these *integrative* processes can be studied experimentally, both at the behavioural and at the neurobiological level; and they can, in principle, be modelled in computational devices, in the form of PDP[11] networks, etc. If the relationship that I have gestured at here between questions of 'consciousness' (or 'phenomenal awareness') and questions of *behavioural integration* has any validity, then, in studying these integrative processes we may, after all, not be so far removed from the scientific investigation of the mechanisms of consciousness.

References

Allport, D. A. (1980*a*). Patterns and actions: cognitive mechanisms are content-specific. In *Cognitive psychology: new directions*, (ed. G. Claxton), pp. 26–64. Routledge and Kegan Paul, Andover, Hants.

Allport, D. A. (1980*b*). Attention and performance. In *Cognitive psychology: new directions*, (ed. G. Claxton), pp. 112–153. Routledge and Kegan Paul, Andover, Hants.

Allport, D. A. (1987). Selection for action: some behavioral and neurophysiological considerations of attention and action. In *Perspectives on perception*

[10] The process of selection for action has also been linked, in some theories, with the process of perceptual integration by which multiple different domains of sensory coding (e.g. different, modular 'feature domains') are bound together in the representation of a given perceptual object (e.g. Kahneman and Treisman 1984; Treisman 1986). This theoretical linkage is controversial, however (e.g. Allport *et al.* 1985). In any case, effective perceptual integration is presumably essential, too, for coherent (non-dissociated) behaviour. Interestingly, failures of perceptual integration, frequently observed, for example, in tachistoscopic experiments (e.g. Styles and Allport 1986), also provoke confused 'phenomenal reports' by the experimental subject ('I don't know whether I *saw* it or not . . .').

[11] Parallel Distributed Processing (c.f. Rumelhart *et al.* 1986).

and action, (ed. H. Heuer and A. F. Sanders), pp. 395–419. Lawrence Erlbaum Associates, Hillsdale, NJ.

Allport, D. A., Tipper S. P., and Chmiel, N. R. J. (1985). Perceptual integration and postcategorical filtering. In *Attention and Performance*, Vol. XI, (ed. M. I. Posner and O. S. M. Marin), pp. 107–132. Lawrence Erlbaum Associates, Hillsdale NJ.

Ballard, D. H., Hinton, G. E., and Sejnowski, T. J. (1983). Parallel visual computation. *Nature*, **306**, 21–26.

Bisiach, E. (1988). The (haunted) brain and consciousness. In *Consciousness in contemporary science*, (ed. A. J. Marcel and E. Bisiach), p. 101. Oxford University Press.

Cheesman, J. and Merikle, P. M. (1985). Word recognition and consciousness. In *Reading research: advances in theory and practice*, Vol. 5., (ed. D. Besner, T. G. Waller and G. E. MacKinnon), pp. 311–352. Academic Press, New York.

Churchland, P. S. (1988). Reduction and the neurobiological basis of consciousness. In *Consciousness in contemporary science*, (ed. A. J. Marcel and E. Bisiach), p. 273. Oxford University Press.

Corteen, R. J. and Dunn, D. (1974). Shock-associated words in a non-attended message: a test for momentary awareness. *Journal of Experimental Psychology*, **102**, 1143–4.

Cumming, G. D. (1972). Visual perception and metacontrast at rapid input rates. Unpublished D.Phil. thesis University of Oxford.

Dennett, D. C. (1978). *Brainstorms*. MIT Press, Boston, MA.

Dennett, D. C. (1988). Quining qualia. In *Consciousness in contemporary science*, (ed. A. J. Marcel and E. Bisiach), p. 42. Oxford University Press.

Forster, P. M. and Govier, E. (1978). Discrimination without awareness? *Quarterly Journal of Experimental Psychology*, **30**, 289–95.

Gabriel, M., Sparenborg, S. P., and Stolar, N. (1986). An executive function of the hippocampus: pathway selection for thalamic neuronal significance code. In *The Hippocampus*, Vol. 4, (ed. R. L. Isaacson and K. H. Pribram), pp. 1–39. Plenum Press, New York.

Johnson-Laird, P. N. (1983). *Mental models.* Cambridge University Press.

Johnson-Laird, P. N. (1988). A computational analysis of consciousness. In *Consciousness in contemporary science*, (ed. A. J. Marcel and E. Bisiach), p. 357. Oxford University Press.

Kahneman, D. and Treisman, A. M. (1984). Changing views of attention and automaticity. In *Varieties of attention*, (ed. R. Parasuraman, R. Davies, and J. Beatty), pp. 29–61. Academic Press, New York.

Norman, D. A. (1981). Categorization of action slips. *Psychological Review*, **88**, 1–15.

Rock, I. and Guttman D. (1981). Effect of inattention on form perception. *Journal of Experimental Psychology: Human Perception and Performance*, **7**, 275–285.

Rumelhart, D. E., McClelland, J. L., and members of the PDP Research Group. (1986). *Parallel distributed processing: explorations in the microstructure of cognition. Two Vols.* MIT Press, Cambridge, MA.

Sloman, A. (1985). What enables a machine to understand? *Proceedings of the International Joint Conference on Artificial Intelligence.*

Sloman, A. (1987). *The space of possible minds.* Cognitive Science Research Reports, No. 28, University of Sussex.

Squire, L. R. and Cohen, N. (1984). Human memory and amnesia. In *Neurobiology of learning and memory,* (ed. G. Lynch, J. L. McGaugh, and N. M. Weinberger), pp. 3–64. Guilford Press, New York.

Styles, E. A. and Allport, D. A. (1986). Perceptual integration of identity, location and colour. *Psychological Research,* **49**, 189–200.

Tipper, S. P. (1985). The negative priming effect: inhibitory priming by ignored objects. *Quarterly Journal of Experimental Psychology,* **37A**, 571–90.

Tipper, S. P. and Driver, J. (1988). Negative priming between pictures and words in a selective attention task: evidence for semantic processing of ignored stimuli. *Memory and Cognition,* **16**, 64–70.

Treisman, A. M. (1986). Properties, parts and objects. In *Handbook of perception and human performance, Vol. 2,* (ed. K. Boff, L. Kaufman, and J. Thomas), pp. 35-1 – 35-70. John Wiley, New York.

Waltz, D. (1975). Understanding line drawings of scenes with shadows. In *The psychology of computer vision,* (ed. P. H. Winston), pp. 19–91. McGraw-Hill, New York.

Weiskrantz, L. (1988). Some contributions of neuropsychology of vision and memory to the problem of consciousness. In *Consciousness in contemporary science,* (ed. A. J. Marcel and E. Bisiach), p. 183. Oxford University Press.

White, C. T. (1963). Temporal numerosity and the psychological unit of duration. *Psychological Monographs,* **77**, No. 12.

Wilkes, K. V. (1988).——, yìshì, duh, um, and consciousness. In *Consciousness in contemporary science,* (ed. A. J. Marcel and E. Bisiach), p. 16. Oxford University Press.

Winston, P. H. (1984). *Artificial intelligence* (2nd. edn.) Addison-Wesley, Reading, MA.

8

Some contributions of neuropsychology of vision and memory to the problem of consciousness

Lawrence Weiskrantz

Each of us will have his or her own idea of what, if anything, is meant by 'consciousness', and what its value might be as a concept, or cluster of concepts, in scientific discourse and theory. But to insist that the value must depend, as a *prerequisite*, on the availability of a precise definition would, I think, be a mistake. Indeed, if we always insisted on precise definitions we all would be speechless almost all the time. Definitions and precise theoretical constructs are the final product, not the starting point of enquiry.

The actual starting point for this chapter derives from certain neuro-psychological observations and reports. In particular, it deals with the reports of patients' 'awareness' of events and capacities in the fields of vision and memory. The problem of 'awareness' and the neurological organization and structure implied by the term were forced upon me by the evidence: they did not emerge from some mysterious vaporous and ill-defined mist, but directly from the empirical evidence of patients themselves. I would like to review some striking disjunctions between verbal reports and patients' actual demonstrable capacities, and then consider some of the implications, within an evolutionary framework. Although I am not providing a definition at this or a later stage, it will become clear, I trust, that I am referring only to a restricted domain of what it means when someone says he 'is (or is not) conscious of something', rather than to embrace an exhaustive or global account of all usages of the term, including some of the usages of some of the other authors in this volume. My main emphasis will be upon the distinction between processing, responding, discriminating—on the one hand— and being able to acknowledge awareness of or to offer a commentary on such functions, on the other hand.

There is nothing inherently strange in our ability to carry out quite complicated tasks without awareness. On the contrary, it could be argued that it is a rare privilege for certain kinds of activities to require awareness. Many motor skills can be performed—are best done, in fact—without reflection. Many bodily processes are sensitive to environmental events without our direct knowledge. The pupillary response, for example, changes continuously in response not only to changes in illumination, but also to the hedonic value of environmental stimuli. And yet no one has a direct awareness of his pupillary response *per se*. There are so many examples of such bodily processes that they probably outnumber considerably those of which we are aware. What *is* strange is finding that there are patients who are *un*aware of events of which we normally expect them—like us—to be aware, even to be vividly aware. Moreover, they are severely disabled by their lack of awareness. There are two categories of such patients I want to consider.

The first are patients with the 'amnesic syndrome'. This condition became best known in recent decades through reports of the patient H M, in whom a surgeon ablated both medial temporal lobes in order to attempt to control an epileptic condition (Scoville and Milner 1957). But in fact such patients have been known to neurologists for a long time through a variety of causes, not only surgery, such as herpes simplex encephalitis, chronic alcoholism, various kinds of poison. Certain features of the amnesic syndrome are present also in pre-senile patients, such as those with Alzheimer's disease, although they are also typically compounded with a host of other cognitive disabilities. I do not wish to discuss the neuropathological question of whether these various aetiologies lead to changes in common, critical sites in the brain, although an argument can be advanced that they do. Nor do I wish to address the question here of whether there may be several different types of amnesic patient, and the complications injected into the field by the study of less than severely affected patients or by associated, but non-obligatory, cognitive deficits. All of these issues have been considered elsewhere (Weiskrantz 1985).

The most striking clinical feature of the amnesic syndrome patient is that he reports not remembering anything new from minute to minute. Moreover, he typically *behaves* in a 'memory-less' way: it is a crippling disease. This is despite there being no necessary drop in intelligence or any necessary deficits in short-term memory, such as the digit span or formally related short-term memory tasks such as the Brown–Peterson paradigm (Brown 1958; Peterson and Peterson 1959), or any loss of well-established skills, such as speech. Thus a patient will show no recognition of having met someone if that person leaves the room and then returns after an interval of just one or two minutes. He cannot learn a list of random paired-associate words even after many trials, nor does he show recognition of lists of words or faces after a minute or two. These features of the amnesic patient's condition are very well known and their occurrence, despite normal short-term memory—taken together with evidence of other patients who are not amnesic but *do* have defective short-term memory—is one of the main pieces of evidence that led to the postulation of multiple memory systems (Baddeley and Warrington 1973; Warrington 1982).

What is less well known, and is still the subject of active current research, is that the patients can be shown to be capable of efficient learning of certain types of tasks, and can demonstrate good retention of them over days or even weeks. The list is now quite long: motor skill learning; visual discrimination learning; cued recall of pictures and words; retention of learning of anomalous pictures in the McGill anomalies test; rule-governed, verbal paired-associate learning and retention; retention of stereoscopic perception of random-dot stereograms; retention of the McCulloch colour-grating illusion; retention of facilitation for solving specific jigsaw puzzles; retention for arranging specific words into specific sentences; classical eyelid conditioning; mathematical problem-solving; and mirror reading (Warrington and Weiskrantz 1982; Weiskrantz 1985 for summaries).

There is considerable theoretical discussion in the current literature of the best descriptive and conceptual characterizations of the examples in this list, but one property that they share is probably necessary, if not sufficient: in none of the tasks in question does the demonstration of learning and memory require the subject to be asked a question directly about what he 'remembers'. Unlike tasks such as free recall or yes/no recognition, where a test inevitably entails such a request, in the tasks listed above the subject is given a cue or stimulus or task and simply asked to provide the appropriate response, such as 'What word fits these initial three letters?' or 'What is the sentence that these random words can form?' or 'Try to keep the spot on the target on the rotating drum'. Evidence of learning and retention is derived directly from measurements of improvement or sustained performance, compared to performance on control tasks on which there has been no experience.

The reason why such a property is important is that it is precisely on questions that directly involve 'memory' that the amnesic patient fails. If he is asked to recite the words that he 'remembers' from a list he was shown a few minutes before, he will perform poorly. If shown the first few letters of each word, however, and asked to complete the word without regard to whether or not he 'remembers' it, he is likely to perform well. This disconnection of the patient's experience of 'remembering' can be shown in striking fashion by interjecting 'memory'-type questions directly into the sequence of a conditioning task that the amnesic subject can learn and retain successfully. He will say he cannot remember anything about the task or what he is doing in the situation, and then immediately afterwards will produce the acquired response to the presented stimulus (Weiskrantz and Warrington 1979).

.While the 'memory-devoid' type of question is probably a necessary feature of tasks yielding successful performance, it is not however a sufficient one. The properties of the cognitive structure itself are also important. Thus an amnesic subject can learn paired-associate words if they are rule-governed (either semantically or phonetically), but not if they are random and unrelated (Warrington and Weiskrantz 1982; Winocur and Weiskrantz 1976), even though the instructions for the two tasks can be identical and there is no explicit reference to the rule. In general, it appears that two general categories of tasks yield successful performance: (a) facilitation through priming, and (b) acquisition of new stimulus–response and stimulus–stimulus relationships of the kind where the through-put from the stimulus can be made directly and with minimal interference to the associated stimulus or response, i.e. can ultimately be made routine and 'automatic'. (Within both of these categories, it may turn out that certain types of events or information are differentially advantaged or disadvantaged; e.g. Schacter *et al.* 1984). All of the tasks listed in the catalogue above fit into one or other (or both) of these categories (Warrington and Weiskrantz 1982).

I shall return later to consider why, despite his impressive list of successful learning skills, the amnesic subject is nevertheless so severely crippled in his everyday life, but for the moment the point to be stressed is the disconnection of the experienced 'memory' for events from the excellent retention of the effects of exposure to those events or situations, for which there is such good evidence. The only person who cannot 'be convinced of that evidence is the amnesic subject himself, because he has no 'memory' of that for which he is showing evidence of retention. He does not act 'from memory', nor can he 'search his memory' if he does not have it.

The second type of patient I wish to describe is affected by lesions in the cerebral cortex at the stage at which visual inputs are normally first received, the striate cortex ('V-1'). There is a well-known and orderly projection of each visual half-field onto each contralateral cerebral hemisphere such that lesions in any particular region of the striate cortex cause restricted regions of 'blindness' in the visual field. Clinically one of the most common visual field defects is one in which a half-field of vision is missing, a 'hemianopia', caused by damage to striate cortex (usually intruding into surrounding cortex as well) in the cerebral hemisphere contralateral to the blind half-field of vision. It was thought for a long time that those regions were absolutely blind. Patients typically say they do not see lights or patterns projected into such a 'blind' region of their fields.

That this was so was becoming increasingly paradoxical, because the retina projects not only to striate cortex, but independently to another five or six, perhaps even more, other targets in the brain by routes that bypass the striate cortex. Therefore, interrupting the input to striate cortex, even though this constitutes the major retinal output, should not block all visual inputs to the brain. Moreover, it has become clear that striate cortex lesions in the monkey, whose visual neuroanatomy is closely similar to that of man, do not cause absolute blindness. Specific regions of the visual field are affected, but the animals can still detect stimuli projected into the affected regions of the visual field. Even if the striate cortex is completely removed in both cerebral hemispheres, the animals can detect visual events, although with reduced acuity, can locate them in space, can discriminate stimuli differing only in orientation in the frontal plane, and even carry out simple pattern discriminations (cf. reviews by Weiskrantz 1972, 1980).

One cannot ask a monkey whether it 'sees' a stimulus or not: one requires it to make a forced-choice discrimination; e.g. if stimulus A, do X, if B, do Y. Or, if and only if A is present, do R. One of the surprising results to emerge recently from testing patients with field defects caused by striate cortex damage is that they, too, if required to respond by forced-choice to visual stimuli projected into their 'blind' fields, can discriminate those stimuli, even though they may fervently deny that they 'see' them. The subject must be persuaded to play a kind of game, like 'if the grating (that you cannot see) is vertical, say 'vertical', and if it is horizontal, say 'horizontal'. Even if you cannot *see* it, have a guess'.

Using forced-choice methods, evidence has been produced that patients can detect visual events in their 'blind' fields, can locate them in space by pointing or by moving their eyes (although with reduced accuracy), can make discriminations between gratings of different orientations with quite good (if not quite normal) accuracy, and can carry out some types of simple shape discriminations. Indeed, as far as one can make direct comparisons, their capacities are of the same order as monkeys in which there is a striate cortex lesion. The first group to have persuaded patients to play the 'respond-by-moving-your-eyes-even-though-you-do-not-see' game was Pöppel *et al.* (1973), but we soon afterwards studied a patient (DB) in whom a similar but larger range of residual visual capacities was seen (Weiskrantz *et al.* 1974), and since then a number of reports, as well as a critique (Campion *et al.* 1983), have appeared (Weiskrantz 1980, 1986).

Despite the demonstration of visual discriminative capacity, the subject characteristically will say that he does not 'see'. He is playing the experimenter's game, and making a forced-choice response such as 'horizontal' or 'vertical' (or 'yes' present, 'no' not present; or 'moving' or 'not moving', or '*X*' or '*O*', depending on the task). Indeed, with the first patient whom we tested, we did not (and still do not) give knowledge of results during the test: when he was told of his impressive results afterwards he expressed open surprise—he thought he was just guessing. We called this phenomenon 'blindsight', which might have been unfortunate because the term has become much better known than the details of the findings. But at least in the sense of connoting a disconnection between a capacity and experience, the term is not entirely misplaced. Interestingly, a closely similar phenomenon has been reported in the somatosensory mode for localization of touch on the hand of a patient with hemi-anaesthesia caused by a cortical lesion in sensory–motor cortex. It has been called 'blind-touch' by Paillard *et al.* (1983). That is, the patient could locate touch stimuli to her skin, but had no awareness of actually being touched.

Patients with occipital lobe damage, including striate cortex but also varying amounts of surrounding tissue, have now been studied in this regard by a number of different investigators. There are various gradations and patterns of results, which cannot be reviewed in detail here (cf. Weiskrantz 1986). Inter-subject variations are not surprising, considering that the striate cortex in man (unlike the monkey) is virtually completely buried, and additional tissue damage, varying in degree and locus depending on the actual details of pathology in each case, is more or less inescapable, whereas in the animal experiments lesions can be placed with some precision. A large number of areas in close proximity to striate cortex in the occipital and parietal lobes appear to be concerned with specialized 'modular' aspects of vision, and hence it is not surprising that in some reports patients are reported to show greater or less residual capacity, say, for example, for orientation or movement discrimination, or some may be able to localize less well than others (cf. Weiskrantz 1980). The important point is not whether all patients are exactly alike, or even what defects are likely to be statistically associated with each other through proximity of lesions, the complex multiple effects of disease, and the influence of testing and treatment procedures. Rather, the most informative findings in neuropsychology concern what *dissociations* are possible, regardless of how often they occur. Even a single demonstration of a highly specific loss, in the absence of other deficits, provides a rich source of inferences about the organization of processing in the nervous system; and instances of 'double dissociation' of two different kinds of capacities allow relatively direct inferences to be drawn about potentially independent processes (Weiskrantz 1968).

There is also bound to be some variability simply because subjects vary in their willingness to venture forward with 'guesses' about events they cannot 'see'—some will even protest that this would be 'lying'. Interestingly, in his experiments on visual masking, Marcel (1983) also found that 'passive' subjects gave more sensitive results than 'active' subjects who 'felt the task to be nonsensical'. And no doubt experimenters themselves will also vary in their flexibility and persistence in offering the types of instructions needed, but, just as in the case of the amnesic patients, the precise form of the instruction can be crucial. Forced-choice methodology is unfamiliar and tedious, especially for the experimenter, and patience may be lost. Perenin *et al.* (1980, p. 608) remark in a report of a patient that they were able to use the forced-choice procedure (with positive effects), 'even if he did not really see anything', but add wrily, 'of course, he often took some time to be persuaded to do so'.

But there is also interesting within-subject variation. As stimulus 'salience' increases, the patient may say insistently that he still does not 'see', but he now has a kind of 'feeling' that something is there. In some cases, if the salience is increased still further, a point may be reached where the subject says he 'sees' but the experience is not veridical. For example, D B 'sees' in response to a vigorously moving stimulus, but he does not see it as a coherent moving object, but instead reports complex patterns of 'waves'. Other subjects report 'dark shadows' emerging as brightness and contrast are increased to high levels. If the conditions are conducive to such non-veridical experiences, the subject may switch out of a pure 'guessing' mode. With D B, for example, Warrington and I found that his performance was sometimes far better if stimulus conditions were kept within the non-salient range and he was merely required to guess (cf. Weiskrantz, 1986).

If it is the case, and this is by no means well established as yet, that there is a consistent transition from 'blind guessing' or 'feeling' to non-veridical 'seeing' (or perhaps even veridical 'seeing'), with variations in stimulus potency along any particular dimension, the question arises whether 'blindsight' is just like degraded normal vision, perhaps near to threshold limits. It has even been argued that it may be based on some remnants of intact striate cortex (Campion *et al.* 1983), although credibility is stretched very far when 'blindsight' cases of complete unilateral hemispheric decortication are considered (Perenin and Jeannerod 1978).

Is the visual capacity of blindsight just like normal vision, but quantitatively degraded? The question is more complex than it might first appear, because normal vision may perhaps change qualitatively as stimuli become degraded. For example, Marcel (1983) has argued that degradation of stimuli (flashed words) by central backward masking affects detectability more severely than it affects semantic processing of the contents of the stimuli, which is opposite to the order of ease of processing one would expect for stimuli well above threshold. Marcel (1983) has also reported preliminary evidence that blindsight patients can process semantic information of 'unseen' words projected into their field defects. But perhaps the clearest evidence of a difference between normal vision and residual vision in blindsight emerges from findings of double dissociations in DB between the two modes. Conditions can be found in which DB can actually *detect* stimuli (gratings) better at a particular locus in his 'blind' field than he can in his intact good half-field. Nevertheless, in those same loci he can discriminate *form* better in his good field than he can in his 'blind' field (Weiskrantz 1980, 1986). And so while there may be a gradation of reported experience with increasing contrast, the sensitivities of blindsight are not the same qualitatively as that of normal vision. The failure of form discrimination cannot be laid simply at the door of decreased detectability, i.e. of decreased salience. The differential bias of the impaired field towards detection and discrimination of localization and orientation, and against form vision, would appear to be in line, in general terms, with 'two visual system' hypotheses (Ingle 1967; Schneider 1967; Trevarthen 1967).

As the subject's attitude is important both in blindsight and masking experiments, the question arises directly about response criteria. Are some modes of vision more conducive to risk-taking than others? It might seem that, in signal detection terms, if d' and β could be determined for blindsight vs. normal vision, then nothing would remain to be determined. These values would not be without interest, but it would be illusory to think that they would provide a final answer. We can measure detection, let us say, in the blindsight field and set the performance level at, say, 90 per cent. We can do exactly the same thing in the intact field, and set the value at 90 per cent. Whether the two fields have different sensitivities or criterion settings does not answer the question why DB resolutely fails to agree that the two suprathreshold stimuli are equivalent. One he says he sees, and the other he says he does not. A signal detection analysis simply fails to capture this difference.

It might be thought, proceeding to a related line of argument, that there are really *two* kinds of forced-choice decisions in the typical blindsight task, namely 'Is this stimulus "*A*" or "*B*"?' and 'Do you "see" or "not see"'? The latter question, in the blindsight field, may be associated with a different criterion level and/or lesser sensitivity than the normal field. This may or may not be true, but it is precisely a restatement of the problem that requires explanation: the restatement does not provide its solution.

Even if different parameters were to emerge from a signal detection analysis for blindsight and normal vision (and from a practical point of view the most compelling cases of blindsight are those in which there are hardly any false alarms for plotting the ROC curves in signal detection analysis), this does not enable one to separate cause from effect. Given that a subject 'sees' in his good field, and must 'guess' in his defective field, different attitudes and response criteria might well be expected to emerge; they could hardly be expected not to emerge. This is not to say that they would not be of interest, but not for solving the problem of their origin.

One should say just a word about a determined, recent attack on blindsight research, if only because not to do so might seem to be avoiding or evading the issue (Campion *et al.* 1983). It is argued that it cannot be concluded confidently whether the striate cortex has been completely damaged in patients. This is true: none of the patients (fortunately) has died yet. When they do die, it will be rather too late to study their vision. But the evidence of monkeys' residual visual capacity in the absence of striate cortex *is* conclusive, and it would also seem highly unlikely that hemispherectomized patients have any intact striate cortex. No matter how the lesion evidence from human patients eventually turns out, the results will be of interest because not all patients show blindsight, and so hopefully the capacity will correlate with the details of the lesions.

But another argument is more serious: namely, that there was an artifact in the studies and the residual vision was really based on stray light affecting the intact regions of the visual fields in some subtle way of which the subject had ill-formed or no 'awareness'. There are several strong lines of evidence against this, which have been reviewed elsewhere (Weiskrantz 1983, 1986) and hence do not require detailed description here. One can mention just one, provided by the natural built-in control, the optic disc in each eye, which *is* truly and absolutely blind because it is where the optic nerve fibres penetrate the retina and hence where there are no receptors. If one directs a spot of light to the optic disc and its size and contrast are appropriately adjusted so that it becomes completely invisible and undetectable by the subject, one has a direct measure of the amount of permissible discriminable diffusion of 'stray light'. Conveniently, there is usually an optic disc within the blindsight field of one of the eyes of the patient. Therefore, one can directly compare his responses to the 'unseen' spot when it falls on the optic disc, and when the same spot is also 'unseen' but falls on a region just adjacent to the disc. When studied this way, DB showed chance performance when the spot was on the optic disc, but very good detection performance when it was off the disc (Weiskrantz 1986, also see Weiskrantz 1983). But in neither location did he 'see' the spot; he reported that he was just guessing. A similar control and result, using a signal detection paradigm for optic disc vs. non-optic disc, has also been reported for another patient by Stoerig *et al.* (1985).

Both of these types of patient, those with the amnesic syndrome and those displaying blindsight, demonstrate quite striking capacities in the absence of the subject's own knowledge. Because one depends so heavily upon a subject's verbal report in clinical assessment it was assumed for many years, in fact, in these types of cases that such residual capacities were absent altogether. The clinician and experimenter can overcome their surprise; but the subject himself is precluded from doing so by his own lack of experience of the phenomena themselves. It is not only the explicit dependence on verbal report of the subject: the very interchange between tester and subject assumes implicitly and richly that the subject will respond to questions about his experience of his knowledge: 'tell me whether you can see the light', or 'tell me which letter you can see', or 'tell me whether you recognize this', and so forth. The subject is being asked not only to discriminate or to retain information, but to say that he knows (or thinks he knows) that he can do so.

There are other similar disconnections from awareness in the clinical neurological field. Tranel and Damasio (1985), for example, have reported intact 'knowledge' of familiar faces by patients with prosopagnosia, as judged by their autonomic nervous system responses, even though these patients (by definition) have no acknowledged awareness of recognizing the faces. And we have already referred to the report of 'blind-touch' by Paillard *et al.* (1983). But the most notable examples are seen with commissurotomy patients (Gazzaniga 1988, this volume). (In this connection, one may say, parenthetically, that blindsight patients are not simply disconnected from access to a verbal capacity. I have tested some of Professor Sperry's patients for their visual capacities, and they are quite different from DB's. Visual parameters such as size did not constrain their response to stimuli directed to their right hemispheres in the way that was so for DB. Moreover, provided they knew the class of stimuli to which they were to respond, their commentaries were qualitatively quite different from DB's: they strongly asserted they 'saw' briefly presented stimuli directed to their right hemisphere, and indeed could offer some description of them. Another argument against a disconnection from a 'verbal' left hemisphere as an explanation is that blindsight has been reported for patients with either left- or right-hemisphere occipital lesions.) A complementary type of mismatch between a capacity and knowledge, perhaps, is seen in patients with Anton's syndrome, who firmly deny that they have a genuine loss of capacity, e.g. blindness.

The blindsight and amnesic patients, despite their striking residual capacities, are nevertheless severely impaired. In both cases there may well turn out to be possibilities of retraining, especially in the case of blindsight (Zihl and von Cramon 1985). With an amnesic patient, so long as he remains amnesic—and this may well be for the rest of his life—such a patient will require continuous custodial care. (Blindsight patients typically have some regions of intact visual field remaining, usually at least a half-field, and so can readily adapt to their deficit. But within their blind regions they are still ill-adapted, and of course cases of bilateral damage, fortunately rare, are grossly incapacitated.) Why should this be? Why should the mere inability to 'know' that one has a capacity, in itself, be so incapacitating?

Lloyd Morgan (1890, p. 375, quoting Mivart) seems to me to have suggested a valuable approach. 'If a being has the power of thinking "thing" or "something", it has the power of transcending space and time. . . . Here is the point where intelligence and reason begins.' To be aware of a visual event is not only to be able to categorize it independently of retinotopic space, to deal with it in canonical form, but to be able to treat that event as an image, to compare it or contrast it with other images, in short to 'think visually' and to guide action relevant to such imagery. Similarly, to 'remember' is to enable one to compare past with present, to reflect, to link separate past events, to order them, and to do so in relation to one's self as a coherent 'thing'. These capacities are precisely the ones that these patients lack.

The biological value of such a capacity of reflection must be beyond dispute, indeed beyond calculation. Much of everyday life can be organized according to automatic routines; even the most extraordinary conversations can become automatic and, in a sense, unconscious, as in verbal exchanges about the weather. Neural capacity for automatic acts need not be larger, or much larger, than the set of elements required for the unfolding of a particular stable act and its input and maintenance, but the penalty for such efficiency is a lack of flexibility outside the bounds of the automatic control. Reflection (in contrast to reflex-ion) frees one from such rigidity of control and brings a vast store of information that 'transcends space and time' to bear upon present action. On the assumption that there is a limited capacity for thought, it would be wasteful in the extreme to assign valuable cerebral capacity to the much simpler circuitry needed for automatic control that otherwise could be engaged in the processes that underlie thought and imaging. It is sometimes argued (Chance and Jolly 1970; Humphrey 1976) that the evolution of 'consciousness' stems from the pressures of social complexity, based on the fact that we are complicated social creatures who benefit from being able to attribute consciousness to other creatures so that we can try to predict their behaviour by role-playing and other forms of thought. The present view is different: social complexity can often be dealt with by automatic routines, and indeed many other creatures aside from man have evolved good ways of dealing with it; e.g. the ant does so very efficiently. Man, in contrast, precisely because he can think about and reflect upon all the social complications and pressures to which he is subject, is probably unique in being at risk in developing maladaptive paranoia (not to mention other forms of socially self-destructive aggressiveness to which mankind seems especially prone). Well-developed paranoia may well be a species-specific feature of humans. It is a penalty, well worth it in the evolutionary balance sheet, of having a capacity to think.

Most would agree, I presume, that evidence from commissurotomy patients, from patients with the amnesic syndrome or with blindsight, have told us something about the organization and logical structure of the nervous system, and they have done so in part because of the disjunction between capacity and 'awareness'. But some may still object to the value of terms such as 'awareness' or 'consciousness' in any such account, or even ultimately in any final account, because of the usual tangles with dualism and causation, among other philosophical issues. It may be that we can ultimately derive operational procedures for trying to decide what kind of nervous organization, or artificial devices, might be said to display 'awareness', but to disallow any assumption of phenomenal awareness in speculating about such a goal seems not only to fly in the face of the virtually universal belief of people in widely varying cultures, but actually to deprive neuroscience itself of information that could not be derived in any other way. The example that seems to me ˙to illustrate this most clearly is that of REM (rapid-eye-movement) sleep. One can observe REM sleep in a wide range of creatures, but, unless one has the opportunity to awaken one who will respond to the question 'Are you dreaming?', one remains ignorant about its correlates and possible function. Neuroscience is better served, I believe, by elevating itself to considering problems of awareness rather than by reducing all phenomena to the level of raw discriminative sequences or other comparable elements. As we have seen, sequences of raw discriminations or of retained facilitations are not enough to serve as an account of the human subject's own behaviour or, indeed, for his own survival.

Having indulged briefly in such exhortation, one cannot stop there: if there is to be a neuroscience or an artificial intelligence of 'awareness' one must rise to the difficult challenge of how to approach the matter beyond just having conversations with subjects. The patients I have discussed seem to be disconnected from a monitoring system, one that is not part of the serial information-processing chain itself, but can monitor what is going on. This disconnection is not the only thing that has happened to such patients, or possibly even the most interesting thing—other issues such as multiple memory systems and 'two visual system' organization are also illuminated by them—but it is what concerns us here. But, at least in the case of blindsight, the first evidence came from animals; it was the fact that monkeys can still carry out visual discriminations, and display control of their eye movements by visual events, that led to a re-examination of the human cases (cf. Weiskrantz 1972 and 1980 for reviews). And so the question arises as to how we would know whether or not the 'monitoring system' is functioning or not in a non-verbal animal. Without such a general and abstract solution the neuroscience of the matter will reach a stalemate.

No one, of course, has the answer as yet, but I speculate that one approach is to put the 'monitoring' response itself under separate control. One approach to this has been developed by Beninger *et al.* (1974), who in effect asked laboratory rats if they knew what they were doing. They allowed the rats to do any one of four possible things that rats do quite frequently and naturally without training: to face-wash, to rear up, to walk, and to remain immobile. But to get food reward, after they had performed any one of these four acts they had to press one of four different levers, so as to indicate what it was they had done. The rats could do that. I have no doubt monkeys could do it. But could a frog, or a pigeon? I doubt that a sea slug or a crab could.

The 'monitoring' lever is, in effect, a 'commentary' lever, and one that runs in parallel with the levers involved in the discriminative capacities themselves. In the present connection, however, it would relate not to reports of a subject's own acts, as in the Beninger study, but to his discriminations or retention of stimuli to which he is exposed. To take the case of blindsight, if *we* had to discriminate between highly distinctive vertical and horizontal gratings we could press one of two keys appropriately to indicate 'horizontal' or 'vertical', and we would also typically have no difficulty in pressing a third key that indicated that we were 'seeing' and not 'guessing' (although it should not be assumed that we would always be accurate under all conditions). This is where we would differ from the blindsight patient, whose third-key response would be 'guessing'. Similarly, the amnesic patient would press one of two keys to discriminate between items to which he may have been previously exposed, but his third key would indicate 'guessing' and not 'remembering'. With more refinement, of course, a confidence rating could be made, as is the practice in some psychophysical procedures. To take another example, a blindfolded paraplegic patient might well display intact spinal reflexes to aversive stimuli applied below the level of the spinal section, but his 'commentary' key responses would obviously and conspicuously not correlate with his reflex responses. But how one might go from the principle of the 'third key' to the actual practice of training a non-verbal or 'non-experiencing' subject makes a challenging 'thought experiment'.

There is a further feature of our adaptability that introduces both a difficulty and a further insight for analysis, namely that we tend to transform as much as we possibly can into automatic routines, even though at the outset they may involve much thought and reflection. Many quite skilled acts, such as are practised by aircraft controllers or pilots, become routine without thought and might even be impeded by thought. Therefore, the status of responses on the 'commentary key' might actually change as acts change from a training phase to a skilled phase. Dickinson (1985) has pursued one approach at the animal level to trying to decide whether a rat is responding 'habitually', i.e. automatically, or not. The strategy is to see whether the animal adapts to a change in the value of the final goal to which his behaviour leads. If the value of the goal is changed outside the situation, will the animal change its behaviour accordingly in the situation, or will it continue inflexibly to reel off its response pattern as before? Dickinson has shown that such a distinction can be made in practice, and has investigated some of the conditions in training that lead to 'habits' or to 'non-habitual' acts. Our well-established skills lack just this feature of a 'commentary'. We do not reflect when speaking that each and every word, or even just one word, was acquired by experience and in this sense is 'remembered'. Nor do we when we drive a car, if skilled, reflect along the lines, 'Ah, yes, I remember that red means stop'. Commentaries are reserved for those sequences and that phase of learning when it is profitable to have them.

'Monitoring' reflects a specific form of neural organization, one which allows, if you will, the third lever to operate in parallel. It is not merely a feature of complexity of neural ganglia *per se*. It may or may not be the fact that such an organization 'emerges' with increasing neural complexity in evolution, but complexity does not by itself entail this organizational feature. There is no reason to think that 'emergence' is an inevitable or sufficient product of increasing complexity, even if a certain level of complexity may be a necessary condition for monitoring. And, in any event, in order to say when something has emerged it is necessary to have some way of identifying it. But the important point is that it is a particular logical structure, rather than complexity as such, that would carry favoured survival value.

We do not yet know how to identify the neural equivalent of the 'third key', but the first stage in the nervous system at which it is likely to emerge, in the case of vision of *objects*, is not the striate cortex but the region where information from the whole of the visual field is united. In the monkey this is thought to be in the infero-temporal cortex (for information about objects; spatial mapping may follow a different anatomical route), which lies several synapses beyond the striate cortex of each hemisphere. The infero-temporal region appears to contain the stage at which information converges from both half-fields of retinal space, which until that point is processed separately in each cerebral hemisphere. The reason for thinking it important in the present context is that we 'image' and 'see' and remember *things* in our visual world in terms of object co-ordinates, *not* in retinal co-ordinates, to use the distinction that Marr (1982) has advanced. It is in the posterior part of this region in the monkey that we have reason to think that the final transformation from retinal to object co-ordinates takes place, and more anteriorly in this region that the storing takes place of prototypical representations of external objects in object co-ordinates (Weiskrantz and Saunders 1984). Even if the suggested identification of this particular localization in the brain turns out to be incorrect, the general logical argument remains about the necessity to achieve representation in object co-ordinates when we think about 'things'. Of course, the achievement of storage in terms of prototypes yields enormous economy for further processing elsewhere in the brain, including those aspects of 'recognition memory' affected in the amnesic syndrome.

Finally, given that we are discussing a type of neural organization, there need be no undue concern regarding the classical problem of the homunculus and the infinite regress. Given that one part of the nervous system has the function of monitoring other processes, there is no difficulty in principle in having hierarchical levels of monitoring, and hence in having varying levels of abstraction, leading, so to speak, to thinking about thinking. The limitation would not be one of logic but of diminishing gains in processing capacity in relation to the increasing neural cost, which would rapidly lead to regression if infinite.

References

Baddeley, A. D. and Warrington, E. K. (1973). Memory coding and amnesia. *Neuropsychologia*, **11**, 159–65.

Beninger, R. J., Kendall, S. B., and Vanderwolf, C. H. (1974). The ability of rats to discriminate their own behaviours. *Canadian Journal of Psychology*, **28**, 79–91.

Brown, J. (1958). Some tests of the decay theory of immediate memory. *Quarterly Journal of Experimental Psychology*, **10**, 12–21.

Campion, J., Latto, R., and Smith, Y. M. (1983). Is blindsight an effect of scattered light, spared cortex, and near-threshold vision? *Behavioral and Brain Sciences*, **6**, 423–86.

Chance, M. R. A. and Jolly, C. (1970). *Social groups of monkeys, apes, and men.* Jonathan Cape, London.

Dickinson, A. (1985). Actions and habits: the development of behavioural autonomy. In *Animal intelligence*, (ed. L. Weiskrantz), pp. 67–78. Oxford University Press.

Gazzaniga, M. H. (1988). Brain modularity: towards a philosophy of conscious experience. In *Consciousness in contemporary science*, (ed. A. J. Marcel and E. Bisiach), p. 218. Oxford University Press.

Humphrey, N. (1976). The social function of intellect. In *Growing points in ethology*, (ed. P. Bateson and R. A. Hinde), pp. 303–18. Cambridge University Press.

Ingle, D. (1967). Locating and identifying: two modes of visual processing. *Psychologische Forschung*, **31**, 44–51.

Marcel, A. J. (1983). Conscious and unconscious perception: experiments on visual masking and word recognition. *Cognitive Psychology*, **15**, 197–237.

Marr, D. (1982). *Vision.* W. H. Freeman, San Francisco.

Morgan, C. Lloyd. (1890). *Animal life and intelligence.* Edward Arnold, London.

Paillard, J., Michel, F., and Stelmach, G. (1983). Localization without content: a tactile analogue of 'blind sight'. *Archives of Neurology*, **40**, 548–51.

Perenin, M. T. and Jeannerod, M. (1978). Visual function within the hemianopic field following early cerebral hemidecortication in man. I. Spatial localization. *Neuropsychologia*, **16**, 1–13.

Perenin, M. T., Ruel, J., and Hécaen, H. (1980). Residual visual capacities in a case of cortical blindness. *Cortex*, **16**, 605–12.

Peterson, L. R. and Peterson, M. J. (1959). Short-term retention of individual verbal items. *Journal of Experimental Psychology*, **58**, 193–8.

Pöppel, E., Held, R., and Frost, D. (1973). Residual function after brain wounds involving the central visual pathways in man. *Nature (London)*, **243**, 295–6.

Schacter, D. L., Harbluk, J. L., and McLachlan, D. R. (1984). Retrieval without recollection: an experimental analysis of source amnesia. *Journal of Verbal Learning and Verbal Behavior*, **23**, 593–611.

Schneider, G. E. (1967). Contrasting visuomotor functions of tectum and cortex in the Golden Hamster. *Psychologische Forschung*, **31**, 52–62.

Scoville, W. B. and Milner, B. (1957). Loss of recent memory after bilateral hippocampal lesions. *Journal of Neurology, Neurosurgery, and Psychiatry*, **20**, 11–21.

Stoerig, P., Hübner, M., and Pöppel, E. (1985). Signal detection analysis of residual vision in a field defect due to post-geniculate lesion. *Neuropsychologia*, **23**, 589–99.

Tranel, D. and Damasio, A. R. (1985). Knowledge without awareness: an autonomic index of facial recognition by prosopagnosics. *Science*, **228**, 1453–5.

Trevarthen, C. B. (1967). Two mechanisms of vision in primates. *Psychologische Forschung*, **31**, 299–337.

Warrington, E. K. (1982). The double dissociation of short- and long-term memory deficits. In *Human memory and amnesia*, (ed. L. S. Cermak), pp. 61–76. Lawrence Erlbaum Associates, Hillsdale, NJ.

Warrington, E. K. and Weiskrantz, L. (1982). Amnesia: a disconnection syndrome? *Neuropsychologia*, **20**, 233–48.

Weiskrantz, L. (1968) Treatments, inferences, and brain function. In *Analysis of behavioral change*, (ed. L. Weiskrantz), pp. 400–14. Harper and Row, New York.

Weiskrantz, L. (1972). Behavioural analysis of the monkey's visual nervous system. *Proceedings of the Royal Society*, **B 182**, 427–55.

Weiskrantz, L. (1980). Varieties of residual experience. *Quarterly Journal of Experimental Psychology*, **32**, 365–86.

Weiskrantz, L. (1983). Evidence and scotomata. *Behavioral and Brain Sciences*, **3**, 464–7.

Weiskrantz, L. (1985). On issues and theories of the human amnesic syndrome. In *Memory systems of the brain: animal and human cognitive processes*, (ed. N. M. Weinberger, J. L. McGaugh, and G. Lynch), pp. 380–415. Guilford Press, New York.

Weiskrantz, L. (1986). *Blindsight: a case study and implications*. Oxford University Press.

Weiskrantz, L. and Saunders, R. C. (1984). Impairments of visual object transforms in monkeys. *Brain*, **107**, 1033–72.

Weiskrantz, L. and Warrington, E. K. (1979). Conditioning in amnesic patients. *Neuropsychologia*, **17**, 187–94.

Weiskrantz, L., Warrington, E. K., Sanders, M. D., and Marshall, J. (1974). Visual capacity in the hemianopic field following a restricted occipital ablation. *Brain*, **97**, 709–28.

Winocur, G. and Weiskrantz, L. (1976). An investigation of paired-associate learning in amnesic patients. *Neuropsychologia*, **14**, 97–110.

Zihl, J. and von Cramon, D. (1985).Visual field recovery from scotoma in patients with postgeniculate damage: a review of 55 cases. *Brain*, **108**, 335–66.

9

Hypermnesia and insight

Matthew Hugh Erdelyi

In attempting to grapple with a concept as inscrutable as consciousness, it is only natural to attempt frontal attacks upon the problem, as have many of the authors in this collection, by asking sensibly direct questions about the construct: What are its different significations? Does it play a causal role? What are the objective criteria by which it may be known? And so forth. There are, however, indirect strategies that might be pursued as well, for, as night vision teaches us, sometimes we look away in order to see better. In this chapter I attempt one such indirect approach: I seek to shed some light on the problem of consciousness by focusing on the unconscious.

Perhaps the tactic is not as perverse as it may strike one at first blush, for I claim (truistically? misguidedly?) that consciousness is a relational concept that has no meaning apart from its complement. We do not, for example, conceive of physicists or chemists as scholars of the unconscious on account of the purported absence of consciousness in the objects'of their inquiry. Consciousness and unconsciousness as concepts may require each other for meaning. It may not be the case, then, that consciousness can only be 'defined in terms of itself' (Angell 1904, quoted by Hilgard 1980), and indeed this may be a formula for conceptual shipwreck.

Now, the problem of the unconscious seems even more impenetrable than that of consciousness: not only is it (like consciousness) not publicly observable, but it is, in addition, not even privately observable. There can be no direct assault upon it. The unconscious may only be deduced from indirect evidence. Of this indirect evidence there are two basic classes, which I have called the two 'paradigms' of the unconscious (Erdelyi 1985). In the present chapter I discuss these two paradigms of the unconscious, underscoring some of their implications for consciousness.

Both paradigms of the unconscious, for example, define unconscious mentation in terms of observed discrepancies in accessibility to consciousness, and thus illustrate the relational nature of the constructs of consciousness and unconsciousness. The methodological problem of response bias is taken up and extended to the conceptual problem of the validity of indicators of awareness. Experimental work is discussed that demonstates that accessibility of memories to consciousness can be enhanced over time (hypermnesia) and that this hypermnesia is dependent upon conscious retrieval effort, without which no hypermnesia occurs. Thus, the activities of consciousness are shown to affect the information accessible to itself—one causal property of consciousness. Since conscious retrieval effort can be deployed (or inhibited) selectively, it follows that we can intentionally determine the relative accessibility to awareness of memories of past events. Because of our propensity to avoid intolerable pain—a peculiar characteristic of consciousness that has far-reaching causal effects—it follows that the contents of our consciousness, as emphasized by psychoanalysis, are subject to defence processes that preclude intolerable contents from consciousness, and result, for episodic memories at least, in forgetting. Thus the contents of consciousness are emotionally biased.

Extending the discussion beyond episodic memory, the distinction is made between manifest (surface, literal) content and latent (deep, figurative) content, which is based on semantic interpretation and results in insight. Manifest contents may be consciously apprehended where latent contents are not. Insight—the concious interpretation of deep meanings—is achieved when unconscious latent contents become accessible to consciousness. Interpretation of the latent contents of events depends upon the contextual surround of these events and is thus subject to psychodynamic distortions through biased sampling of the contextual ecology. Real-life stimuli, it is argued, typically result not from the degeneration of the physical stimulus (through brief exposure, low illumination, masking, etc.) but from the management of the contextual surround. Consciousness-raising, as it has evolved in psychoanalysis, focuses not so much on the recovery into consciousness of hitherto inaccessible memories or inputs (hypermnesia), but on the semantic interpretation or re-interpretation of complex life events which yields access to hitherto inaccessible latent contents (insight).

The recovery or hypermnesia paradigm

The simplest paradigm of the unconscious, interestingly, is the least direct. This is the *recovery* or *hypermnesia* paradigm, which is realized when a subject who cannot access some information at time 1, t_1, manages to access it at a subsequent time, t_2. Formally, the paradigm is defined by the inequality

$$\alpha_2 > \alpha_1,$$

where α_2 is accessibility of information to awareness at t_2 and α_1 is accessibility to awareness at t_1. Note that the unconscious is not defined in its 'own terms' but rather with reference to consciousness/awareness.

The logic of this consciousness-raising paradigm is straightforward enough. If in the absence of further external information, $\alpha_2 > \alpha_1$, the recovered information, especially if it involves episodic memories, must have come from some unconscious buffer.

It is not that easy to implement the paradigm in a methodologically satisfactory fashion. Consider a study by Haber and Erdelyi (1967) in which an attempt was made to recover subliminal tachistoscopic inputs through the agency of free associative fantasy. The subjects in the experimental group were presented the complex picture stimulus shown in Fig. 9.1 for 500 msec. After exposure to the stimulus, the subjects were required to provide a recall (α_1), in the form of an exhaustive, labelled drawing, in which everything was to be included, short of outright guessing. Thereupon the subjects (who were tested individually) reclined in a comfortable armchair and free-associated aloud for 40 minutes, at the end of which period they were asked to

Fig. 9.1. The stimulus presented for 500 msec by Haber and Erdelyi (1967).

produce a second recall (α_2) of the original stimulus. One of the control groups was treated identically except for the free-association task; subjects in this group engaged in a sensori-motor task for 40 minutes instead. Another control group, instead of being exposed to the tachistoscopic stimulus, copied the α_1 of a yoked experimental counterpart, then free-associated for 40 minutes, and then tried to produce an improved post-fantasy reproduction, 'α_2'. The results were straightforward; the free-association subjects, by a variety of scoring criteria, improved significantly in post-fantasy recall relative to pre-fantasy recall $(\alpha_2 > \alpha_1)$, whereas the control subjects failed to do so. An example of pre- and post-fantasy recall is presented in Fig. 9.2.

The experiment suggests that, because of the hypermnesic property of free association, the subliminal became supraliminal; i.e. the unconscious became conscious. It should be noted also that the free associations of the experimental subjects reflected stimulus contents (relative to the yoked controls, who did not see the stimulus) that were never actually accessed in either α_1 or α_2, implying that free associations index registered information that was never consciously accessed. This point will be returned to below. It will be observed that in the example (Fig. 9.2) there were not only more correct recalls ('hits') but also (it appears) more incorrect recalls ('false alarms'). Thus it does not necessarily follow that accessibility had improved, since it is possible that the free-associating subjects merely adopted laxer response criteria (β) after fantasy production than before fantasy production, thereby producing both more hits and more false alarms. To resolve the methodological issue, a series of experiments were undertaken involving both recall and recognition indicators of accessibility (Erdelyi 1970). One technique introduced was that of *forced recall*, in which both experimental and control subjects were required (forced) to generate an equivalent number of responses, guessing if necessary. Another tack was the application of signal detection procedures to recognition memory. Every study converged on the same conclusion: although the original Haber and Erdelyi (1967) results were readily replicable, the recovery effect was shown to be entirely due to response bias shifts. Thus subjects were not *recovering previously inaccessible memories* but only *reporting previously unreported memories*.

The response-bias interpretation of the foregoing data are worth considering in connection with some of the phenomena mentioned by Bisiach (1988), Weiskrantz (1988), and others (all in this volume). From the experimental subjects' reports and behaviour in the Haber and Erdelyi (1967) recovery study, it seemed clear that the subjects had no notion that they possessed more information than that which they produced in α_1: indeed, they insisted that they had produced all that

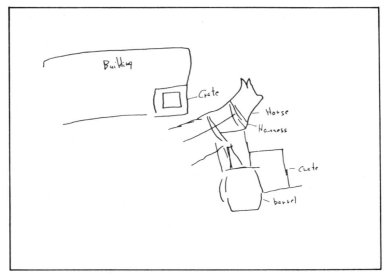

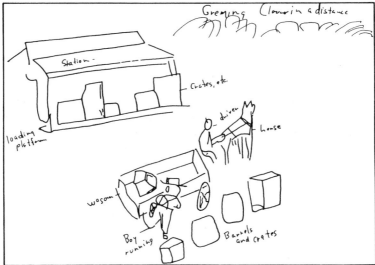

Fig. 9.2. Pre- and post-fantasy recall of subject HJ.

they could possibly produce 'short of outright guessing'. Moreover, as already noted, the experimental subjects' free-associations contained stimulus information never recovered in either recall trial. The question arises—and it is no longer clear to me whether it is a methodological or conceptual issue—whether we should identify accessibility with the sensitivity parameter, d', or with response bias, β. In experimental psychology (including my own work) no such question has been raised, at least not until very recently (see Bisiach 1986; Bowers 1984; Holender 1986): β has been thought of as merely a *response* bias, having nothing to do with consciousness/accessibility; it is d' that has been taken to index consciousness. My problem is that certain putative phenomena of behaviour without awareness—blindsight, for example—seem to represent drastic versions of what is commonly dismissed as β effects: the subject claims not to see, but when he is forced to respond—to guess, to grasp, and so forth—his performance is above chance, or even normal. So the question arises whether 'blindsight' and cognate phenomena represent a pathology of consciousness or of response bias (β). One possible argument against the response-bias interpretation is that the subject, after all, adopts normal criteria for responding to stimuli impinging on the intact perceptual field. Nevertheless, it is not clear, from a purely formal standpoint, how the issue is decidable, for it could be argued that blindsight is precisely a bias (β) pathology localized in the defective field.

The issue that intrigues me here is not so much the methodological one as the conceptual one. Should β—the subject's criterion for deciding what he sees/remembers/is aware of—be, after all, the criterion of consciousness, and not d' (which one would obtain from forced responding, guessing, etc.)? If we were to adopt β as a criterion of consciousness, literally hundreds of rejected experiments on subliminal perception would require re-evaluation. We would have to commit ourselves, also, to espousing the unpalatable view that response pay-offs alter awareness. (Discomfort over such problems, probably, has led some (e.g. Merikle and Cheesman 1986) to require 'qualitative' differences between conscious and unconscious processes as well, a suggestion that has something arbitrary about it and may not be altogether lacking in ambiguity). We see here one of the frustrating aspects of the problem of consciousness: not even in the most technical context of measurement is it clear what constitutes the criterion of consciousness. The problem amounts ultimately to the validity of our indicants of awareness, and there is no reason why indicators of awareness should be more tractable than indicants of other complex constructs, such as intelligence (Erdelyi 1986).

Let us adopt, as I have conservatively adopted in the past, a strict approach to the definition of accessibility/awareness and identify it with d' or some homologue. Can it not be demonstrated that subjects are truly capable of recovering subliminal inputs or forgotten memories, as opposed to merely relaxing response criteria for them? Despite many years of effort, I have never been able to demonstrate d' recoveries in the realm of subliminal perception. Only in the context of memory paradigms, where the targets of recovery are materials that were once in consciousness and were subsequently forgotten (subliminal memories rather than subliminal percepts), was I finally able to demonstrate bona fide recovery or hypermnesia effects. A review of this experimental programme can be found in Erdelyi (1984). I will here focus on the most powerful demonstration of hypermnesia, that of Erdelyi and Kleinbard (1978). The subject in study 1 (Jeff Kleinbard) was exposed to a list of 40 simple pictures (e.g. pipe, Christmas tree, snake, boomerang, feather, etc.) for five seconds each and then asked to try to recall all 40 items (in writing), guessing if necessary up to 40 responses. Thus the subject was forced to produce 40 non-repeated responses in all, for which he was given seven minutes. After the end of the first recall effort (which was collected by the experimenter), the subject was asked to try again to recall as many items as possible, guessing if necessary to generate 40 non-repeating responses. The procedure was repeated several times in the laboratory, after which the subject was dismissed but asked (without his foreknowledge) to generate several recall efforts every day for the next week, in the manner in which he had been tested in the laboratory. Figure 9.3 presents Kleinbard's retention function (right panel), alongside that of the classic Ebbinghaus retention function (plotted from Ebbinghaus 1964, p. 76). It will readily be seen that in contrast to Ebbinghaus's *amnesic* function, Kleinbard's is *hypermnesic*. Kleinbard obviously had a large store of inaccessible memories that he was eventually able to recover into consciousness. [This result is readily replicable with naïve subjects, e.g. see Erdelyi and Kleinbard (1978, study 2).]

What accounts for the difference between the two functions? The stimulus (nonsense syllables in the case of Ebbinghaus, pictures in the case of Erdelyi and Kleinbard) is certainly a factor but not the only one, since the forgetting functions can be obtained with pictures as well. The retention measures are different (savings vs. recall), but this is of little import since Ebbinghaus functions are readily obtained with recall and recognition, as well as with savings. The use of multiple recall trials in the Erdelyi and Kleinbard study as opposed to a single trial at different time intervals (for different items but for the same type of materials) is not the essential factor either, since hypermnesia can be obtained with a single test procedure as well (Shapiro and Erdelyi 1974).

The critical difference (cf. Erdelyi 1977, 1984) appears to be the deployment of *consciousness* upon the target items in the hypermnesia studies—i.e. conscious review of and search for the target items— in contrast to the intentional withdrawal of consciousness from the stimuli in the Ebbinghaus-type situation. The crucial factor, then, is conscious thinking (reviewing, searching, retrieving, attending) in the

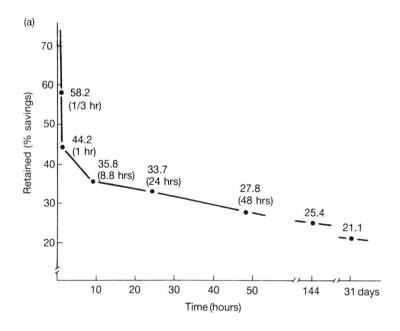

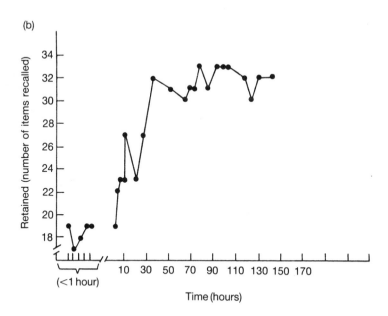

Fig. 9.3. (a) Decrease of retention (amnesia) over time (adapted from Ebbinghaus 1964). (b) Increase in retention (hypermnesia) over time (Erdelyi and Kleinbard, study 1, 1978).

guise of multiple recall trials and silent concentration/think intervals between trials. It has been clearly shown by Roediger and Payne (1982) (see also Erdelyi 1977, 1984) that prevention of conscious retrieval effort eliminates hypermnesia. Actually, absence of consciousness in general probably eliminates *both* hypermnesia and forgetting in the recall of pictures and words (Erdelyi 1977).

The implication of the foregoing is that in normal subjects *consciousness* is a prerequisite for hypermnesia or forgetting—at least for certain types of materials and memory measures. Thus the voluntary deployment of consciousness (thinking/concentration/retrieval effort, etc.) toward or away from some memory complex on a chronic, long-term basis will dramatically determine the memory's accessibility to consciousness. Here then is a possible causal aspect of consciousness: consciousness in part determines what is accessible to itself, which in turn has causal effects upon behaviour. Not all memories obey this rule. Motor responses or conditioned responses may actually increase over time with no conscious review or practice. But the deployment of conscious retrieval effort is what drives, at least in part, the fate of episodic accessibility for verbal and pictorial materials.

Figure 9.3 shows us that we have, in effect, an 'up-function' and a 'down-function' for conscious retention. Since we can partially determine (through selective deployment of consciousness) the extent to which the up- and down-functions operate for a particular memory complex, it follows that we have considerable control over the accessibility of our own memories. And since, moreover, our motives (interests, fears, wishes, etc.) affect this deployment, it would seem obvious that emotional factors, by biasing conscious deployment, bias what is accessible to consciousness. This processing bias, as I have argued elsewhere, constitutes a basic cognitive mechanism of defence processes (Erdelyi 1984, 1985).

The dissociation paradigm

Whereas the hypermnesia paradigm involves a discrepancy between the same indicator of accessibility assayed at different times, the dissociation paradigm arises from the observed discrepancy between two different indicators that are observed concurrently. Formally, the dissociation paradigm is defined by the inequality

$$\varepsilon > \alpha,$$

where α, as before, is an indicator of information accessible to consciousness, and ε is an indicator of information that is available (but not necessarily accessible). The paradigm entails the simple notion that if the amount of information that is available (ε) is greater than the information that is accessible to consciousness (α), then the inaccessible but available information is unconscious. Once again the unconscious is defined relative to the conscious and not in its own terms.

Most of the experimental literature on unconscious processes involves the dissociation paradigm. In the Marcel (1983a) type of experiment on semantic activation without awareness, for example, the indicator of conscious discrimination is null ($\alpha = 0$), but another indicator, as in priming effects, is positive ($\varepsilon > 0$); hence $\varepsilon > \alpha$. Similarly, in the Haber and Erdelyi (1967) study, in the absence of conscious recall on either α_1 or α_2 ($\alpha = 0$), free associations (ε) nevertheless reflected available stimulus information ($\varepsilon > 0$). This paradigm is also pervasive in the clinical literature. For example, Breuer and Freud's (1895, p. 7) classic formula, 'hysterics suffer mainly from reminiscences', conveys the notion that the hysteric's symptoms reflect active memories ($\varepsilon > 0$) which are not, however, accessible to consciousness ($\alpha = 0$). Split-brain phenomena are the pure-culture version of the dissociation paradigm in that they feature a dissociation not just of the mind but of the brain itself.

In real-life situations, including psychotherapy, the dissociation paradigm tends to involve indicators that are more complex than those encountered in the laboratory. Usually it is not a single measure or event that indicates availability but a set of events. Thus in real-life situations the dissociation paradigm tends to assume the form

$$[\varepsilon] > \alpha,$$

where $[\varepsilon]$ represents a set of events, $\varepsilon_1, \varepsilon_2, \varepsilon_3, \ldots, \varepsilon_n$, which, *taken together*, indicate some information or emotion that is available (but not necessarily accessible) to consciousness.

I wish to consider here a complex form of complex indicators, hence termed 'hypercomplex indicators' (Erdelyi 1985), which involves not merely the summation of separate events, but their interaction. Hypercomplex indicators arise from the fact that the meaning of an isolated event, ε, is not necessarily the same as the meaning of the selfsame event given a set of other events, $\varepsilon_1, \varepsilon_2, \varepsilon_3, \ldots, \varepsilon_n$, or $[\varepsilon]$; that is,

$$\varepsilon \neq \varepsilon \,|\, [\varepsilon].$$

Put in more familiar terms, an isolated event, ε, may acquire new, indeed radically different, meaning when considered in a particular context, $[\varepsilon]$. The distinction, variously referred to in the contemporary literature as that between surface and deep meaning, reference vs. sense, semantics vs. pragmatics, etc, turns out to be the seminal distinction of psychoanalysis between manifest content (surface meaning, ε) and latent content (deep meaning, $\varepsilon \mid [\varepsilon]$), which is the key to the understanding of all depth-psychological materials, conscious or unconscious, including dreams, psychotic thinking, symptoms—and jokes.

As an illustration of the manifest-latent content distinction, consider the following Groucho Marx witticism, which caused him some trouble. He was interviewing a man on his television programme *You Bet Your Life* and had the following exchange with him:

Groucho: How are you?
Man: Fine.
Groucho: Well, tell me, are you married?
Man: Yes; I've been married for nine years.
Groucho: Gee, that's swell! Do you have any kids?
Man: Yes, nine already and the tenth is on the way.
Groucho: Wait a minute! Ten kids in nine years?
Man (sheepishly): Well, I happen to love my wife very much.
Groucho: Well, I love my cigar too, but I take it out of my mouth once in a while!

Clearly, it is not necessary to be a Freudian to understand the joke and its 'Freudian symbols'. Those who 'get' the joke—a vast majority of people —understand that the significant content of the retort is not 'cigar' and 'mouth' but 'penis' and 'vagina'. Thus

$$\varepsilon_i = \text{cigar, but } \varepsilon_i \mid [\varepsilon] = \text{penis,}$$

and

$$\varepsilon_j = \text{mouth, but } \varepsilon_j \mid [\varepsilon] = \text{vagina.}$$

Not 'getting' the joke is layman terminology for not interpreting, and therefore not having insight into the latent content of the retort, without which of course there is nothing witty about it. Note that the context, $[\varepsilon]$, is all important for the interpretation. The retort taken by itself, without the preceding context, fails to generate the latent meaning.

There are several important points underscored by the illustration. One of them is that it is a mistake to assume, as psychology has done throughout most of its lifespan, that meanings inhere in specific stimuli, or indeed that stimuli are palpable, materially defined. From a strict physical standpoint there is no stimulus 'penis' or 'vagina' in the retort, even if these are the critical psychological stimuli. As obvious as this might have become in post-behaviourist psychology, it still has not begun to be 'worked through' in modern cognitive psychology. I find it remarkable that virtually all research on subliminal stimulation is still carried on at a crude physicalist plane, with tachistoscopes and other stimulus degraders serving as modulators of the perceptibility of manifest stimuli. The usual concern lies with durations, luminances, masking patterns, noise background, and so forth. All this, in my view, misses the really interesting realm of subliminal perception, which deals with latent as opposed to manifest stimuli—with insight rather than sight —and which is modulated not by tachistoscopes and the like but by the manipulation of context. Thus it is possible to vary the contextual environment to produce different degrees of 'seeing' (insight) of the latent content (Erdelyi 1985).[1]

An important feature of 'deep' stimuli, which has critical psychodynamic implications, is that significant portions of the context reside in the observer. For example, the interpretation of the Groucho Marx joke depends not only on the textual context provided but also on the subject's knowledge of sex, without which the witticism is meaningless. The observer, moreover, contributes not only his store of knowledge for the interpretation of deep stimuli but also, critically, determines the operative context by deciding what part of the external and internal information domain to bring to bear upon the interpretation. The selective and constructive aspect of insight—the detection of deep meanings—is overwhelming compared to the acknowledged role of selectivity and construction in the processing of manifest events. Consequently, decisions about what aspects of the potential context to sample determine at a fundamental level what is actually 'seen' or 'remembered' consciously. It is at this level, and not at the manifest one traditionally pursued by experimental psychology, that defence processes can be clearly observed. Thus the pervasive type of 'perceptual defence' in real life concerns not the failure to see some briefly flashed word, but rather the failure to 'see' the deep meaning of complex events. The usual type of perceptual defence is not some laboratory version of psychogenic blindness, which is a rare phenomenon indeed, but the ubiquitous failure of *insight* into deep stimuli.

In short, because of its physicalist bias, psychology has unwittingly studied defence phenomena in the recalcitrant realm of the manifest— ε—when the ubiquity of defence processes occur in the realm of the latent— $\varepsilon|[\varepsilon]$. Because of the proclivity of consciousness to avoid

---

[1] According to Marcel (1983*b* and 1988, present volume) the very act of conscious apprehension is an act of conceptual categorization in terms of constructs accessible to consciousness; thus central masking, in his view, does not physically degrade the stimulus but rather disrupts the act of categorization essential for awareness. Marcel's position, if I understand it, is that all conscious experience, whether of manifest or latent contents, constitutes acts of insight. This undoubtably is true, and points to the fact that the manifest/latent content distinction is but a convenient simplification. Stimuli, strictly, are not all-or-none deep but more-or-less deep, being based on lower- or higher-order categorizations.

pain—here is another peculiar feature of consciousness—tendentious sampling of the context occurs which produces desirable $\varepsilon|[\varepsilon]$ without disturbing the perception or memory of any particular ε. Thus the dissociation paradigm of the unconscious occurs, in the case of hypercomplex indicators, when

$$\varepsilon|[\varepsilon] > \alpha.$$

The manifest content, ε, may be entirely conscious, but the latent content, $\varepsilon|[\varepsilon]$, is inaccessible.

It is difficult to provide in manageable form evidence for latent contents that are inaccessible to consciousness; i.e. into which the subject has no insight. Unlike in the case of the Groucho Marx joke, where the latent content was (one may assume) quite conscious, unconscious latent contents are 'deeper' and require often massive contextual background to be persuasively revealed. I shall here present a proto-clinical example (from Erdelyi 1985, pp. 98–100)[2] which is at least suggestive of several such subliminal latent contents:

The Case of J.

Preamble

A close friend, J. had recently completed his Ph.D. studies in an abstruse field of science and won an appointment to one of the country's top universities. After his first year, which was a brilliant one, J. took a year's leave of absence to accept a post-doctoral fellowship at one of the leading laboratories in his area of speciality. J. returned a changed person. Emaciated physically, he was withdrawn and sullen. He drank heavily and experimented with drugs. He appeared chronically anxious, suffered from a variety of tics, and had lost a great deal of his social grace. He was singularly unsuccessful with women. His performance in his department deteriorated rapidly, and despite efforts by his colleagues to help him, he was not able to complete his term of contract. Subsisting on his savings for several months, he led a desultory, aimless existence. Eventually he found himself a job at a small bookstore, which he managed to hold down despite some initial frictions with his boss, Mrs. W., and the other clerk at the store. He seemed content with this simple existence, though he had horrendous long-distance arguments with his widowed mother, who was outraged at his having 'squandered' his career in science. With his new job, J.'s drinking abated somewhat, and he gave up, for the most part, the drugs that he had been taking. Even so, he was unusually taciturn and became progressively more isolated socially; also, he stopped visiting his mother.

---

[2] Certain facts of this case have been omitted or altered to protect the privacy of the individual discussed.

I feared greatly for my friend. It was around this time that I gave him a long-distance phone call to find out how he was managing.

Text of phone conversation with J.

His answer to my conventional 'How are you?' was melodramatic.

'*I have finally done it! I have flipped!*'

'What do you mean?' I asked him anxiously.

'*Right in front of my apartment. The sidewalk was a sheet of ice* [it was winter]. *I flipped full circle and broke my leg. The doctor said it was a classic skier's fracture. My whole leg is in a cast.*'

I was astonished. Could he really have been missing the idiomatic sense of 'flipped'? It was not impossible since he was an immigrant and spoke an accented English. But it seemed unlikely.

J. proceeded to tell me that after an initial tendency '*to minimize the whole incident—so many people break an ankle* [sic?], *big deal!*'—he was '*only now coming to realize how ill* [he] *truly was.*'

I was again startled by his mode of expression. Was he alluding, after all, to a psychological crisis? Finally I suggested that he might be depressed, a common reaction.

J. replied that he had been depressed for a while but that his most notable psychological experience was an '*inexpressibly violent attack of guilt after the accident.*'

At this point, J. shifted to another matter, the impending visit of his older brother from _____, his first to the United States. (This older brother, J.'s only sibling, was very different from J. He was handsomer, married, and the father of two children; also he was internationally famous in his own specialty of science, the purpose of his visit being, in fact, to deliver a series of invited colloquia at American universities.) J. related how his mother had been pestering him day and night over the phone about going to _____ Airport to meet his brother, despite the fact that J. lived in another city and had no car. He now matter-of-factly pointed out that he would not be able to meet his brother. Thereupon, J. launched into a remarkable tirade. Screaming into the phone, he declared that *his brother was after all a grown-up man with a family; he had children, a wife; surely he would know how to find a taxi to the City; as far as he was concerned, his mother could rant and rave and he would not care in the slightest; his brother would surely not be the first person in the history of* _____ *Airport to arrive unmet; nor would he be the last; he could, like countless others, find the wit to find himself a taxi; and if his mother couldn't see this, well then she could just as well go fuck herself! And so could his brother. By God, he had a broken leg and he was in no condition to go meeting anybody at* _____ *Airport!*

I asked whether his mother was still nagging him about making the trip to the airport.

He barely replied (she wasn't) and launched into a new diatribe.

Eventually, I asked J. about the 'excruciating guilt' he had previously mentioned.

He explained that he *so hated to overburden his poor associates at the bookstore who, with him out, would have to do all kinds of extra work; they were really very nice people, and he just didn't want to push his own work upon their shoulders.* Indeed, upon his return from the hospital, he had called up to say that he would come in to work despite his cast, but his boss, Mrs. W., had vetoed this, assuring him that she and the other clerk could manage at the bookstore, that he need have no concerns about his salary or his job, and that he should rest two weeks before returning to work. J. had protested, but she was firm.

Despite this, J. said, *'I felt incredibly guilty the whole day; it was unbearable, overwhelming! I felt so guilty about those two doing all my work.'*

I finally asked J. whether part of his guilt might not be connected to his incapacity with respect to the *other two people* of whom he was telling me—his *mother* and his *brother.*

To this I received a curt and vehement response: *'Most certainly not! I have suffered almost two nervous collapses from her nagging, and I'll be damned if I'll suffer another one! They have nothing to do with it.'*

After a brief new outburst of vituperation about his mother and brother we shifted to a few trivia and concluded the conversation.

One potential latent content into which J. has no insight is associated with the manifest statement:

$$\varepsilon = \text{'I have finally done it! I have flipped!'}$$

The contextual backdrop suggests, however, that J.'s more telling communication is:

$$\varepsilon \,|\, [\varepsilon] = \text{'I am in a psychological crisis' (or some such)}.$$

Note that J. denies the interpreted latent content, which means either that he has no insight into the deep meaning of his own utterance or that the author, as J. believes, has produced a faulty interpretation. Similarly, when J. alludes to

$$\varepsilon = \text{'those two' [in the bookstore]}$$

he seems really to be stating, despite his disavowal,

$$\varepsilon \,|\, [\varepsilon] = \text{my mother and brother}.$$

There is no absolute way of resolving the issue of whether the $\varepsilon \,|\, [\varepsilon]$ extracted by the author is a true latent content for which J. has no insight or whether it is simply non-existent. The point to some extent is empirical. It is possible to submit the material to independent judges and evaluate the consensus among them. (For the Groucho Marx joke, the consensus is extremely high; for the case of J. it is high, but less so.) Consensual validation seems to be an unsatisfactory criterion, yet it is ultimately all that we have—not just in the realm of insight but also in that of sight.

But this is not really the interesting question—whether the stimulus is *really there*. No stimulus is really 'there', as we experience it; perceptual awareness is always an interaction phenomenon—the sound of two hands clapping—of something out there with something within us. The interesting point is that in the realm of latent content much more variability exists among people in what meanings/insights they extract from some input, because they can sample the contextual ecology tendentiously, often for the purpose of defence, to avoid insights that would be intolerably painful. Thus, if we take the author's interpretation of 'ε = "those two"' to be veridical, then we may treat J.'s unawareness of '$\varepsilon|[\varepsilon]$ = my mother and brother' as a form of 'perceptual' defence (perceptual in the sense not of conscious sight but of conscious insight). It is this type of tendentious processing—biased insight brought on by biased context selection—that characterizes the defence mechanisms of relatively normal individuals (who do not, for example, have literal positive or negative hallucinations).

The main purpose of insight therapies such as psychoanalysis is consciousness-raising in the sense of insight rather than of manifest sight or memory. Insight is brought about by a gradual restructuring of the subject's operative context set, not by the sharpening of perception or memory. This type of consciousness-raising requires tremendous cognitive effort and must take into account the emotional factor of the tolerability of the target insight. Thus what consciousness is willing to tolerate and what it is willing to exert itself upon determines substantially its own content.

Acknowledgements

This work was supported in part by grant no. 6-64345 from the City University of New York PSC-CUNY Research Award Programme, by grant no. MH 19156 from the National Institute of mental Health, US Public Health Service, and by a grant from the Institute for Experimental Psychiatry.

References

Bisiach, E. (1986). Through the looking-glass and what cognitive psychology found there. *Behavioral and Brain Sciences*, **9**, 24–5.
Bisiach, E. (1988). The (haunted) brain and consciousness. In *Consciousness in*

contemporary science, (ed. A. J. Marcel and E. Bisiach), p. 101. Oxford University Press.

Bowers, K. (1984). Being unconsciously influenced and informed. In *The unconscious reconsidered*. (ed. K. S. Bowers and D. Meichenbaum), pp. 227–72. John Wiley, New York.

Breuer, J. and Freud, S. (1895) *Studies on Hysteria*. Translated by A. Strachey and J. Strachey. In *The Standard Edition of the Complete Psychological Works of Sigmund Freud*, Ed. J. Strachey, Vol. 2. Hogarth Press, London, 1955.

Ebbinghaus, H. (1964). *Memory*, (trans. H. A. Ruger and C. E. Bussenius). Dover, New York.

Erdelyi, M. H. (1970). Recovery of unavailable perceptual input. *Cognitive Psychology*, **1**, 99–113.

Erdelyi, M. H. (1977). Has Ebbinghaus decayed with time? Paper presented at the Eighteenth Annual Meeting of the Psychonomic Society, Washington, D C. Available from M. Erdelyi, Brooklyn College, City University of New York.

Erdelyi, M. H. (1984). The recovery of unconscious (inaccessible) memories: laboratory studies of hypermnesia. In *The psychology of learning and motivation: advances in research and theory*, (ed. G. Bower), pp. 95–127. Academic Press, New York.

Erdelyi, M. H. (1985). *Psychoanalysis: Freud's cognitive psychology*. W. H. Freeman, New York.

Erdelyi, M. H. (1986). Experimental indeterminacies in the dissociation paradigm of subliminal perception. *Behavioral and Brain Sciences*, **9**, 30–1.

Erdelyi, M. H. and Kleinbard, J. (1978). Has Ebbinghaus decayed with time? The growth of recall (hypermnesia) over days. *Journal of Experimental Psychology: Human Learning and Memory*, **4**, 275–89.

Haber, R. N. and Erdelyi, M. H. (1967). Emergence and recovery of initially unavailable perceptual material. *Journal of Verbal Learning and Verbal Behavior*, **6**, 618–28.

Hilgard, E. R. (1980). Consciousness in contemporary psychology. *Ann. Rev. Psychol.*, **31**, 1–26.

Holender, D. (1986). Semantic activation without conscious identification in dichotic listening, parafoveal vision, and visual masking: a survey and appraisal. *Behavioral and Brain Sciences*, **9**, 1–66.

Marcel, A. J. (1983*a*). Conscious and unconscious perception: experiments on visual masking and word recognition. *Cognitive Psychology*, **15**, 197–237.

Marcel, A. J. (1983*b*). Conscious and unconscious perception: an approach to the relation between phenomenal experience and perceptual processes. *Cognitive Psychology*, **15**, 238–300.

Marcel, A. J. (1988). Phenomenal experience and functionalism. In *Consciousness in contemporary science* , (ed. A. J. Marcel and E. Bisiach), p. 121. Oxford University Press.

Merikle, P. M. and Cheesman, J. (1986). Consciousness is a 'subjective' state. *Behavioral and Brain Sciences*, **9**, 432.

Roediger, H. L. and Payne, D. G. (1982). Hypermnesia: the role of repeated testing. *Journal of Experimental Psychology: Learning, Memory, and Cognition*, **8**, 66–72.

Shapiro, S. R. and Erdelyi, M. H. (1974). Hypermnesia for pictures but not words. *Journal of Experimental Psychology*, **103**, 1218–19.

Weiskrantz, L. (1988). Some contributions of neuropsychology of vision and memory to the problem of consciousness. In *Consciousness in contemporary science*, (ed. A. J. Marcel and E. Bisiach), p. 183. Oxford University Press.

10

Brain modularity: towards a philosophy of conscious experience

Michael S. Gazzaniga

Thinking about thought has a long and distinguished history. It is what philosophers do and it is one of the more difficult human mental activities. To have a philosophy of something is to try to have an understanding of how one comes to know what it is one knows about the subject in question. What makes that so difficult is that philosophical thinking demands working out relations between two different levels of analysis. A philosophy of conscious experience would require knowledge about the content and mechanics of conscious experience and the process by which consciousness was gained. A science of conscious experience, by comparison, is much easier. Students of that enterprise need only worry about representational systems independent of the stuff that sustains those representations and independent of a consideration of how they came to know the ideas expressed in the representation. There are a million things to know within that framework, and there should be little mystery why many scientists investigating the nature of human consciousness are reluctant to engage in the additional chore of considering a philosophy of conscious experience. To achieve the goal of having a philosophy of conscious experience would require, it seems to me, the philosopher of conscious experience to know about the brain. From my perspective, understanding brain logic will illuminate these traditional epistemological issues. I will take as my assignment, therefore, the outlining of what I feel are important brain facts, facts a philosophy of conscious experience ought to incorporate.

My message is that in order to know how we come to know what we know we must learn how and why the human brain seems reflexively given over to the process of generating hypotheses and explanations for events (information) it contacts. Why must self-produced actions, thoughts occurring suddenly in our conscious experience, mood swings, externally occurring events, and other perturbations of the status quo be interpreted? Why does the human species not content itself with simply recording internal and external actions and leave the possible causal aspects of the experience in question alone? Why must we come to lay claim that we know why an event occurred and gradually transform this thought into a firmly held belief? What is it about us that finds us always transcending a literal contact with the environment and insisting on an interpretive view of events? Our species does all of this with a vengeance. Even fatalists from religious or whatever background have asked the question 'Why?' That they have accepted a stock answer handed to them by a religious system or whatever does not negate the reality of their original question. The interpretation we give to the question 'Why?' starts with analysing such things as simple hand movements, simple moods, and builds into larger culturally based phenomena. It is everywhere in our species, and I think that there are some provocative clues now available from brain research that shed light on this phenomenon.

My approach will be to outline a series of findings on human patients who have undergone brain bisection and to argue that the findings suggest that human brain architecture is organized in terms of functional modules capable of working both co-operatively and independently. These modules can carry out their functions in parallel and outside of the realm of conscious experience. The modules can effect internal and external behaviours, and do this at regular intervals. Monitoring all of this is a left-brain based system called the interpreter. The interpreter considers all the outputs of the functional modules as soon as they are made and immediately constructs a hypothesis as to why particular actions occurred. In fact the interpreter need not be privy to why a particular module responded. None the less, it will take the behaviour at face value and fit the event into the large ongoing mental schema (belief system) that it has already constructed. If a module effects a behaviour that is dissonant with the belief system established through prior interpretive actions, the behaviour will tend to change the belief system. [One quickly sees the importance of not exposing such a brain to an environment that would encourage certain modules to act in a way counter-productive to the current belief system (Gazzaniga 1985).]

It is important to understand how the concept of modularity is being applied in this context. Modularity refers to functional units that can produce behaviours and trigger emotional responses. This differs from the concept of modularity as it is commonly used in cognitive science (Fodor 1983; Kosslyn 1983). In that setting it refers to the identifiable components that are part of the mechanism of specific mental functions such as mental imagery or language. I will discuss modularity from this standpoint as well. The concept of modularity has also been used in the neurosciences where specific brain structures and areas are identified as carrying out specific functional activities. This is yet another way the term is used, and I will also report on studies relevant to this framework.

Overall, my view, driven by brain studies to be outlined below, argues that our species generates what it thinks it knows as a consequence of a special human brain capacity that insists on interpreting events. This interpretive system, which appears tied to the capacity to make inference, is based in the left hemisphere of most humans, and I believe, as will be made clear, that it is found in the distribution of the left, middle cerebral artery. What is not yet determined is whether the kinds of inferences that can be made about a set of events (data) are limited. Does our species's interpretive system have a very finite repertoire of algorithms upon which to draw when inferring the meaning of the data it is considering? I am convinced mine does, and so, for those who believe human inventiveness is infinite, I shall now proceed to report the observations that find my interpreter making the foregoing claims. Perhaps others will process the data differently.

If the brain is organized in the way I suggest, it is easy to imagine that rich dissociations would be potentially demonstrable in patients with brain damage or disconnections. For independently functioning modules, brain damage ought to eliminate their participation in overall cognitive activities. For modules working in co-operation, brain damage or disconnection ought to yield impaired performance in the remaining brain system. In short, studying the neurological patient should illuminate how modular the brain is in the overall construction of our apparent conscious unity.

It should not come as a surprise to discover that particular dissociations are stumbled upon somewhat unexpectedly and rather routinely. The idea of brain modularity is sufficiently undefined at this time that the cataloguing of phenomena that will someday comprise the total picture of modularity will go on for years. The first dissociation I would like to discuss demonstrates the extent to which non-conscious processes can influence behaviour. I will describe this set of experiments in detail as illustrative of our general testing procedures.

Speech without conscious awareness

We are now able to show how a non-conscious system can discreetly govern an overt behaviour (Gazzaniga *et al.* 1987). Until recently, the view that functional brain architecture incorporates non-conscious parallel processes has been more commonly asserted than demonstrated. In the experimental psychological literature on normals, modern studies have focused on how subliminal stimuli are effective in facilitating subsequent perceptual and semantic judgements (Marcel 1983). Studies on the neurological patient have claimed that information presented in a blind field of vision produced by occipital lobe damage can be useful in generating manual and ocular responses (Pöppel *et al.* 1973; Weiskrantz *et al.* 1974), although there is now evidence that not all lesions in primary occipital cortex allow for dissociations between spoken and manual responses under conditions of a forced choice response (Holtzman 1983). Additionally, patients with extinction on double simultaneous stimulation following parietal lobe damage are able to have the information presented in the extinguished field influence cognitive judgements (Volpe *et al.* 1979).

In a new study (Gazzaniga *et al.* 1987), the results suggest that non-conscious processes can control overt behaviour. Specifically, it is possible to show how the right hemisphere can set up a left-hemisphere-specific response without the left hemisphere being able to consciously access the information inserted by the right brain. In short, the findings suggest that response behaviours can be set up and carried out without conscious awareness of the elicited behaviour prior to its occurrence.

The experimental demonstration of the existence of such systems was made possible by means of tests conducted on a patient with a particular brain condition—the existence of an MRI[1]-verified full callosal section. Case JW is a 32-year-old male who underwent staged surgery in 1979. Post-operation, JW evidenced the capacity to comprehend language in both his right and left himispheres. However, JW can produce speech only from his language-dominant left hemisphere (Gazzaniga *et al.* 1984).

For present purposes I will report on two sets of experiments. The first detailed the capacity of the left hemisphere to name visual information strictly lateralized to the right hemisphere. The second set of experiments was directed at ascertaining the left hemisphere's awareness of the knowledge it possessed. The initial period of each trial was common to all conditions: JW was instructed to fixate a point on the centre of the screen and, when the experimenter ascertained that accurate fixation had been achieved, a single digit was presented for 150 msec 6° directly to the left or right of the fixation stimulus. Two seconds then elapsed and a tone sounded. Conditions differed with regard to the sequence of events following the tone.

In the first study demonstrating the crossed speech phenomenon, a target digit was either a '1' or a '2', and JW's task was to report verbally the digit that appeared after the tone. In this test JW's left hemisphere

---

[1] Magnetic Resonance Imaging.

could accurately verbalize which of the two stimuli appeared in either visual field. In a second study the subject was required to substitute a difficult word-associate for the simple numbers being flashed. In this complex naming condition, which was identical in all respects to the simple naming condition, except for the response that was required, J W was instructed to respond 'indescribable' when the '1' appeared and to respond 'indestructable' when the '2' appeared. In this test, as in the first, JW was able to respond accurately through speech to stimuli presented in both visual fields.

In the final experiment in this set, the target digit '2' was substituted by '9' in the left visual field only. This condition was identical in all other respects to the simple naming condition carried out in the first experiment. Thus either a '1' or a '2' appeared in the right visual field, whereas either a '1' or a '9' appeared in the left visual field. In this test J W verbally responded to the left visual field '9' as if it were a '2', and overall remained highly consistent in naming the stimuli presented in both fields.

Taken together, these experiments show that J W can name from the left hemisphere information presented to the right half-brain. The next set of experiments examines whether or not J W's left hemisphere is aware of this capacity.

In the first experiment in this group, J W was required to make an interfield comparison using a pointing response. Here, at the tone, two vertically aligned digits (1/2 or 2/1) appeared either in the same hemifield as the target (within-field) or in the opposite hemifield (between-field). J W was required to point with his right hand to the digit that matched the target. The results were clear. J W was able to match the sample information to the target number only when the sample information was presented to the same hemisphere that initially saw the information. Thus a stimulus presented to the right brain such as a '1' could be matched by the right brain but not by the left.

Similar results were seen in another experiment that required J W to make an interfield comparison (same/different judgement). For this experiment, all target digits (either '1' or '2') appeared in the left visual field and, at the tone, a probe digit (either a '1' or a '2') appeared for 150 msec in the right visual field. J W was required to report verbally whether the target and probe digits were the 'same' or 'different'. As in the preceding experiment, J W was unable to carry out a comparison task requiring between-field matching.

These results clearly show that J W's left hemisphere was not able to consciously access the information made available to the speech response system of the left hemisphere. It would appear that the information transmitted to the left hemisphere from the right hemisphere establishes a response readiness for one of two possible outcomes. This readiness is established outside the realm of conscious awareness. In follow-up studies, confidence judgements were taken from the left-hemisphere for both the left and right visual field presentations. When the stimuli appeared in the right visual field, the accuracy and the ratings were high. When the stimuli appeared in the left visual field, the accuracy was high but the confidence ratings were low; to borrow terminology established for other visual studies, it could be called 'blind-naming'.

The dramatic finding that a response system such as speech can be prepared and be capable of functioning without the left hemisphere being aware of the information the speech system possesses is consistent with the view that non-conscious processes can be active in the production of behaviours. In this particular set of experiments the split-brain patient is useful in revealing how such organizational features of human cognition operate.

It is not clear how the information presented to the right hemisphere is transmitted to the left hemisphere. Simple cross-cueing strategies seen in other tests do not appear active in J W (Gazzaniga and Hillyard 1971). If cross-cueing were active, the left hemisphere would be aware of the nature of the information and would be able to respond in the between-hemisphere comparison tasks. As a result it would appear that the transfer of information is internal and neural in nature as opposed to the external strategies these patients can use. Consistent with this view are recent observations using evoked potentials on case J W (M. Kutas, S. A. Hillyard, and M. S. Gazzaniga 1985, unpublished observation). It was shown that early responses in the visual evoked potential were different for the '1' as opposed to the '2' when these were flashed in the left visual field. Clearly the information is being encoded in a differential way, and information is transmitted to the left speech system. The route of transmission is unknown, but is presumably subcortical.

This demonstration of how an unconscious process can directly influence a behaviour is also reminiscent of other examples where the triggering stimuli were emotional in nature. I have previously reported a series of studies that showed how stimuli presented to the disconnected right hemisphere can influence the emotional state of the left hemisphere (Gazzaniga, 1970), even though the left hemisphere is not cognizant of the stimulus that triggered the emotional response in the right half-brain. Recently, LeDoux *et al.* (1984) have developed an animal model for studying such conditioned emotional responses and have begun to be able to track the actual neural pathways involved in an auditory conditioned response. These pathways course through the mid-brain and never reach cortical systems. LeDoux can selectively block the learning of the conditioned response by lesioning the pathways in the mid-brain. These startling findings suggest a reason why our emotional life seems so out of control to our cognitive processes. Conversely, we can begin to see a brain mechanism that explains why we construct a story at the conscious level for emotional responses triggered independently at the mid-brain level. Cortical mechanisms have to construct a theory as to why there is a felt state since the brain systems triggering the emotional state do not have direct, neural access to cortical processes. As in the split-brain patients, where we can examine these kinds of mechanisms experimentally, the emotional tone set up by one disconnected half-brain can be transmitted through mid-brain systems over to the other. It remains for the other half-brain to interpret the meaning within its ongoing cognitive context, whether it be positive or negative.

The experimental psychological and brain sciences are now able to move beyond the claim that non-conscious processes influence behaviour and are able to isolate the systems and study them directly. The realization that processes accessible to conscious experience represent only part of the overall process by which consciousness functions complicates the task of identifying all the players in the game of coming to know how we know things. At the same time it liberates us from viewing conscious experience as the product of a totally accessible rational process.

Whole brains and half-brains: modular entities and interactions

A major tenet in human brain science is that there are specialized functions within the cerebral cortex. It has long been known that for most humans the left brain is specialized for language and speech. More tentative but still widely believed is the claim that the right hemisphere is specialized for certain non-verbal skills such as facial recognition, spatial location, and other non-verbal skills. Thus neurologists have been aware for years of the 'modular' nature of the brain and have argued convincingly that any model of brain function that assumes there is something like a unified cognitive mechanism is in error. General problem-solving devices and other highly integrated views of how cognition works (Newell and Simon 1972) simply will not do.

One of the main challenges of human brain science is to attempt to isolate subfunctions of the cognitive system and to assess whether or not there is any biological validity to the cognitive constructs that are proposed. The need to put more ornaments on the modular tree is obvious. In my laboratory, at the Cornell University Medical College, work guided somewhat by the basic idea of modularity continues, and we continue to unearth many puzzling phenomena, some of which identify new modules and some of which speak to how other features of brain organization emphasize the modules' integrative action.

Studying the split-brain patient allows the examination of one half-brain independent of the major influences of the other half-brain. This is both revealing and masking with respect to our overall objective of identifying modular systems of the cerebral cortex. I will first review what I take to be the most revealing studies. Later we shall see how the study of disconnected brains masks how the two half-brains normally interact to produce what appear to be specialized functions.

Dissociating language and cognition

There are a few split-brain patients who possess language in both the left and the right half-brain. Most patients only possess language in the left hemisphere. The few who possess language in both allow one to examine how the introduction of language to the right hemisphere augments the overall cognitive capacity of that brain structure. To understand the continuum in which these observations are made, it is necessary to know what right hemispheres without language can do and also what I mean by the right-hemisphere possessing language.

In the past, split-brain patients without right-hemisphere language have not been studied as extensively as they might have been. After their post-operative assessment, which quickly reveals the cognitive state, or, better, the lack of cognitive state, in the right brain the patients as a rule have not been followed up. If the right brain is largely unresponsive to the processing of simple or complex stimuli, interest wanes in further delineating the patient's status. None the less, in the several patients we have studied in more detail a disconcerting picture emerges. The range of responsiveness to patterned stimuli ranges from none to the capacity to make simple matching judgements above chance. In the patients with the capacity to make perceptual judgements not involving language, there was no ability to make a simple same/different judgement within the right brain when both the sample and match were lateralized simultaneously. In other words, when a judgement of sameness was required for two simultaneously presented figures, the right hemisphere failed. Also, these same patients were unable to carry out tasks calling upon specialized right-hemisphere skills such as facial recognition tests. This profile is commonly seen in patients of all kinds, including patients of similar and sometimes greater overall intelligence than those who possess some right-hemisphere language.

This minimal profile of capacity stands in marked contrast to the patients with right-hemisphere language. The right brain of these patients is most responsive, and their overall capacity to respond both to language and non-language stimuli has been well catalogued and reported. Perceptual judgments are made easily, as are same/different judgments of all kinds. In some tests, such as facial recognition tasks, these patients demonstrate a superior performance in the left visual field. It was these kinds of findings that led me to believe that the presence of language in the right brain made the difference, and that, moreover, language processes were central to good cognitive operations (Gazzaniga 1980, 1983). It was upon this foundation that we planned a series of experiments that tried to elucidate further the cognitive capacities of the right brain in these patients with language. Our hope was to correlate varying cognitive capacity with the varying competence of language skills within the group of patients who possessed language.

The East Coast series of patients we studied included case JW, who understood language and who had a rich right-brain lexicon as assessed by the Peabody picture–vocabulary test as well as by other special tests (Gazzaniga *et al.* 1984). At the same time, JW has no capacity to speak out of the right brain. Our hope of contrasting JW's performance on a set of cognitive tasks with a more robust right-brain language system was made possible by studies on cases VP and PS. These patients were able both to understand language and to speak from each half-brain. Would this extra skill lend a greater capacity to the right brain's ability to think?

The different language capacities of the two groups consisted primarily of the capacity to speak, to understand some syntactic relationships, and to comprehend sentential strings. JW's right hemisphere is able to understand only individual words as evidenced by his capacity to pick word-associates (Gazzaniga *et al.* 1984). Thus if the word 'fruit' is presented to the right hemisphere, the left hand is able to point to the word 'apple' out of a list. In this manner superordinate and subordinate relations can be tested, as can synonyms/antonyms and so on (Sidtis *et al.* 1981). However, while JW's right hemisphere performs well at picking associates, the right hemisphere appears to have little insight into what it is doing. If a series of words are presented that vary in their categorical quality, such as 'fruit', 'apple', 'hardware', 'hammer', and so on, the right hemisphere is unable to judge whether or not one word is superordinate to another. Thus while the left brain has no problem judging which of the words is a 'category' vs. 'no category' word, the right brain performs barely above chance. It appears that possessing the capacity to correctly pick words that have a relationship to one another does not mean that the responding system knows why it is doing so.

The literalness or concreteness of the right hemisphere can be ascertained from another simple test. If a word, such as 'book', is spelled backwards, so that the right hemisphere sees the string 'koob', it cannot sort it out such that it can pick a simple associate. The left hemisphere under the same tachistocopic conditions has no problem with the task.

Of course, in all tests demonstrating lack of function the possibility always exists that the task was not understood. Yet in these tests as in all of our tests, the task is explained, then demonstrated, and then, under free-field conditions, practice trials are run. In most cases the response mode is identical to that in other tests that are completed with success.

At the same time, JW's right brain seems relatively at ease in understanding simple requests. Quite remarkably, if a picture of something, for example a horse, is flashed to the right brain, the left brain will typically speak out and say that it saw nothing. The examiner can then say things like, 'Don't draw what it is; draw what goes on it'. The patient might say something to the effect of 'What are you talking about; I didn't see anything'. Then the left hand will pick up a pencil and draw a saddle. In this particular case, JW drew an English saddle, a sketch that would appear ambiguous if you did not know the context. JW said that he did not know what he had drawn. He was then asked to draw a picture of what was flashed. The left hand then drew a horse and, after completing the picture, JW grinned and said, 'That must be a saddle'.

JW's right hemisphere, however, has failed to reveal any understanding of syntax. Thus while case VP's right hemisphere can appreciate functors, so that the difference between the triplet 'playing the field' and 'the playing field' is easily detected, JW's cannot. JW has poor overall sentential understanding as compared to VP. For example, if a sentence is read aloud, such as 'He forgot to water his new plant', and is then followed by the question 'Was it . . . ?', again read aloud except for the last missing word, and then the missing word 'dry' was flashed to either the left or right brain, JW scored at chance with the right hemisphere and 87 per cent with the left. VP, on the other hand, scored 77 per cent with the right brain and 92 per cent with the left. Clearly, VP possesses far better language skills that does JW. The exact specification and nature of the difference remains for future research to identify.

None the less, it turns out that the right hemispheres of both patients are poor at making simple inferences (Gazzaniga and Smylie 1984). For example, when shown two pictures one after the other, such as a picture of a match and a picture of a wood pile, the right hemisphere cannot combine the two gnostic elements into a causal relation and choose the proper result; i.e. a picture of a burning wood pile as opposed to a picture of a wood pile and a set of matches. In other testing, simple words were serially presented to the right brain. The task was to infer the causal relation that obtains between the two lexical elements and to pick the answer from a list of six possible answers printed and in full view of the subject. A typical trial would consist of words like 'pin' and 'finger' being flashed to the right brain, with the correct answer being 'bleed'. While the right hemisphere when tested separately could always find a close lexical associate of all the words used, it could not make the inference that 'pin' and 'finger' should result in the answer 'bleed'. All of this goes on, of course, after the right hemisphere has been shown how to do the task under free-field conditions and several examples have been presented. The successful completion of a task under these conditions must mean that the left hemisphere was controlling the response. Still, the right hemisphere was free to inspect and watch how the task was done.

With VP the tests were pursued and further simplified. Instead of two words being flashed to either the left or right visual field, one word was spoken and the other was then lateralized to either the left or right brain. Thus the word 'pin' would be spoken, followed by the word 'finger' being flashed. This simplification seemed to make no difference. The right hemisphere remained poor at carrying out the task.

When another test was administered that reduced still further the cognitive demand, both VP and JW performed poorly with their right hemisphere. This test assessed the capacity of each half-brain to identify a common attribute between two different words. Again, in order to simplify the testing procedure, one word was spoken and the other flashed to either the left or right brain. Thus the word 'fire-truck' was spoken and the word 'elephant' was flashed to either the left or right brain. There were four words in full view of the subject: 'size', 'colour', 'speed', and 'texture'. The right brain scored above chance at about 50 per cent (chance being 25 per cent) for both patients, while the left was near perfect. When the right hemisphere was correct on this task it picked a likely associate to the flashed word.

Similarly, the right hemisphere performs poorly on simple mathematical problems. When flashed a series of numbers and asked to perform simple addition, subtraction, multiplication, and division, the right hemisphere can identify the stimuli but performs poorly on carrying out the required computation.

Finally, the right hemisphere in these three patients also proved to be poor at solving a spatial reasoning task. In this task, a geometric shape with a specific design is lateralized to either the right or left brain. Placed in full view in front of the patient are four other figures, one of which fits exactly into the geometric stimulus that had been lateralized to form a perfect square. The task is exceedingly simple. Yet, while the left hemisphere found the task easy to solve, the right hemisphere performed poorly.

From all these tests we conclude that perhaps language is a data structure, a system called upon to label and express the computations of other cognitive systems. The language system itself is not able to perform cognitive activities, such as inference. In this regard, it is helpful to remember that patients in the early stages of Alzheimer's disease are frequently quite intact with respect to language, but are unable to solve the simplest problems or make the simplest inferences. At the same time, it is also apparent that the presence of language of the kind described in the right hemisphere in these patients correlates with a half-brain that is capable of carrying out many more mental activities than a right hemisphere without language. Understanding this fact is the objective of some new, ongoing studies of ours that compare performance of patients pre- and post-operatively on tasks that draw upon lateralized skills.

Bihemispheric interactions

Over the past 25 years of split-brain research it has been difficult to isolate what the potential costs to cognition might be by having the human cerebrum divided in two. Many earlier studies have shown that there is no change in reaction-time response to simple discriminations (Gazzaniga and Sperry 1966); in the capacity to form hypotheses (LeDoux *et al.* 1977); and in verbal IQ (Campbell *et al.* 1981). There have been some reports that negative effects can be registered on memory function (Zaidel and Sperry 1974), whereas others have not confirmed this concern (LeDoux *et al.* 1977). There are data suggesting that hemispheric disconnection actually allows each half-brain to function without perceptual interference from the other, and thereby confers on the whole brain a super-normal capacity to apprehend perceptual information (Gazzaniga 1970; Holtzman and Gazzaniga 1985). While most prior studies have been carried out in the belief that each half-brain is a functioning, independent system that operates no differently when separated than when connected, new studies are beginning to challenge this original view. The old working assumption was based on the kind of behavioural profile seen in the split-brain cases who possess language in each hemisphere. Each hemisphere seemed capable of responding in its own way to a wide variety of stimuli.

But then there are the other cases in whom right-hemisphere performance after surgery was poor to non-existent. Prior to surgery these epileptic patients performed normally on so-called right-hemisphere tasks. The question became whether these abilities were locked in to the silent right-hemisphere after disconnection from the dominant left half-brain. We plan a large study on this issue, and are encouraged by some early results. Consider the cases EB, DR, and LL. Each is interesting in a different way.

Case EB is a 23-year-old woman who has suffered from epilepsy since the age of 12. Prior to callosal surgery she underwent a right occipital resection in Montreal, with the aim of removing her epileptic focus. This produced a left-visual-field scotoma. Her seizures were not brought under control, and, at the age of 21, she underwent posterior section of the corpus callosum. Prior to this surgery she was examined on a number of tests including the nonsense wire-figure test of Milner and Taylor (1972). In brief, this task requires the matching of irregularly shaped wire figures. Four are placed in front of a subject, out of view. One of the figures is placed in the hand and then removed. Moments later the subject is required to find a match from the group of four. This task is believed to tap into right-hemisphere specialized systems, and case EB was able to perform the task with either hand. It appears that her intact callosum assisted in distributing the information arriving in her left brain from the right hand over to the specialized system in the right hemisphere. Or that, at least, is how we have come to think about these kinds of things.

After the posterior half of the callosum was cut, EB, in typical split-brain fashion, was unable to name objects placed in the left hand. The fibres crucial for the interhemispheric transfer of tactile information had been severed, and, as a result, what the right hand knew the left did not. She also proved to be a patient without right-hemisphere language. While she was able to find points of stimulation on the left hand by touching them with the left thumb, thereby demonstrating good, right-hemisphere, cortical somato-sensory function, she was unable to retrieve with the left hand objects named by the examiner. This task is managed easily by patients with right-hemisphere language. Most importantly, however, EB could no longer perform the wire-figure task with either hand.

Since EB could perform the task pre-operatively, it seems clear that the right hemisphere had the capacity to contribute to solving this kind of task when it was connected to the left. Disconnected from the left, it appears unable to function. This kind of finding suggests that the left hemisphere may normally contribute certain executive functions to specialized systems in the right brain. What was thought to be one module actually is the product of the interaction of at least two modules, each located in a different brain area.

The same general finding was seen in the pre- vs. post-operative scores on the block design test for the two other patients, cases DR and LL.

This test is considered to be a right-hemisphere task, although there are several reports suggesting that left-hemisphere damage can also cause deficits in its performance. Both of the patients underwent full callosal section. Case LL had had a right temporal lobe resection prior to the callosal surgery, carried out in an effort to control his epilepsy. His callosal surgery was performed in two stages. Case DR, a 39-year-old woman, had her surgery carried out in one operation. Prior to the callosal surgery of these patients, their performance on the block design subtest of the WAIS (Wechsler Adult Intelligence Scale) was fast and accurate, well within the normal range. This simple test requires arranging four to nine, red and white coloured blocks in a pattern that matches a picture of arranged blocks. The pre-operative tests were carried out with the right hand. Post-operatively, neither the left nor the right hand of either case could perform the task with ease. The time it took to solve the simplest patterns doubled, and completion of the more difficult patterns was simply not possible. Case LL revealed no other right-hemisphere function except for the capacity to locate with the left hand a point of light flashed in the left visual field. He was unable to carry out with the right hemisphere the simplest match-to-sample test using pictorial or verbal stimuli. Yet it appears that, pre-operatively, the right hemisphere when connected to the left participated in the solving of the block design problem. Also, since the post-operative scores for the left hemisphere were also lowered, the left hemisphere obviously benefited from processes located in the right half-brain.

Case DR had a more responsive right hemisphere. She was able to carry out visual match-to-sample tasks for lateralized visual, but not verbal, stimuli within her right hemisphere. Yet when two geometric shapes were presented sequentially to the right hemisphere, she was unable to make a same/different judgement, thereby indicating that her right hemisphere was not capable of simple problem-solving. At the same time, there is evidence that DR's right hemisphere understands some simple nouns. Yet, even with the far greater capacity to process information within the right brain, neither the left nor the right hand could perform the block design test as well as the right hand had performed it pre-operatively. Here again we see evidence that the normal contribution of the right hemisphere to solving such tasks can be realized only when it is connected to the left. And again, the left hemisphere was also benefiting from right-brain participation prior to the operation.

For future surgeries, a more comprehensive pre- and post-operative battery of tests is planned. However, when the evidence to date is taken together it suggests that there are dissociable factors active in what look to be unified mental activities. In short, one can begin to envisage that there are something like executive controllers that are active in manipulating the data of specialized processing systems. These controllers tend normally to be lateralized in the left brain, and, when the right brain becomes isolated from their influence, the specific functions of the right brain become hard to detect when tested alone.

Windows on cognition: the incomplete callosal section

Deducing the nature of cortical organization and how it relates to cognitive representational systems is a task that receives assistance from many quarters. The newest opportunity for us grows out of the recent observation that MRI studies of the human brain allow for determining the extent of callosal disconnection actually achieved during split-brain surgery. Before this technology was available, reports on the extent of surgery on all patients relied on the accuracy of surgical notes. Now, MRI is able to verify or correct the surgical claims (Gazzaniga *et al.* 1985).

For present purposes, two paradigmatic cases will be reviewed. Case JW was discovered to have a complete callosal section with the unapproached anterior commissure remaining intact. JW does not transfer any perceptual information between the hemispheres. Colour, pattern, and brightness information can not be cross-compared, thereby leaving but one observation of interhemispheric integration. Under conditions of sustained stimulation some crude spatial information can be integrated between the disconnected half-brains (Holtzman 1984). In any event, a spared anterior commissure has not yet proved capable of transmitting any useful cognitive or perceptual information between the hemispheres. It seems reasonable to assume that it transfers something, yet studies to date have failed to identify what this might be.

Case VP, however, allows for different insights. MRI revealed sparing of fibres both in the splenium and in the rostrum of the corpus callosum. Sparing in the splenium suggests the possibility that visual pattern and colour information might transfer between the two hemispheres. Yet in test after test in VP there is no such indication. It is too soon to tell whether the failure of simple transfer reflects regional differences of function within the splenium or whether it is related to the number of fibres spared, or both.

Further tests on VP have revealed a most remarkable interaction that was not seen in JW. The task required VP to judge whether or not two words, one presented to each visual field, rhymed. There were four conditions: the words (a) did not look or sound alike, (b) looked alike but did not sound alike, (c) sounded alike but did not look alike, or (d) both _ looked and sounded alike. VP is able to judge correctly whether the words rhyme only when the words both look and sound alike. Such a finding suggests the highly specific way in which the cortex encodes information. It appears that the visual system, which is still marginally interconnected by some fibres, can send some kind of verifying signal that is useful if information has already been transmitted through another modality. Without that bit of redundancy in the system, the information transferred appears to be of no use.

The left-brain interpreter

I have outlined a picture of brain function that reveals its apparent modular organization. The functioning modules do have some kind of physical instantiation, but the brain sciences are not yet able to specify the nature of the actual neural networks involved. It is clear that they operate outside the realm of awareness and express their computational product to the motor system directly. Catching up with and assessing what the brain is doing seems to be a function of an interpretive module residing in the left hemispere. I think it need not always be in the left, but that is where it is for most humans. To watch the interpreter work under strict experimental conditions is most dramatic.

We first revealed the phenomenon using a simultaneous concept test. The patient is shown two pictures, one exclusively to the left hemisphere and one exclusively to the right, and is asked to choose from an array of pictures placed in full view in front of him the ones associated with the pictures lateralized to the left and right brain. In one example of this kind of test, a picture of a chicken claw was flashed to the left hemisphere and a picture of a snow scene to the right hemisphere. Of the array of pictures placed in front of the subject, the obviously correct association is a chicken for the chicken claw and a shovel for the snow scene. PS responded by choosing the shovel with the left hand and the chicken with the right. When asked why he chose these items, his left hemisphere replied: 'Oh, that's simple. The chicken claw goes with the chicken, and you need a shovel to clean out the chicken shed'. Here, the left brain, observing the left hand's response, interprets that response into a context consistent with its sphere of knowledge—one that does not include information about the left-hemifield snow scene.

It is interesting to note that, although the patients possess at least some understanding of their surgery, they never say things like, 'Well, I chose this because I have a split-brain and the information went to the right, non-verbal hemisphere'. Even patients who are brighter than PS, based on IQ testing, view their responses as behaviours emanating from their own volitional selves, and as a result, incorporate these behaviours into a theory to explain why they behave as they do. Certainly one can imagine that at some future point a patient might be studied who chose not to interpret such behaviours in terms of an overlying psychological structure that prevented the response. Or one can imagine a patient learning by rote, as it were, what a 'split-brain' is all about, and why, therefore, a certain behaviour most likely occurred. With that understood they may well not offer an explanation.

There are occasions in which a patient, having trouble controlling his left arm due to a transient state of dyspraxia, will tend to write off anything this arm does under the direction of the right brain, thereby making the foregoing test inappropriate for demonstrating the phenomenon. In such situations, a single set of pictures is presented and only one hand is allowed to make the response. Thus, in this test the word 'pink' is flashed to the right hemisphere and the word 'bottle' is flashed to the left. Placed in front of the patient are pictures of at least 10 bottles of different colour and shape. When this test was run on JW, on a particular day when he kept saying that his left hand was doing what it wanted to do, he immediately pointed to the pink bottle with his right hand. When asked why he had done this, JW said: 'Pink is a nice colour'.

Another example of this phenomenon, of the left brain interpreting actions produced by the disconnected right brain, involves lateralizing a written command, such as 'laugh', to the right hemisphere by tachistoscopically presenting it to the left visual field. After the stimulus is presented, the patient laughs and, when asked why, says: 'You guys come up and test us every month. What a way to make a living'. In still another example, if the command 'walk' is flashed to the right hemisphere, the patients will typically stand up and begin to leave the testing van. When asked where he is going, the left brain says: 'I'm going into the house to get a Coke'. However you manipulate this type of test, it always yields the same kind of result (Gazzaniga 1983).

There are many ways to influence the left-brain interpreter. As already mentioned, we wanted to know whether or not the emotional response to stimuli presented to one half-brain would have an effect on the affective tone of the other half-brain. In this particular study, we showed under lateralized stimulus-presentation procedures a series of film vignettes that included either violent or calm sequences. In these studies we used an eye-tracking device which permits prolonged lateralization of visual stimuli while the eyes remain fixated on a point (Holtzman 1984; Gazzaniga and Smylie 1984). The computer-based system keeps careful track of the position of the eyes so that if they move from fixation the movie sequence is electronically turned off. For example, in one test a film depicting one person throwing another into a fire was shown to the right hemisphere of patient V P. She reacted: 'I don't really know what I saw; I think just a white flash. Maybe some trees, red trees like in the fall. I don't know why, but I feel kind of scared. I feel jumpy. I don't like this room, or maybe it's you getting me nervous.'

As an aside to a colleague, she then said, out of my earshot, 'I know I like Dr Gazzaniga, but right now I'm scared of him for some reason'. Clearly, the emotional valence of the stimulus has crossed over from the right to the left hemisphere. The left hemisphere remains unaware of the content that produced the emotional change, but it experiences and must deal with the emotion and give it an interpretation.

The same kind of phenomenon is observed when more neutral stimuli are presented, such as scenes of ocean surf, nature walks, and the like. The patient becomes calm and serene. Taken together, these examples show that both covert as well as overt responses are interpreted, and confirm and extend earlier experiments carried out on the California patients (Gazzaniga 1970).

The kind of thing we see in these special patients and under these kinds of laboratory conditions can be related to many everyday experiences. Consider how often we go to bed in a good frame of mind (or the opposite), only to awake feeling depressed and cranky (or the opposite). The cognitive data structure, which is to say the facts about our life, has not changed during the night, so why the change in mood? Could it be that a set of prior memories has become activated and has unleashed biochemical mechanisms that give rise to a specific mood state? The idea here is that the left-brain interpreter would try to make sense out of these feelings, and may well and somewhat gratuitously attribute a cause for them to otherwise innocent concepts also existing in the conscious realm at that time.

In the patients studied to date, the interpreter has been most demonstrably represented in the left hemisphere. A natural question to consider is whether or not the right hemisphere also has an interpreter, or, more likely, could develop one if disconnected from the left. The intact brain would not need two such modules. For a right brain that cannot talk, it would be difficult to gain evidence of such a function. Yet such right brains, when examined non-verbally, as on the inference tests described above, do not perform well. This would suggest that they do not have a functioning interpreter. For the patients with bilateral speech, no clear evidence has yet appeared that the right hemisphere carries out interpretive functions. In the standard tests used to elicit the phenomenon, each hemisphere can quite simply state what it saw and, when asked why, can tell you why. And, as in the patient whose right hemisphere can understand but not produce language, even a right hemisphere that can speak cannot carry out the simple inference tasks.

The idea of a central and crucial mental module or system that interprets the behaviours and activities of other modules comes up in a number of other contexts. Recent developments in the study of amnesia have suggested that patients frequently suffer from so-called 'source amnesia' (Schacter 1986; Schacter *et al.* 1984). Patients remember the content of a prior event but not its source. What they commonly do under a task demanding identification of the source is to make up a source which they quickly believe to be real. In hypnosis research, the interpreter is commonly seen at work when a subject explains why a post-hypnotic suggestion has been carried out (Hilgard 1977).

Conclusions

It is difficult to imagine Descartes insisting today that the only truly knowable subject would be mathematics. It clearly is knowable since the mathematician starts by fiat, setting up the hypothesis to be examined. Yet in the twentieth century there are factual systems being generated all the time that in effect are fiats for the mind's interpreter to play on. Those of us in brain science assume that one day all the characters in the brain play will be known and, as a consequence, that the interpreter will also have its data. In the foregoing I reviewed some new facts in what I assume to be not a hopeless objective, the task of specifying the operating characteristics of the human brain.

The picture that emerges from studying damaged brains, whether they be by the elegant disconnection process used in the control of epilepsy, or by focal lesions produced by stroke or tumour, all point to a brain model that is heavily committed to parallel processes that are co-active in our conscious lives. Their function proceeds, as it must, and as do most other physiological processes, outside of our awareness. Corralling all of these activities and making sense of them appears to be the function of special processes present in the left brain of humans. This function, the interpretive function, works on the products of the modular activities to build a schema that can explain the logic behind all of the ongoing activity that results in a behaviour. Behaving, alas, becomes a powerful determinant in what we come to believe as true.

Acknowledgements

This work was aided by USPHS (United States Public Health Service) grant no. NS-17778-04.05; by the McKnight Foundation; and by the Alfred P. Sloan Foundation.

References

Campbell, A. L., Bogen, J. E., and Smith, A. (1981). Disorganization and re-organization of cognitive and sensorimotor functions in cerebral commissurotomy: compensatory roles of the forebrain commissures and cerebral hemispheres in man. *Brain*, **104**, 493–511.

Fodor, J. (1983). *The modularity of mind*. MIT Press, Cambridge, MA.

Gazzaniga, M. S. (1970). *The bisected brain*. Appleton–Century–Crofts, New York.

Gazzaniga, M. S. (1980). The role of language for conscious experience: observations from split-brain man. In *Motivation, motor and sensory processes of the brain, progress in brain research*, Vol. 54, (ed. H. H. Kornhuber and L. Deecke), pp. 689–96. Elsevier, Amsterdam.

Gazzaniga, M. S. (1983). Right hemisphere language; a twenty year perspective. *The American Psychologist*, **38**, 525–37.

Gazzaniga, M. S. (1985). *The social brain*. Basic Books, New York.

Gazzaniga, M. S. and Hillyard, S. A. (1971). Language and speech capacity of the right hemisphere. *Neuropsychologia*, **9**, 273–80.

Gazzaniga, M. S. and Smylie, C. S. (1984). Dissociation of language and cognition: A psychological profile of two disconnected right hemispheres. *Brain*, **107**, 145–53.

Gazzaniga, M. S. and Sperry, R. W. (1966). Simultaneous double discrimination response following brain bisection. *Psychonomic Science*, **4**, 261–2.

Gazzaniga, M. S., Holtzman, J. D., Gates, J., Deck, M. D. F., and Lee, B. C. P. (1985). MRI assessment of human callosal surgery with neuropsychological correlates. *Neurology*, **35**, 1763–66.

Gazzaniga, M. S., Smylie, C. S., Baynes, K., Hirst, W., and McCleary, C. (1984). Profiles of right hemisphere language and speech following brain bisection. *Brain and Language*, **22**, 206–20.

Gazzaniga, M. S., Holtzman, J. D., and Smylie C. S. (1987). Speech without conscious awareness. *Neurology*, **35**, 682–685.

Hilgard, E. R. (1977). *Divided consciousness: multiple controls in human thought and action*. John Wiley, New York.

Holtzman, J. D. (1984). Interactions between cortical and subcortical visual areas: evidence from human commissurotomy patients. *Vision Research*, **24**, 801–13.

Holtzman, J. D. and Gazzaniga, M. S. (1985). Enhanced dual task performance following callosal commissurotomy in humans. *Neuropsychologia*, **23**, 315–21.

Holtzman, J. D., Volpe, B. T., and Gazzaniga, M. S. (1984). Deficits in visual–motor control despite intact subcortical visual areas, (abstract) *Neurology*, **34**, 187.

Kosslyn, S. M. (1983). *Ghosts in the mind's machine: creating and using images in the brain*. W. W. Norton, New York.

LeDoux, J. E., Risse, G., Springer, S., Wilson, D. H., and Gazzaniga, M. S. (1977). Cognition and commissurotomy. *Brain*, **100**, 87–104.

LeDoux, J. E., Sakaguchi, A., and Reis, D. J. (1984). Subcortical efferent projections of the medial geniculate nucleus mediate emotional response conditioned to acoustic stimuli. *Journal of Neuroscience*, **4**, 683–98.

Marcel, A. J. (1983). Conscious and unconscious perception: an approach to the relations between phenomenal experience and perceptual processes. *Cognitive Psychology*, **15**, 238–300.

Milner, B. and Taylor, L. (1972). Right hemisphere superiority in tactile pattern-recognition after cerebral commissurotomy: evidence for nonverbal memory. *Neuropsychologia*, **10**, 1–15.

Newell, A. and Simon, H. A. (1972). *Human problem solving*. Prentice-Hall, Englewood Cliffs, NJ.

Pöppel, E., Held R., and Frost D. (1973). Residual visual capacities in a case of cortical blindness. *Cortex*, **10**, 605–12.

Schacter, D. L. (1986). The psychology of memory. In *Mind and brain: dialogues in cognitive neuroscience*, (ed. J. E. LeDoux and W. Hirst), pp. 189–214. Cambridge University Press, New York.

Schacter, D. L., Harbluk, J. L., and McLachlan, D. R. (1984). Retrieval without recollection: an experimental analysis of source amnesia. *Journal of Verbal Learning and Verbal Behavior*, **23**, 593–611.

Sidtis, J. J., Volpe, B. T., Holtzman, J. D., Wilson, D. H., and Gazzaniga, M. S. (1981). Cognitive interaction after staged callosal section: evidence for a transfer of semantic activation. *Science*, **212**, 344–6.

Volpe, B. T., LeDoux, J. E., and Gazzaniga, M. S. (1979). Information processing of visual stimuli in an extinguished field. *Nature*, **282**, 722–4.

Weiskrantz, L., Warrington, E. K., Sanders, M. D., and Marshall, J. (1974). Visual capacity in the hemianopic field following a restricted occipital ablation. *Brain*, **97**, 709–28.

Zaidel, D. and Sperry, R. W. (1974) Memory impairment after commissurotomy in man. *Brain*, **97**, 263–72.

11

Integrated field theory of consciousness

Marcel Kinsbourne

The problem

A credible theory of the biological role and brain basis of consciousness must take into account the actual structure and organization of the cerebrum, which is the acknowledged chief organ of the mind. To date, formulating such a theory has been handicapped by too restrictive a model of central nervous functioning. This prevalent view has been characterized by Pribram (1971, p. 9) as holding 'that each cortical point . . . is specialized for a unique function. The integration necessary to account for behavioral and psychological processes is . . . accomplished by . . . permanent associative connections between neurons'. Though endorsing local functional specialization, I find it necessary to qualify the emphasis on associations by noting two overriding design characteristics of cortex:

1. Its network characteristic (Sholl 1956). Cerebral neurons, directly or indirectly, are all interconnected. No anatomically demonstrable preferential pathways exist within the cortical neuropil [the feltwork of axons and dendrites within which neuron-to-neuron communication largely occurs (Pribram 1971)].

2. Its operation in terms of excitation–inhibition balances. There being no circumscribed (anatomically 'encapsulated') modules in cortex, the programming of mental operations is protected from interference by inhibitory surrounds (Walley and Weiden, 1973; Cook 1986). Cortico–cortical connections, the U-shaped fibre tracts that sweep from grey through white and back to grey matter, all carry impulses in both directions (Pandya and Kuypers 1979). They are therefore suited to the modulation of excitation–inhibition balance. The brain basis of a given mental operation can flexibly be insulated from other cortical activity, or integrated with it. I will argue, under the headings of twelve propositions, that, to the extent a mental operation is selectively insulated, it is automatic, and outside awareness; to the extent it is co-ordinated

with other mental operations, it contributes to the 'integrative field of awareness'. Each proposition is controversial, and the evidence for each is circumstantial. My effort is confined to showing that individually the propositions are plausible and, collectively, they offer a coherent working hypothesis about the brain basis of consciousness.

Propositions

The cortical network is specialized for selective representation

At the neural level of organization, mental operations proceed through a series (or parallel set) of transformations of central representations. The prevalent notion that representations are 'processed' implies that the cortical network is divisible into substrates for representations and into processors that operate upon them. There is no independent support for such a subdivision of the cerebral neural networks. It is simpler and sufficient to suppose that each cortical area represents input within the constraints of its neural specialization and current state of interrelationships with other areas (including activating and inhibitory influences). Some representations are available as contributions to the field of conscious awareness and, if attended, enter it.

The term representation is usually applied to accounts of experience. But mental life also features intentions. Are intentions representations? They are the representations of potential outcomes. The planning function of the frontal lobes is to programme activity that will approximate the external status quo to the represented intention. The well-known Sokolovian model of match–mismatch between anticipated and experienced input (mismatch eliciting an orienting response) is equally applicable to the match between the *intended* and the *realized* state of affairs. In both cases an input pattern is matched to a schema. Mismatch will lead to an orienting response, with a shift of the focus of attention from the overall goal to the instrumental acts, which may have to be reprogrammed.

Awareness is a composite of multiple, coincident representations

Outdated notions of cognition as based on information flow from processor to processor lead naturally to a hierarchical model of conscious functioning, with some master area—the left angular gyrus (Geschwind 1970)—nominated to be the site of a conscious overview of input and of a conscious formulation of intentions. Yet no documented focal cerebral lesion has ever deprived its victim of attention and intention, leaving unconscious operations in place. Rather than suggesting the existence of what James (1890) ironically termed a 'pontifical neuron' that knows and determines all, the evidence points to a quite different model, of parallel representation (Merzenich and Kaas 1980). In accord with this view, I propose that the conscious domain is not the product of an all-or-nothing vantage point, but is a multi-componential field. Clinical illustrations derive from cerebral syndromes in which awareness is restricted or divided. The issue in each case is not 'Can such and such be experienced?' but, rather, 'Is the individual aware that such and such is no longer being experienced?'

A distinction is necessary between whether we experience something and whether we are aware of whether we are experiencing something. Certain manifestations of damage to the posterior part of the cerebrum illustrate this point. One is the condition following destruction of the visual cortex. The patient can see little or nothing (is 'cortically blind') but acts and speaks as if he could see as well as ever. The victim of cortical blindness not only sees nothing; he does not experience the fact that he is not seeing anything. His conversation is rich with allusions to visual experiences which conflict with physical reality. His confabulations about visual percepts represent his attempts to use his previously acquired knowledge about visual events in the face of the sudden loss of the ability even to imagine what it is like to have a visual experience. A similarly striking restriction of awareness occurs in neglect/denial syndromes in which the left side of the patient's body is no longer experienced. In these situations the patient ignores the left side of his body and fails to intentionally use it. Statements he makes about his left side are oblique and inferential. They represent attempts to draw upon his fund of knowledge about what it must be like, and reluctant attempts at that, as the lack of support from direct experience is perplexing. No doubt patients' failure to execute intentional movements and plans directed leftward following certain right frontal lesions are similarly to be classified as a suspension of attempts to represent the intended outcomes of leftward action. Organizing activity leftward ceases to be an option within the patient's repertoire of intentions. He is therefore unaware of the fact that he is not performing such acts. Nor can patients with any of the above syndromes be made aware by explanations of what they do not realize spontaneously. One may deny what he is told, another may appear to acquiesce. But the information supplied is not meaningful to the patient. His subsequent activities ignore the deficit, as before.

Some characteristics of behaviour in the split-brain state can be understood in this light. Such patients have had the interconnection between their cerebral hemispheres surgically severed. Each hemisphere retains its relation to the brain stem, independently of the other. The patient's behaviour when functioning through one disconnected hemisphere suggests that he is unaware that his control over ipsilateral sensory–motor facilities has become grossly restricted. Nor does he exhibit the bewilderment and distress at this loss of control that one might expect. Disconnected from the relevant areas of representation, he simply fails to represent that loss within awareness. The picturesque rationalizations that the left (verbal) hemisphere has repeatedly been reported to produce when bodily actions occur under right-hemisphere control (Gazzaniga 1988, this volume) are not unlike those that a patient with unilateral neglect might generate relative to the left side of his body.

Specialized cortical areas represent, on an ongoing basis, the state of each sensory field. They represent both the presence of input and its absence (a homogeneous Ganzfeld, the silence of an anechoic chamber). Absence as well as presence is positively entered into awareness. When the representing area itself is destroyed, absence of input within the relevant domain is no longer represented centrally at all.

Awareness is based on continually revised representation

The individual represents aspects of input from the external and the internal environment, variously disassembling, assembling, and transforming them. Disassembling is shifting to a more local (restricted and detailed) view, termed focal attention. Assembling is a shift to a more global overview, with sacrifice of detail, termed diffuse attention. Transforming is imaging changes in the state of things. A rough analogy is in vogue between this shifting, representational activity and the discontinuous operation of certain machines, which process input, represent the product (for the view of the operator or for further processing by another part of the machine), and process again. There is no reason to suppose that such a jerky discontinuity characterizes the human control systems. To say that processing occurs over time, in a manner of which the individual is unaware (Lashley, 1958), misses the point. We do not experience *how* we represent. The representations of the moment *are* our experience at that moment. [See Shallice (1988, this volume) for a fuller discussion.] The neural network, responding to the flux of input, changes state (i.e. pattern of activity). Models using parallel distributed rather than symbolic computational architecture seem suited to this concept of cerebral functioning (McClelland and Rumelhart 1986).

Neural activity proceeds concurrently at various levels of organization. Awareness characterizes the level of organization at which the brain represents physical change, represents previously experienced change, and represents intended action and its consequence, all in relation to each other. A machine with multiple channels of parallel but independent information flow would not simulate awareness. But if the channels transmit information that is interrelated to form an overall configuration, its elements having differential access to response mechanisms, then an aspect of awareness has been simulated.

Awareness in diffuse and in focal attending

What is represented in consciousness has often been confounded with what can be reported in particularly detailed ways. (Hence the over-emphasis on language when consciousness is considered.) Appearances that are elaborately and analytically encoded (in focal attention) are subsequently more accessible [in episodic memory (Tulving 1983)]. Hence those contents of awareness that exist when a subject is in a state of focal (effortful) attention have been emphasized, to the point that theorists have equated the boundaries of subjective experience with those of what one is aware of when focally attending (Johnson-Laird 1983). But, as Sperling (1960) showed in his 'partial report' experiments, and as is intuitively obvious, vastly more is experienced at a glance than can be communicated by explicit serial report (or subsequent recollection). Sperling's emphasis was on the identification of items in terms of differential location, which entails focal attending. Although focal attention may restrict awareness to a few task-relevant cues, one does not spend one's waking life in a state of relentlessly focused attention. People more often grant their attentional mechanisms wide range, opening their awareness to the experience of a diverse, if incompletely specified, environment.

Diffuse and focal attention both subserve the function of integrating information. But in the first case the integration is parallel, in the second sequential. During diffuse attention various percepts are placed into spatio–temporal relationship to one another. When attention is focal the precise combination of a limited number of features is specified and related to representations of previous experiences in the observer's mind. Focal attention is selective. It limits the perceptual field, and the set of comparisons that are of interest. When that set is incompletely specified, and response uncertainty prevails, I suggest that this calls for much inhibition of information travelling to (Skinner and Yingling 1977) or within the connected cerebral network, and is therefore experienced as effortful. In contrast, diffuse attention is relatively free of such constraints. Inhibition is not called for, and therefore this state is not experienced as effortful.

A task is difficult when the set of relevant cues has not been fully specified. Awareness of present and prior context is potentially useful as the subject tries one and then another way of accomplishing the task. When the range of cue utilization has been narrowed by repeated trials, and this practice has perfected the performance, the task becomes easier and requires less conscious attention. Shallice (1988, this volume) remarks that in dual-task performance the subject may be unaware of the easier action sequences.

Up to a certain degree of specification, perceptual analysis proceeds in parallel (Treisman and Gelade 1980), creating a rich environment of which we are fully conscious. Pre-attentive processes (Neisser 1967) supply parallel representations to consciousness. Only when more specific differentiation is required do the limitations of focal attention come into play. This is not because further differentiation involves qualitatively different mental operations. It is a matter of degree only. Because the cerebrum is a neural network, focal activity will tend to diffuse. For fine differentiation a high degree of inhibitory surround is needed—the cognitive masking of Walley and Weiden (1973)—which necessarily excludes other uses for the inhibited areas of brain. There is no known stuff in the brain that corresponds to attentional resource (cf. Navon 1984). Rather, there are different ways of using the finite neural network. When wide areas of the network are involved in one mental operation (in terms of patterned activation and the surrounding inhibition) other operations are deprived of the necessary cerebral functional space (Kinsbourne and Hicks 1978). In other words, multi-purpose cerebral computing space can be used either for a wide-ranging but shallow encoding, or for a single but difficult mental operation.

An example is afforded by the panoramic visual field. This yields diverse and structured information instantly to the glance, without any laboured successive extraction of information. A rich auditory field (for instance, of orchestral music) is similarly readily apparent. Yet when we focally attend, panoramic information slips from awareness. If one were to take literally the limited resource metaphor, one would have to suppose a trade-off between resource for focal analysis and resource for the definition of the ambient field. One would assume that in focal attention the attended field constricts because of a shortage of attentional resources; they have been siphoned off to fuel attention at its focus. My alternative suggestion is that it is because the periphery has become irrelevant and potentially interfering that the range of the attended field constricts. The field constriction could be a by-product of the hypothesized inhibitory surround that safeguards the precision of the activity of the active cerebral area.

I have discussed elsewhere, and in more detail, reasons to suppose that

the neural network itself is the 'limited resource' (Kinsbourne 1981)
How limited it is depends on the particular task-combination and on the
level of practice. The fashionable parallel/serial distinction for mental
operations does not characterize qualitatively different processes but
dichotomizes what is a matter of degree. If one mental operation draws
upon all of the applicable neural network, the next mental operation
needs must wait.

Awareness permits general application of specific mental operations

Rozin (1976) has pointed out that, whereas sophisticated transformations may characterize specific adaptive behaviours in many otherwise
behaviourally unsophisticated species, it is only in mammals and
outstandingly in man that such programmes became accessible for
general application to diverse problems as they arose. Special purpose
programmes—Fodor's (1983) modules—could become available for
more general purposes through genetically determined replication.
Alternatively, their constituents could be dismantled, making the logic
generally available. Subjecting the mental activity in question to conscious analysis would be an efficient way of accomplishing this, as, once
it is formulated in awareness, the logic in question becomes available for
unlimited application. By means of what adaptation does the mammalian brain, and notably the human brain, become able to subject
automatic mental representations to deliberate analytic scrutiny?
Presumably intermediary representations within an information-
processing sequence are isolated for selective attention. This would
appear to call for a sophisticated, selective inhibitory capability.

   The mediating role of awareness in rendering special purpose programmes generally accessible may have contributed greatly to the
adaptive advantage that led to its evolution. It pre-adapts the organism
to a virtually unlimited range of unforeseen contingencies.

Consciousness is not qualitatively different from other properties of nervous systems

The properties of things describe what happens when things interact. A
thing cannot have an absolute property (*in vacuo*) any more than a
sensation can have an absolute quality (see Dennett 1988*a*). A
thing may weigh upon another thing to a certain extent, interrupt rays
of light or reflect them in a certain way, but in these and countless other
ways the specific context with which the thing interacts enables it to
manifest properties. Similarly, even an ostensibly simple stimulus such
as a primary colour renders a subjective impression that is determined
by its context (of other inputs and of the mental state of the perceiver).

All properties are interactive. Just as figure could not exist without ground, so some context with potential for interaction is always needed to enable anything to manifest any of its properties. Thus it is a property of neurons to generate awareness when they interact. Consciousness is a property of neurons, although the neuron is not conscious. Properties that manifest when components of a system interact have been called 'emergent' (Bunge 1977). Emergence is no more or less mysterious than the organization of the physical world in general, and to characterize consciousness as emergent (e.g. Sperry 1984) in no qualitative way distinguishes it from everything else in the world. Certainly, the conscious functions of neural systems are neither more nor less 'emergent' than their unconscious functions (e.g. autonomic control). Nor is it any more perplexing that certain forms of neuronal organization give rise to conscious experience than that other organizations underlie processes of which we are not aware. In so far as emergence is based on a *constellation* of events, each of which has a finite probability of occurrence, some emergent properties may depend on an unlikely conjunction of events, and thus be rare. Awareness, and particularly self-awareness, may be highly improbable emergent properties. But this would not characterize these properties as *qualitatively* different from other properties that emerge in organized cell populations in the animal body.

Self-consciousness, focused attention, and the brain

Of all mental states, intense concentration is perhaps the least suitable for elucidating the most elaborate level of conscious experience, self-consciousness. Not only is the concentrating subject unaware of all but the most intense stimuli extraneous to those concentrated on, he may even be unaware of himself as a concentrating agent. The focus of attention is so narrowed that self-consciousness itself is among the excluded facets of experience. The self is a construct. It is incorporated in experience only at those times when the focus of attention permits.

According to the multi-componential field model that is here being considered, the construct of the self should have a definable neural basis, like any element capable of entering into consciousness. It is not yet possible to specify this based on the usual analysis by clinical deficit, since focal damage that documents loss of self-consciousness, while conscious processes remain intact, has not been described. (Perhaps we have not asked patients the right questions.) But one clue implicates the temporal lobes. This is the phenomenon of depersonalization, as seen in temporal lobe epilepsy, in schizophrenia [which allegedly involves dominant temporal mechanisms (Gruzelier and Flor-Henry 1979)], and occasionally in intact and sane people.

Temporal lobe epileptic phenomena comprise a set of experiences dissociated from their customary referents—*déjà vu, jamais vu*, various emotions. One such experience can be described as autoscopic. The patient has a sense of himself, separate from his physical person. The patient is, momentarily, 'beside himself' (Baldwin 1962). Depersonalization can be construed as a paroxysmal activation of self-consciousness separate from what the self is constructed to organize: the individual's stream of consciousness.

Orienting brings the remembered as well as the present into awareness

Patients with unilateral neglect of space act as if oblivious of the left side of configurations they attend to, whether they be centred or to the right or to the left of midline (Kinsbourne 1970). The right-hemisphere lesion has not simply destroyed the ability to represent either immediate or remembered experience to the left of centre; the neglect tendency affects representations of both sides of space. It is not that the representation *per se* is incomplete. Bisiach and Luzzatti (1978) have demonstrated that, viewing a scene from opposite directions in his imagination, the patient with neglect mentions items from each perspective that he omits from the other perspective. In each case an intact representation is brought to awareness incompletely (defective on the left of the observer's mind's eye). Even normal subjects, if their attention to briefly exposed stimuli is manipulated so as to bias it to one end of a line, can be prone to 'neglect' the other end. Reuter-Lorenz *et al.* (1990) briefly exposed transected lines to the left or right of fixation. Subjects judged whether the transections bisected the line. A cue at one end of the line (e.g. a square) biased attention so as to induce a systematic bias in the bisection judgement. This suggests that consciously representing is a constructive process comparable to representing input acquired by orienting across the lateral plane (Kinsbourne 1970, 1987; Posner *et al.* 1984). But it is not necessary to suppose that an internal 'scanner' makes passes over an existing representation, which may be incomplete (as Bisiach and Luzzatti assume). The neglect state, as we have discussed, generates directionally based representations, and this is sufficient to account for the clinical phenomenology, without postulating scanners within the brain (and representations within the scanner, leading to an infinite regress). Orienting across a scene from memory need not be regarded as a two-stage process: (i) representing, (ii) scanning. The representation itself could be directionally constructed and therefore enter awareness directionally.

Imagery generation does appear to differ from direct perception in one relevant respect. It seems to be generally deliberate or conscious. Thus it might be more vulnerable still to cortical deficit than the comparable direct percepts—and apparently is, according to Farah (1984). Posner *et al.* (1985) find certain distinctions between external and internal visual scanning mechanisms, which could correspondingly reflect the automatized nature of the former, but not the latter.

Left hemisphere lends depth, right lends breadth, to awareness

The simultaneous nature of diffuse attention and the successive nature of focal attention suggest a division in terms of complementary hemispheric specialization (Kinsbourne 1982). It takes left-hemisphere damage to impair the depth of conscious analysis, but right-hemisphere damage to restrict the extent of the conscious field. By depth I mean relation of present to previous (and prospective) relevant experiences. Breadth describes the extent of current experience. Contrast simultanagnosia (left posterior damage) with visual-spatial agnosia (right posterior damage). We have interpreted the left posterior deficit as one of serial identification (Kinsbourne and Warrington 1962). On account of this lesion it takes longer to identify visually each in turn of simultaneously presented, briefly exposed items. The patient is nevertheless aware of the number of items that are simultaneously present [and possibly their enduring influence could be demonstrated by non-episodic means—e.g. priming (Marcel 1983)]. Correspondingly he remains able to subitize (estimate visual numerosity) efficiently (Kinsbourne and Warrington 1962). In contrast, the visual–spatial agnosic can attend to only one object at a time. As he does so, he loses awareness of everything else. For him, identifying concurrently present items is hardly the issue. He is not even aware of them (although a moment earlier they might have been the focus of his pathologically constricted attention). Adjacent and even overlapping forms slip from awareness. The function of maintaining a structured perceptual field is impaired. One form, perceived, blocks both the current impact and the recent memory of the others, so that the patient cannot relate the one in view to the existence, let alone the locations, of others. Relative to a depth vs. breadth dichotomy, the left posterior lesion impairs serial matching of single items to sample in memory store. The right posterior lesion impairs the simultaneous perception of multiple objects in a structured perceptual field.

Intuition is a fallible guide to reality

A major theoretical problem relative to consciousness has been to reconcile the intuition of a unique and unitary self with neurobiological reality. Where in the brain is the self located (represented)? If different sectors of brain can independently support voluntary action, are there, in the split-brain case, two selves (Bogen 1969)? If there are, must they always have been there (even before the commissurotomy)? If so, was one held prisoner by the other (Puccetti 1973)?

Those debates present the intuition of unity as the given, with which our theories have to be reconciled. This is not necessarily reasonable. If the concept of self evolved, it did so on account of adaptive advantage, not because it reflects some objective truth. The concept of self reifies the organizing activity of a cybernetic device that incorporates its history ('experience') into the basis for its actions. It is the construct around which are organized impressions and intentions that reach awareness. If the brain is divided saggitally—or coronally, or in any other manner (Kinsbourne 1982)—with the result that some mental operations proceed out of contact with others, this degrades the adaptive use of the self-concept. But if it implies a multiplication (or unveiling) of selves, it does so in no more profound a sense than the division of a communication network results in two distinct (albeit more limited) networks.

This sceptical view of what is intuitively obvious has general applicability. Intuitions may or may not correspond to what is 'really' the case. Intuitions presumably evolved by selection, like other aspects of behaviour, because they steer actions in adaptive directions under given circumstances. Any philosopher can easily invent situations other than those on which a given intuition was based, for purposes of which the intuition would be useless or misleading. The overreliance on intuition is apt to lead to philosophic shipwreck.

One case in point is the intuition that the self is indivisible (Dennett and Kinsbourne, in press).

Awareness is surgically divisible

When a normal person does two unrelated things at the same time, different parts of the brain subserve the two performances. But, unless one activity fully pre-empts attention (Shallice 1988, this volume) the subjective experience is that of a single person doing two things. The unifying effect of awareness relates both activities to the same construct of the self and thus to each other. This has obvious utility in that the individual can at all times determine the pay-off for maximizing performance on the one vs. on the other activity (Shallice 1978). Response conflict is observably set up, and adaptively resolved. Were there no interconnections to resolve response conflict, the organism could not reconcile its responses to complex stimulus situations, different aspects of which call for conflicting performance. Instead, each performance would be separately driven, oblivious of the other, if not even obtruding on or invalidating it. When a person does something of which he is unaware, he cannot integrate it with his intended behaviour, and it may indeed contradict the latter. Thus the twisting fingers of the lying witness contradict his lies, as surely as his dry mouth and wild polygraph excursions do so. Such a dissociation is found in the split-brain (commissurotomized) individual. By default, the split-brain situation dramatically illustrates the role of awareness in adaptive behaviour. The anatomical disconnection divides the normally unified field of awareness into two fields, not co-ordinated with each other, yet overlapping in what they represent. Each hemisphere supports its own independent representation across the whole range of conscious experience but is depleted of data to which only the other hemisphere is privy. The 'centre of narrative gravity' (Dennett 1988b) shifts from moment to moment, from one to the other of the separated hemispheres. This makes it at least theoretically possible that each hemisphere, employed in a task that does not fully engage it, has attention to spare to relate its ongoing experience to its self-consciousness. Being concurrently aware of different things the two hemispheres may have different experiences. The following example from our research illustrates this (Trevarthen 1974).

Split-brain patients were simultaneously presented with one incomplete form just right of the fixation point and another one just left; for instance, three sides of a square open centrally to the right and two sides of a triangle open centrally to the left. The patient reports a complete square (picking it out with the right hand) and a complete triangle (picking it out with the left hand). The patient discerns no contradiction in identifying two totally different objects in the same location in the visual field. The discontinuity in the field of awareness precludes the two hemispheric representations being matched against each other and therefore they need not be reconciled with each other. The very separation that limits the performance to a single hemisphere's neural substrate permits another performance to occur concurrently based on the other hemisphere. In special circumstances, even contradictory judgements can be made relative to the same point in space without internal conflict (Kinsbourne 1975). Two tasks based on different hemispheres can be done as efficiently as each alone, and independently of each other (Holtzman and Gazzaniga 1985). Note that disconnection does not in general increase the organisms's overall capacity; rather, it bestows an advantage in the special case in which the intact operator would suffer interference from response conflict (Holtzman 1984). The split-brain subject processes information poorly with either hemisphere (presumably for lack of neural computational space) and is particularly poor in dual-task performance when both tasks are based on one hemisphere (Kreuter *et al.* 1972). In practice, parallel independent processing is not likely to happen much of the time. Not only is the free-moving organism likely to experience much the same overall input bilaterally, but the disconnected hemispheres constitute an unstable system competing for resources (Teng and Sperry 1974; Kinsbourne 1975; Holtzman and Gazzaniga 1982). For instance, in a concurrent finger-tapping–speaking task, a mistake by the vocal left hemisphere regularly brought the finger-tapping controlled by the right hemisphere to a halt (Kreuter *et al.* 1972). When split-brain subjects responded manually to the presence of a target letter in either half-field, response in the left half-field would be withheld (two cases) or be nondiscriminative (one case) (Kinsbourne 1975). The verbally activated left hemisphere pre-empts ascending activation, and, in the absence of the normal equilibrating function of the corpus callosum (Kinsbourne 1974), the right hemisphere effectively becomes a non-participant in the task.

Activity in one of separated hemispheres has been shown to unconsciously bias responses programmed by the other or to unconsciously guide them (Trevarthen and Sperry 1973; Holtzman 1984). It is specifically and selectively awareness that is split. It follows that the two disconnected hemispheres can concurrently be aware of different things. It is normal for them to receive different inputs, but the unified awareness of the integrated brain permits the construction of a more complete representation incorporating information processed by both hemispheres. Intrahemisphere disconnections can also separate response systems, so that different ones respond differently or even contradictorily to the same situation. Moreover, the anatomically dissociated brain is by no means the only case in point for dissociated consciousnesses. In hysterical and in hypnotic states a single individual can be host to two or more perspectives on events that seem to run their course out of contact with each other, and even do so concurrently (Hilgard 1977). Must one of these be automatic [for instance, a perspective based on the right hemisphere, as Eccles (1966) supposed]?

A narrow range of awareness suffices for specific acts of discrimination or execution (left hemispheric); automatized, the act calls for little more. The relational function of the right hemisphere, however, can usually not be automatized, as the set of possible alternative arrangements of external things (episodes) is so great that each episode is likely to be to some extent novel. Goldberg and Costa (1981) advocated the view that the right hemisphere is specialized for novelty detection. If so, its functions, more than those of the left hemisphere, would appear to invoke awareness (Sperry *et al.* 1979). If at any time a hemisphere works like an automaton, it is the left [see Gazzaniga (1988, this volume) for a contrary opinion].

While there is no evidence for anything like two, continuous, independent streams of consciousness in the split-brain patient, as sometimes alleged (Sperry 1966), that *in principle* he can house at the same time two versions of awareness experiencing different events remains a possibility.

Two awarenesses can coexist

Neither of the two separated hemispheres houses the 'real self'. Representations in each hemisphere continue to be related to the construct of the self, as they did before the operation. Experience continually reshapes the self-concept. If that construct is differently informed by the two hemispheres at the same time, the disconnection precluding the individual from being aware of any incongruity, each hemisphere's activity must engender a somewhat differently delineated self. If by some hypothetical arrangement the flow of information to the two disconnected hemispheres could be kept separate and disparate over an appreciable period of time, presumably the construct of the self would be shaped by a different cumulative experience within each hemisphere. Also, in so far as the cognitive strengths and emotional proclivities of each hemisphere are not the same, processing of, and response to, the onrush of events would to some extent differ, further amplifying discrepancies in the experience of the two hemispheres. This dramatic circumstance does not indicate that the organism housed two distinct personalities (one perhaps suppressed by the other) prior to the callosal section (Bogen 1969; Puccetti 1973). The intact individual's unified awareness precludes this and blends the different skills and styles of the two hemispheres into the unified behavioural repertoire of the individual. The lesson for consciousness of the split brain is quite a different one. It further demonstrates the distributed nature of the brain basis of awareness. Awareness is a property of neural networks, not of any particular locus in the brain. Were it possible further to subdivide the human brain, there is no reason in principle why several or many independently aware neural systems might not result. They would not be as perfectly equipped with a full range of input and output facilities as each hemisphere is, but that is a matter of detail, not principle. The only theoretical limit on the size of the set of consciousnesses would be the (as yet unknown) level of system complexity below which neurons are not able to actualize their potential for generating representations of which the individual is aware.

Conclusions

Awareness describes the unified field within which representations of experienced input, re-experienced (recollected) input, and goals of contemplated action are interrelated. Its scope ranges from global, encompassing a wide multi-modal but incompletely specified field, to focal, restricted to a figure scrutinized in analytic detail. Self-awareness is conditional on the availability of attentional resources. In contrast, representations outside the conscious field are relatively isolated, lack relationship to context, and cannot be recovered through episodic recollection. The ability to interrelate a wide range of information, past, present, and anticipated, lends flexibility to behavioural control. This presumably explains the evolution of awareness as a response of the mammalian brain to selection pressures. At the neural level awareness is based on the concurrent activities of distributed loci in the differentiated, cortical neuronal network. Depletion in the brain of such loci restricts the range of conscious experience; the subject is unaware of the restriction. Disconnection of areas of cerebrum can leave the network capable of generating representations within awareness that co-exist but can no longer qualify each other.

References

Baldwin, M. (1962). Hallucinations in neurologic syndromes. In *Hallucinations*, (ed. L. J. West), pp. 77–86. Grune and Stratton, New York.

Bisiach, E. and Luzzatti, C. (1978). Unilateral neglect of representational space. *Cortex*, **14**, 129–33.

Bogen, J. E. (1969). The other side of the brain. II. An appositional mind. *Bulletin of the Los Angeles Neurological Society*, **34**, 135–62.

Bunge, M. (1977). Emergence and the mind. *Neuroscience*, **2**, 501–9.

Cook, N. D. (1986). *The brain code*. Methuen, Andover, Hants.

Dennett, D. C. (1988*a*). Quining qualia. In *Consciousness in contemporary science*, (ed. A. J. Marcel and E. Bisiach), p. 42. Oxford University Press.

Dennett, D. C. (1988*b*). Why everyone is a novelist. *Times Literary Supplement*, September, 16–22.

Dennett, D. C. and Kinsbourne, M. (in press). Time and the observer: The where and when of consciousness in the brain. *Behavioral and Brain Sciences*.

Eccles, J. C. (1966). Conscious experience and memory. In *Brain and conscious experience*, (ed. J. C. Eccles), pp. 314–44. Springer, Berlin.

Farah, M. J. (1984). The neurological basis of mental imagery: a comparative analysis. *Cognition*, **18**, 245–72.

Fodor, J. A. (1983). *The modularity of mind*. MIT Press, Cambridge, MA.

Gazzaniga, M. H. (1988). Brain modularity: towards a philosophy of conscious experience. In *Consciousness in contemporary science*, (ed. A. J. Marcel and E. Bisiach), p. 218. Oxford University Press.

Geschwind, N. (1970). The organization of language and the brain. *Science*, **170**, 940–44.

Goldberg, E. and Costa, L. D. (1981). Hemisphere difference in the acquisition and use of descriptive systems. *Brain and Language*, **14**, 144–73.

Gruzelier, J. and Flor-Henry, P. (1979). *Hemisphere asymmetries of function in psychopathology*. Elsevier, Amsterdam.

Hilgard, E. R. (1977). *Divided consciousness: multiple controls in human thought and action*. John Wiley, New York.

Holtzman, J. D. (1984). Interactions between cortical and subcortical visual areas: evidence from human commisurotomy patients. *Vision Research*, **24**, 801–13.

Holtzman, J. D. and Gazzaniga, M. S. (1982). Dual task interactions due exclusively to limits in processing resources. *Science*, **218**, 1325–7.

Holtzman, J. D. and Gazzaniga, M. S. (1985). Enhanced dual task performance following callosal commissurotomy in humans. *Neuropsychologia*, **23**, 315–321.

James, W. (1890). *Principles of psychology*. Holt, New York.

Johnson-Laird, P. N. (1983). *Mental models: towards a cognitive science of language, inference and consciousness*. Cambridge University Press.

Kinsbourne, M. (1970). A model for the mechanism of unilateral neglect of space. *Transactions of the American Neurological Association*, **95**, 143–5.

Kinsbourne, M. (1974). Lateral interactions in the brain. In *Hemispheric disconnection and cerebral function*, (ed. M. Kinsbourne and W. L. Smith), pp. 239–59. Thomas, Springfield, Il.

Kinsbourne, M. (1975). The mechanism of hemispheric control of the lateral gradient of attention. In *Attention and performance V*, (ed. P. M. A. Rabbitt and S. Dornic), pp. 81–97. Academic Press, London.

Kinsbourne, M. (1981). Single channel theory. In *Human skills*, (ed. D. H. Holding), pp. 65–90. Wiley, Chichester, Sussex.

Kinsbourne, M. (1982). Hemispheric specialization and the growth of human understanding. *American Psychologist*, **37**, 411–20.

Kinsbourne, M. (1987). Mechanisms of unilateral neglect. In *Neurophysiological and neuropsychological aspects of spatial neglect*, (ed. M. Jeannerod), pp. 69–86. Elsevier, Amsterdam.

Kinsbourne, M. and Hicks, R. E. (1978). Functional cerebral space: a model for overflow, transfer and interference effects in human performance. In *Attention and performance VII*, (ed. J. Requin), pp. 345–62. Erlbaum, Hillsdale, NJ.

Kinsbourne, M. and Warrington, E. K. (1962). A disorder of simultaneous form perception. *Brain*, **14**, 235–46.

Kreuter, C., Kinsbourne, M., and Trevarthen, C. (1972). Are deconnected cerebral hemispheres independent channels? A preliminary study of the effect of unilateral loading on bilateral finger tapping. *Neuropsychologia*, **10**, 453–61.

Lashley, K. S. (1958). Cerebral organization and behavior. In *The brain and human behavior, Research Publications, Association of Research on Nervous and Mental Disorders*, No. 36, (ed. H. C. Solomon, S. Cobb, and W. Penfield). Baltimore, MD.

Marcel, A. J. (1983). Conscious and unconscious perception: an approach to the relations between phenomenal experience and perceptual processes. *Cognitive Psychology*, **15**, 238–300.

McClelland, J. L. and Rumelhart, D. E. (ed.) (1986). *Parallel distributed processing: explorations in the microstructure of cognition*. MIT Press, Cambridge, MA.

Merzenich, M. M. and Kaas, J. H. (1980). Principles of organization of sensory–perceptual systems in mammals. *Progress in Psychobiology and Physiological Psychology*, **9**, 1–42.

Navon, D. (1984). Resources—a theoretical soup stone? *Psychological Review*, **91**, 216–34.

Neisser, U. (1967). *Cognitive psychology*. Appleton–Century–Crofts, New York.

Pandya, D. N. and Kuypers, H. G. J. M. (1979). Cortico–cortical connections in the rhesus monkey. *Brain Research, Amsterdam*, **177**, 176–82.

Posner, M. I., Walker, J. A., Friedrich, F. J., and Rafal, R. D. (1984). Effects of parietal injury on covert orienting of visual attention. *Journal of Neuroscience*, **7**, 1863–74.

Posner, M. I., Choate, L., Rafal, R. D., and Vaughan, J. (1985). Inhibition of return: neural mechanisms and function. *Cognitive Neuropsychology*, **2**, 211–28.

Pribram, K. H. (1971). *Languages of the brain*. Prentice Hall, Englewood Cliffs, NJ.

Puccetti, R. (1973). Brain bisection and personal identity. *British Journal of Philosophy and Science*, **24**, 339–55.

Reuter-Lorenz, P. A., Moscovitch, M., and Kinsbourne, M. (1985). Hemisphere control of spatial attention. *Brain and Cognition*, **12**, 240–66.

Rozin, P. (1976). The evolution of intelligence and access to the cognitive unconscious. *Progress in psychobiology and physiological psychology*, **6**, 245–80.

Shallice, T. (1978). The dominant action system: an information-processing approach to consciousness. In *The stream of consciousness*, (ed. K. S. Pope and J. L. Singer), pp. 117–57. Plenum, New York.

Shallice, T. (1988). Information-processing models of consciousness: possibilities and problems. In *Consciousness in contemporary science*, (ed. A. J. Marcel and E. Bisiach), p. 305. Oxford University Press. John Wiley, New York.

Sholl, D. A. (1956). *The organization of the cerebral cortex*. John Wiley, New York.

Skinner, J. E. and Yingling, C. D. (1977). Gating of thalamic input to cerebral cortex by nucleus reticularis thalami. In *Attention, voluntary contraction and event-related cerebral potentials, progress in clinical neurophysiology* (ed. J. E. Desmedt), pp. 30–69. Karger, Basel.

Sperling, G. (1960). The information available in brief visual presentations. *Psychological Monographs*, **74**, (11, Whole No. 498).

Sperry, R. W. (1966). Brain bisection and mechanisms of consciousness. In *Brain and conscious experience*, (ed. J. C. Eccles), pp. 289–313. Springer, New York.

Sperry, R. W. (1984). Consciousness, personal identity and the divided brain. *Neuropsychologia*, **22**, 661–73.

Sperry, R. W., Zaidel, E., and Zaidel, D. (1979). Self-recognition and social awareness in the deconnected hemisphere. *Neuropsychologia*, **17**, 153–66.

Teng, E. L. and Sperry, R. W. (1974). Interhemispheric rivalry during simultaneous bilateral task presentation in commissurotomized patients. *Cortex*, **10**, 111–20.

Treisman, A. and Gelade, G. A. (1980). Feature-integration theory of attention. *Cognitive Psychology*, **12**, 97–136.

Trevarthen, C. (1974) Functional relations of disconnected hemispheres with the brain stem, and with each other: monkey and man. In *Hemisphere disconnection and cerebral function*, (ed. M. Kinsbourne and W. L. Smith), pp. 187–207. Charles C. Thomas, Springfield Ill.

Trevarthen, C. and Sperry, R. W. (1973). Perceptual unity of the ambient visual field in human commissurotomy patients. *Brain*, **96**, 547–70.

Tulving, E. (1983). *Elements of episodic memory*. Oxford University Press.

Walley, R. E. and Weiden, T. D. (1973). Lateral inhibition and cognitive masking: a neuropsychological theory of attention. *Psychological Review*, **80**, 284–302.

12

Consciousness in science and philosophy: conscience and con-science

Richard L. Gregory

'If you can't explain it—deny it' is one strategy for dealing with embarrassing questions such as 'what is consciousness?' But few of us now follow the behaviourists in *denying* consciousness; even though we cannot account for why we have sensations. Matter, also, is deeply mysterious. Physicists cannot say what electrons are made of. But as matter is not *private* we can talk about material objects (and abstract objects, such as numbers) in ways denied us for discussing experience. And it is hard to draw effective analogies (which are essential for useful explanation) from objects to mind, as material and other objects are public but mind is uniquely private.

Wittgenstein, in his remarkable book *Philosophical Investigations* (1953), gives a deep, fascinating discussion on whether we can compare sensations. The concern is whether it is possible to generalize from one's own sensations to other people's; or to animals, or intelligent machines, or (even more unlike us) to 'conscious stones'. In *Philosophical Investigations* (1953, Vol. I. para. 293) Wittgenstein states this famous argument: 'Suppose everyone had a box with something in it: we call it a "beetle". No one can look into anyone else's box, and everyone says he knows what a beetle is like by looking at *his* beetle. Here it would be quite possible for everyone to have something different in his box.' The 'beetles' are of course sensations, such as colour or pain. Wittgenstein concludes that other people's sensations can only have significance for us in terms of behaviour. He was not, however, a behaviourist—for he did not claim that behavioural accounts are adequate to describe, for example, pain. He allowed that there are sensations, though he denied that it is possible to describe sensations, as it seemed to him impossible to generalize language from the single case of one's own private experience to anyone else's. In his terminology of 'language games', developed for talking about shared objects and events, *private* 'language games' are not possible; so we cannot talk of experience, even though we can discuss external objects which, perhaps indirectly, we experience in perception. So according to Wittgenstein, although through all our lives from childhood we assume that other people feel pain, and taste wine much as we do, language can 'model' (or 'picture') states of the world but not states of mind. Wittgenstein (1953, Vol. I, para. 282) concludes: '"But doesn't what you say come to this: that there is no pain, for example, without *pain-behaviour*?" It comes to this: only of a human being and what resembles (behaves like) a living human being can one say: it has sensations; it sees; is blind; hears; is deaf; is conscious or unconscious.'

We may note that this emphasis on behaviour refers to the entire organism, though the neurosciences are interested in the roles of specific neural processes and tend to attribute consciousness to specific structures and functions of the nervous system. Recently, the Oxford philosopher Kenny (1984) has defended Wittgenstein's position on this, commenting:

This dictum is often rejected in practice by psychologists, physiologists and computer experts, when they take predicates whose normal application is to complete human beings or complete animals and apply them to parts of animals, such as brains, or to electrical systems . . . . I wish to argue that it is a dangerous practice which may lead to conceptual and methodological confusion (page 125).

His point is that if a brain could not think without its body, its sensory and motor systems, and so on, we cannot, without risk of confusion, say that 'a brain thinks'; only that 'a person thinks'. But are we so confused? It is usual to say that a table supports a book without reference to the floor, or to the ground the table stands on, and this is hardly confusing. It is confusing, however, to say 'the eye sees', because in an important sense it is the brain that does the seeing, rather than the eye which feeds it with neural signals. The importance, it seems to me, is not normal language usage so much as what one takes to be important conceptually, or factually. It could indeed be that the (often useful) role of philosophers as *guardians of semantic inertia* can block progress in understanding by overstern warnings of impending confusions. Indeed, although their strictures based on linguistic usage derived from the past are sometimes very important, analytical philosophy is too often overcautious— unable to break its hold on the past to take the necessary plunge when new facts are discovered or new techniques made available (Gellner 1959). Science is far more brave!

Is Wittgenstein overcautious in not allowing us to talk about consciousness apart from guilt *behaviour*, thinking *behaviour*, pain *behaviour*, and so on? Certainly we often wish to talk of others' *sensations*. Thus when we see, or think we see, someone in pain we are not concerned only with his behaviour. We believe we know what an injured person feels, even though he is lying still. Clearly it is this that concerns us, or anaesthetists would be out of work. Perhaps Wittgenstein's semantic caution makes him a Red Queen ordering our heads to be cut off—preventing us speaking of ourselves or appreciating sensations in others.

It is tempting to think that we must be appreciating others' sensations when we accept their testimony as observers of external objects. Could we have 'objective' knowledge without 'subjective' experience of observers, including scientists? A way to think about this is to imagine co-operating, non-conscious computers finding out about the world from signals provided by their own detectors or sensory systems and sharing and comparing each other's data stores and hypotheses of the way things are. Could all this be non-conscious? If so, what does our consciousness *do*?

It came as a jolt, whose shock waves have not yet settled down, when the founder of modern studies of perception, Hermann von Helmholtz first spoke in 1866 of perceptions as *conclusions of unconscious inferences* (see Southall 1962). This was not a popular notion at the time as it threatened the assumed necessity of consciousness for moral decisions for praise or blame. Thus a man can hardly be brave if he does not experience fear; or be praised for a discovery if he did not consciously set out to find it or realize the significance of what he had found. Helmholtz's sensationless mental processes of the nineteenth century became the sensation of the twentieth century—in Freud's unconscious mind. Helmholtz questioned conscious intention leading to perception (on which judgements depend), while Freud questioned the status of motives—as being largely unconscious and affected by unconscious forces. The moral dilemmas are highlighted (one might say cartooned) in artificial intelligence (AI), which may claim to encompass brain intelligence (BI) though leaving out consciousness, and, if implicitly, rejecting moral responsibility for machines—even though the brain is accepted as a machine. AI has not yet spelled out how the brain machine can be different, so that human behaviour is praiseworthy and blameworthy. (Unless, indeed, brain machines are conscious, though bright computer machines live in the dark of no consciousness—and so no conscience. Hence the study of these issues are dubbed *con-science*! But can we do effective experiments on consciousness? A problem is that, as consciousness is private, they really need to be done on oneself. An inadequate example now follows.

Switching consciousness with anaesthesia

Philosophers draw distinctions and parallels with words: scientists identify causes or origins of phenomena (when possible) by switching them on and off. This immediately suggests that anaesthetics should be useful for investigating the brain-dependence of consciousness. With this in mind, some years ago I had myself anaesthetized, under controlled conditions in an operating theatre, with various tests and simple experiments carried out by my colleague John Harris, while very slowly I lost consciousness, under the influence of Ketamine. The results are described more fully in *Odd perceptions* (Gregory 1986, pp. 198–208):

The initial effect of the Ketamine was violent instability and breaking up of the visual world. There was very soon loss of pain sensations; though other sensations remained and some were enhanced. There was marked synaesthesia (senses normally distinct becoming linked, so touch stimuli would produce brilliant colours). Here are some of the findings, as recorded on tape during the experiment:

1. *Stereopsis*
   During the infusion of Ketamine (which was selected as muscle tone remains, so it was possible to speak to the tape recorder), stereopsis (for Julesz random dot stereo pairs) was still present—though the visual world was in violent unstable motion. This was surprising, and suggested that the early stages of visual processing were not affected by the Ketamine.

2. *Colour vision*
   All the Ishihara colour-dot test plates were shown before infusion. 20 minutes into infusion plates 1–17 and plates 22–25 were shown. Of the 21 responses obtained, 4 were different from the pre-infusion responses. Of these, 1 was probably due to JH mishearing RLG's speech (74 confused with 24). Not too much weight should be given to the other errors, since they involve visual confusion such as 2 with 7. Misperception of the Ishihara plates was minimal, if it occurred.

3. *Visual acuity*
   This was measured crudely before and during infusion, with the kind of Snellen chart which presents a continuous passage of prose, and where the type size of each line or block of several lines is progressively reduced. Before infusion, the chart was held at a distance of about three feet, and the type size noted at which there were beginning to be reading errors. 12 minutes after infusion started, a second test suggested difficulty with a type size 25% larger. But this measurement depended on ability to read aloud fluently; which was impaired. Acuity did not seem to be greatly reduced from normal, in spite of the violent visual instability.

4. *Visual Recognition*
   11 minutes after infusion I could immediately recognise a picture of the Queen. At 25 minutes I described a scene of Mohammed Ali with children accurately, but failed initially to identify Ali. And failed to recognise a picture of

David Owen, (then Foreign Secretary) either during infusion or much later. In general, visual recognition seemed good, as I could name the people in the room, and also most of the visual illusion test figures which were presented.

5. *Auditory Recognition*

Four familiar voices (of colleagues) were recognised (at 21 minutes): three within 5 seconds, the fourth within 15 seconds. There was no evidence of impairment of auditory recognition.

6. *Distortions and Changes in Pattern Vision*

These were apparently one of the more striking effects of the Ketamine. 6 minutes after infusion, I said of the Zollner illusion, that it was very odd and moved about. It was so odd that I could not say whether the normal illusory distortion was present. At 16 minutes the Necker cube was presented: it looked 'wibbly' and as though made of plasticine. It was coloured and shimmery; not at all like the usual cube, as the parts were somehow separate. It did not reverse in depth. At 17 minutes Boring's mistress-wife figure appeared as a sequence of features, that could have belonged to one or the other percept, as they flashed in and out very fast, quite unlike the usual perceptual reversals of the entire figure, normally appearing as a young or an old woman. Percepts were intermittent, as though the stimulus was being turned on and off. At 18 minutes the Mach card was presented. Like the Necker cube, it did not reverse in depth. At 33 minutes (after atropine and physostigmine) the ceiling and the Zollner illusion figure appeared as before, the latter looking totally unlike its usual appearance—as though being strobed, with about two waves running along it per second. This was rather like binocular rivalry. At 35 minutes the Necker cube appeared as before, with no reversal of depth. At 36 minutes the Boring figure appeared as before, though with more irrelevant details, such as the texture of the paper.

Pattern and object vision were clearly abnormal; yet visual abilities such as colour and stereopsis were hardly affected. Minutes after infusion, when I moved my eyes the wall appeared to move with them, especially at the beginning of the eye movements. The entire world jazzed about most disconcertingly. I am amazed that I could do these tests so well (especially fusion of the Julesz random dot stereo pairs) in spite of the continuous violent movements and actual dislocations of the visual world.

7. *Changes in Proprioception and 'Body Image'*

At 4 minutes I could accurately describe how my legs were crossed, but had to think hard how to do so. I had a sinking feeling, and felt that my middle had gone hollow, like plasticine. At 12 minutes, my Achilles tendon was squeezed, with a lot of force, to produce bruising which lasted several days; but it didn't hurt at all. So although touch remained, pain was lost. I did however experience more and more unpleasant sensations, together with a 'distancing' from the world, and I felt almost as though I was floating.

At 14 minutes, I was asked to try to bite my tongue, but didn't know where my tongue or mouth were, and was not able to do so (fortunately!).

8. *Auditory Disturbances*

I often reported a buzzing sensation in my head. This was most unpleasant. The tape reveals hammering and banging which was going on in a nearby room;

but the loud buzzing was quite different from these, and different also from the infusion pump.

9. *Time Perception*

On one occasion (18 minutes) I said that events took longer in my world. On listening to the tape playback, this sounded faster than the original events had seemed. (But what of playbacks of ordinary events, such as a conversation? Would this seem faster, or slower, than the original? This could make an interesting experiment on the perception of time. It should be done before we can interpret the change of time under anaesthesia—which seemed rather paradoxical since, although each event was drawn out, I am not sure that time seemed to go slower than usual.)

10. *Tactile Recognition*

There was impairment of recognition by active touch, though touch sensitivity seemed normal. I could name three objects (clothes brush, comb, cotton reel) which were presented before as well as during infusion. However, this took much longer under the drug. And although some aspects of sensation were unimpaired (e.g. I could feel individual bristles on a brush) somehow the object didn't feel as it usually did. And when an object was both seen and touched, there was a kind of mis-match between the visual and the tactile world.

11. *Synaesthesia*

I experienced (and reported) that stroking the bristles of the brush, or having my hand stroked by a comb when my eyes were closed, gave brightly coloured red and green images. This was the most dramatic effect. One really did enter another world: a world of brilliant supersaturated colours and shapes quite unrelated to what I was touching. I suppose one's inhibitory processes were failing to isolate the senses, and touch stimuli (though why only touch?) triggered dream-generating mechanisms.

So, what did we find out about consciousness—or unconsciousness? Not very much; but this was just one journey to unconsciousness. There are many other routes. Other anaesthetics have different physiological and psychological effects; and of course ordinary sleep is different, for going to sleep is generally pleasant, but this Ketamine journey was highly unpleasant. In general, it was as though inhibitory barriers between the senses, and the central store of images and associations, started to melt so that the mind and consciousness lost integrity. Surely there is much of interest to be found out by progressively switching off the brain—especially if we could record brain activity occurring through various anaesthetic journeys to unconsciousness.

Brain language and the self

The Scottish philosopher David Hume in 1740 held that the self is *made of sensations* and that it is 'a bundle or collection of different perceptions' (see Selby-Bigge 1978). He toyed with the notion that the self's continuity through gaps of consciousness is due to sensations being embodied in a mind substance, much as external objects are often thought to hold together and be maintained by continuity of an underlying substance. But Hume later came to reject the notion of such underlying matter—both for the object world and for self—writing it off as a conceptually vacuous explanation. Hume then came close to saying that self, for each of us, is an hypothesis, based on fleeting experiences. On this account the ultimate status of consciousness and of matter may seem similar and equally mysterious—which seems to make the status of consciousness less of a bother, as at least we are not alone in our puzzle. Perhaps, indeed, science is simply unable to answer questions below a certain 'depth'. Hume's questioning of the status of substance and of cause (he did not actually deny cause) remain of key importance. He rejects the *will* as an *anima* outside the physiology of organisms, describing it: 'By the *Will*, I mean nothing but *the internal impression that we feel, and are conscious of, when we knowingly give rise to any new motion of our body, or new perception of our mind*' (see Selby-Bigge 1978). For Hume the will is not a cause of action, but just a link in a chain (of which we are somehow aware) which is physical as much as mental. Like any other link, it may equally be thought of as caused (whatever 'cause' means) by past events or as causing later events. In considering how we may know these things, Hume states that we no more know by experience the inner workings of the mind than we know, without experiment and hypothesis, the processes of physics.

Hume's emphasis on *'new'* motion and *'new'* perception may possibly reflect a deep characteristic of consciousness (though it is not clear that this is what he intended): that we are primarily aware of novelty, or *surprise*. Surely, indeed, we are hardly if at all aware of completely predicted actions or events. Awareness is primarily of discrepancies between predictions and what then happens. Thus while driving a car there are few sensations of what is happening (at least as we recall the situation) until something unpredicable occurs; then one is suddenly and acutely aware. The surprise may be surprising *absence*, as for the non-ticking of a clock. Actually I have two chiming clocks in my sitting-room and I am seldom aware of them striking the hours. I shall indeed hazard the postulate that: 'Consciousness is always associated with some surprise'.

Now, following Shannon's theory of information (Shannon and Weaver 1949) we accept that amount of information is directly (logarithmically) related to surprise. Surely it is highly suggestive that just as *information* is given by *surprising* events (or surprising non-events), just so is the awareness of *consciousness* related to *surprising* occurrences, or absences. In both cases—for gain of information and for the awareness of consciousness—the surprise of discrepancies from prediction seems to be the key.

Surely this emphasis on *surprise* for information-bearing [and the amount of information represented in consciousness is small (Mandler 1982)] raises a general problem for neurophysiology; for it implies that it is not just neural activity (which can be measured by normal physical means) but also surprising *absence* of activity that counts. But how is this to be measured? The probability (or improbability) of signals (or words) is a matter of how often each occurs from the possible set or *ensemble* from which they are selected. But the surprise of messages —such as 'Your aunt has won the Nobel Prize'—cannot be assessed without knowing their significance and context. Thus if you knew that your aunt had been nominated, this would enormously reduce the surprise, and the amount of information, of the message. But if you did not know about Nobel Prizes (or aunts) it would be meaningless and of no surprise or information value. The surprise and the information are limited to those who have the necessary ensemble of possibilities for selection by the message. So messages are meaningless, secret, except to owners of the necessary ensemble of possibilities for selection by the message. Is this a clue to why consciousness is private to each brain-owner?.

Surprise, and also information, belong to *descriptions*. An event may be unusual but it cannot be surprising except to an observer with knowledge of what might be. Although an event may *provide* information it cannot *be* information, or be information-bearing except to a perceiving memory. But descriptions require some kind of language (a biological or a computer language), so, surely, it is helpful to think of mind and consciousness as associated, not so much in terms of the physical (or physiological) brain activities, but more specifically in terms of some kind of private 'brain language', written and read by neural processes (which may be open to physiological recording) but depending on some organized data base or knowledge, and formal (grammatical) rules for the messages to have significance.

The biologist John Young says, in his *Programs of the brain* (1978):

> If the essential feature of the brain is that it contains information then the task is to learn to translate the language that it uses. But of course this is not the method that is generally used in the attempt to 'understand the brain'. Physiologists do not go around saying that they are trying to translate brain language. They would rather think that they are trying to understand it in the 'ordinary scientific terms of physics and chemistry'.

This is so now; but will we continue to see brain activity in the same way if we find a Rosetta Stone for reading its signals as a language?

Surely neuroscience will take Young's advice and desert physics and chemistry for a while to translate and read brain language—what could be more exciting? Then, perhaps, we may read another's thoughts —even to knowing his sensations—almost, if not quite, at first-hand. But this can only be first-hand—and this is a highly reductionist suggestion—if the brain language normally reads itself and if we can read it as though it is the language of our own brain. If possible, this would release the beetles from Wittgenstein's boxes.

How could this come about? There are some clues. Given that information is closely related to surprise, it is highly suggestive that the peripheral nervous system, at least, responds especially to *changes* of stimulation. As Adrian so well described, in *The basis of sensation* (1928):

> It is easy to multiply instances of sensations fading owing to the adaptation of their receptors to a constant environment. We cease to be aware that our clothes are touching our bodies as soon as we put them on. This may be due partly to the diversion of our attention to more interesting topics, but even if we try to focus it on the body surface we find that there is little to feel as long as we do not move. . . . The fact that the receptors can be moved about in relation to the external world enlarges their scope enormously. To gain information about the environment there is no need to wait for it to change. . . .

Adrian, of course, established beautiful correspondences between this loss of continuous (and non-information bearing) peripheral neural activity to loss of consciousness for continuous stimuli.

In perhaps something of the same vein, Barlow (1977) says of recording optic nerve signals:

> What comes to be more prominently represented in the optic nerve message is also more prominent subjectively, and similarly for features like prey and predators that are important for an animal. It is fascinating to see how information about the factors known to trigger various forms of behaviour is preserved in the optic nerve message, whilst information about other aspects of the retinal image is discarded. It is also fascinating to have an opportunity of studying in the retina the physiological mechanisms that achieve some degree of specificity for these behavioural releasers.

Barlow (1977, p. 259–72) calls these trigger features an '"Alphabet" of one or two dozen symbols, each replicated in many places to cover the whole visual field'. Comparing this with the letters of a human language, he considers that (at least in the periphery) brain language is much simpler, as 'Letters and words are used in a mutually exclusive fashion, whereas nerve fibres (with some exceptions) are not'. Taking this beyond the optic nerve, to recorded brain activity in the primary visual cortex, and then beyond where it is no longer clearly related to

retinal stimuli, Barlow supposes that this, for the physiologist, confusing activity, may be distributions of descriptions to secondary regions
—as 'words' and 'sentences'. The snag is that we do not know the brain
language at this stage, which is where one might expect it to use
combinatorial tricks of written language, and so be very difficult to
elucidate what is going on by single-cell recording.

Should philosophers take acount of this kind of experimental data
and scientific speculation, and technical puzzles set by lack of data?
Wittgenstein (cf. Ayer 1985, p. 106) was surprisingly sceptical of
psycho–physical (mind–brain) laws, saying: 'No supposition seems to
me more natural than that there is no process in the brain correlated
with associating or with thinking; so that it would be impossible to read
off thought-processes from brain-processes' (Wittgenstein 1967, para.
608). Is this why Wittgenstein thought that we can only talk of *sensation behaviour*—as, because brain activity may not be closely related,
we could not even hope to see the 'beetles' in our friends' 'boxes' by
recording from their brains?. If so, Wittgenstein's position is based on an
empirical doubt, that mind may not be based on brain processes, which
only a few neuroscientists now share. If, indeed, mind and consciousness *are* directly related to brain processes (and brain-stored data or
knowledge), which is the prevailing view, there seems to be no reason in
principle why we should not compare 'beetles'—private sensations
—quite apart from overt behaviour. Allow that we can concentrate on
specific neural activity [in spite of Kenny's (1984) objection to singling out
particularly involved anatomical regions or processes]; then why should
not brain recordings be read so that we know another's sensations?

This would be different from Wittgenstein's *behavioural* links, for
there need be no overt behaviour and the recorded brain states should be
far more closely related (even to identity) than behaviour or language to
sensations. Already, evoked potentials and recordings of local cerebral
blood flow allow us to identify processes and to see which regions of the
brain are specially active during private bouts of mental arithmetic,
dreaming, thinking, hearing, or seeing. These are not visible to anyone
else by overt behaviour; but they are made visible by these new non-
invasive techniques for recording brain activity. If such new techniques
allow us to bridge minds, as surely they begin to do, this would not be
the first time that technology has broken through limits to what is set as
possible by philosophers. All through the history of science new techniques have raised and answered new questions. When informative they
are necessarily surprising, and so upsetting as they have not been
predicted and so cannot have been warned against. This might suggest
why it is that technology is ignored by analytical philosophers even
though it has generated so much of our knowledge.

Are illusory surfaces given by unconscious computations?

As we have said, Helmholtz saw perceptions as conclusions of un-conscious inferences. He effectively suggested that these inferences generate (conscious) sensations. If we identify these consciousness-generating inferences, or computations, then surely AI and BI will meet. With this end in view, let us consider in necessary detail some suggestive perceptual phenomena—such as the fascinating sensations of illusory contours (Schumann 1904, cited in Woodworth 1938; Kanizsa 1955, 1979). Many of these might better be called illusory surfaces bounded by illusory borders; as we see surfaces that have a small brightness difference from the surrounding background with clear borders. I have suggested that they are perceptually postulated to account for surprising gaps (Gregory 1972). The brightness difference increases (by up to about 12 per cent) as the probability that the gaps occur by chance decreases. The suggestion is that the surface is inferred by Bayesian-like estimates of the probability that they are truly gaps (and so there is no occluding surface) or that there is more probably a nearer eclipsing or occluding surface responsible for the gaps (Harris and Gregory 1973; Gregory and Harris 1974). Is this awareness of an (illus-ory) surface created by the assessed probability of a surface? I suspect that it is, and that this is so for *all* sensations and perceptions. This is a particularly useful case to consider, as these surfaces and edges are clearly not simply 'stimulus-driven' but depend on whether there 'should' on the evidence be a surface. On this view of the matter, though this is an illusion (which shows up what is happening), it sets the paradigm for understanding all perception and sensation.

Where, specifically, in the brain do illusory surfaces and contours originate? The electro-physiological evidence is that there is no single-cell activity corresponding to the activity recorded when a true edge is passed across the corresponding receptive field, as recorded in the primary cortex, area 17 (Baumgartner *et al.* 1984; Sillito *et al.* 1982). There is, however, recorded activity related to the illusory contour in the higher level visual region of area 18 (Baumgartner *et al.* 1984). If, indeed, we can regard these visual surfaces as computed from prob-abilities, then can we say the same for *all sensations*? Experience with robot vision shows that signalling the presence, and precise position, of edges is difficult and requires elaborate summing and assessing of evidence. Much as for the illusory contours, each edge needs to be a *postulate*—which may or may not be correct. And when the odds are really odd, robots will, like us, be fooled into illusion. Does this mean that they will be conscious? Only if such inference from surprise is a *sufficient* condition for awareness.

The notion that sensations are given by unconscious computations fits very well the recent physiological work of Zeki (1977, 1986) on cortical 'colour regions', which appear to create colour sensations not simply according to input stimuli but rather by the kind of computations Land suggests in his 'retinex theory' (Land 1983). Land's striking experiments show that perceived colour depends on estimates of surrounding luminances and not simply on the relative activities of the three kinds of cone cells in the retina. It is by these computations, Land suggests, that objects maintain the same colour though the lighting changes: so colours are seen as belonging to objects though they are generated within us.

Could a man-made machine compute sensations—and so be conscious? Could it experience colours and pains? Certainly, computers can handle symbols brilliantly, so if they had the appropriate computations to perform, why should they not also see red and feel pain? It is hard to deny this as a possibility without having to evoke some special *substance* for consciousness; but this (like phlogiston as an account of heat) is hardly a useful explanation for it does not bridge across to anything else we know about. If claimed for consciousness, it asserts that sensations are not only private but unique beyond analogy with anything we know. Possibly this is so; but to assume it is to give up attempting an effective explanation. It is holding back from plunging into a structured hypothesis. When science is not immersed in hypotheses (however philosophically wild and woolly) it is nothing, and gets nowhere. But hypotheses should be tested. How would we recognize that a sensory computer is conscious?

We may for a start (though this is not adequate) see whether it *understands*. A now familiar machine which handles our language is a *word-processor*. But, surely, it is not party to our thoughts as we write. It obeys commands to print, save, delete, and so on; but we have no reason to believe that it understands the text—if only because it cannot spell appropriately to meaning. Now on the edge of technology lie machines for typing to dictation. Will they be able to spell? In a language of consistent spelling there would be no problem; but in English it is important to select the appropriate spelling, according to the intended meaning and the context. Thus a person who is able to spell correctly words which sound alike, such as 'key' and 'quay', must understand their meanings in the current context: 'I can't find the key for the quay' makes sense; but hardly 'I can't find the quay for the key'. But these would be indescriminably typed out by a present-day dictation machine. Such an error would betray lack of understanding (for A I or B I) of intended meaning; for the context must be understood to select the right spelling. This is where puns can be so suggestive. For us, 'The piece of Cod, which passeth all Understanding' is funny: will it ever be funny for a machine? Or will we have to rationalize our spelling to match their dumbness? As a matter of fact it is already possible for machines to select correct spellings for a limited set of contexts, and computer systems having rich data-bases as in hospitals or travel agencies can answer questions usefully. So in some sense they understand: at least if they work semantically 'top–down' from context rather than syntactically through 'bottom–up' rules. These can be distinguished experimentally, so it is an entirely possible test for whether a machine understands. A lack of understanding suggests that it does not have the necessary access to stored knowledge for the kinds of inferences we suggest are associated with consciousness.

This, however, is within language. The significant tests of understanding—and the building up of understanding—surely come from interacting with the world. [The principle objection to Searle's (1980, 1984) Chinese room argument for denying understanding in computers is that even babies would not gain understanding of symbols if brought up without interaction with objects and events of the world (Gregory 1987).]

A few years ago, demonstrating understanding would have been sufficient for claiming consciousness; but as AI advances, criteria both of intelligence and consciousness recede—they lie always just beyond current attainment. This shows, indeed, that we do not accept purely behavioural criteria for consciousness (though we may for intelligence), as most people are loath to call a machine conscious *whatever* its behavioural performance. There is, however, another conceivable kind of test for consciousness in a non-human candidate, if sensation is regarded as given by some kinds of inference or computations. This will depend on establishing just which, and preferably how, certain computations produce consciousness. And it may assume that *how* they are carried out (by protoplasm or silicon) is unimportant. So what we need is an adequate general hypothesis—in which certain kinds of computations are seen to produce sensations and perceptions. This might be similar to seeing how the armature coils of a dynamo, revolving in a magnetic field, produce electricity; or how a car engine, from its interacting parts, produces power; or how a sewing-machine produces stitches. It is, indeed, characteristic of machines that their parts produce surprising results—which can only be understood by immersion in hypotheses.

From con-science to science?

Will new technologies serve neuroscience for reading brain language, and so for understanding the role of consciousness? Will they show us why colour and pain seem so *immediate* though sensations depend on elaborate mechanisms and chains of *inferences*? For, given that mind is generated by brain processes, there would seem to be no conceptual objections to the possibility of reading a brain as a book. And if we do come to read brain activity, could we find out why some inferences, or other brain processes, are *unconscious* though others are *conscious*? We may hazard the guess that it will turn out to be *surprise*—departures from prediction, requiring memory cross-index search and rapid sensory update—that is the key to consciousness. If so, it will not be sufficient to decode brain language; it must be read in the context of what is memory-stored and organized as predictive hypotheses in the other mind. But when we read a book we only have a propositional account of the author's thought, experience, perception, or whatever: we do not tap his consciousness. So why should we gain direct experience of another's consciousness by reading his brain? The suggestion is, perhaps, that it will set up in our brain just the same experiences and sensations.

This is very much the situation discussed by Wisdom in *Other minds* (1952; especially IV, pp. 80–113). Wisdom considers the status that might be accorded to a claimed telepathic sharing of another's sensations. (He is not claiming that telepathy is possible.) Here Wisdom contrasts measuring colour blindness with a standard test (deriving numbers from a dial) with, by looking into his mind on a screen, experiencing another's colour vision by telepathy. This, however, Wisdom will not allow, for 'no telepathic knowledge of what is in Smith's mind will do as direct knowledge of what is in his mind, for such knowledge is always still too little and finally too much'. Thus he explains: 'Besides the fact that any telepathy consists in predicting sensation from sensation and is thus (a) indirect and (b) still confined to the knower's sensations, there is the point that only if the telephathic sensations are *extremely* like those they are used to predict is there any inclination to speak of the facts of telepathy as directly seeing into the mind of another'. And a little later: 'You can't have direct knowledge of a feeling, a sensation, without having that feeling or sensation. . . . But if when and as Smith or any other man felt depressed we all felt depressed then there would be no point in Jones or anyone else saying "Smith feels depressed". It would do as well to say "A depression this morning".' But is this (or Wisdom's other beautifully expressed arguments, in which he tries to show that such questions are symptoms of deep confusion) a vital objection to any claim of sharing sensation through bridging brains? I am not sure that it is. Thus although we might only accept telepathy (if we could ever accept it) on the basis of behavioural prediction, our supposed brain language recordings (if these can ever be accepted) might be based on correlations with one's own experience. At least there might be far deeper and richer predictions than for the telepathy case. And does it matter if the subjective experience is shared—to become in this sense objective? Would this not be the needed move from con-science to science?

Such intimate brain reading may sound like science fiction, but this is no argument for why it should not become fact. Far more important is whether arguments such as Wisdom's (which I have only touched upon here) preclude it on logical grounds. This is an extremely difficult issue; but I incline to the view that, as neurological techniques become more powerful, this will be seen as an empirical matter, to be solved by experiment, though we can only dimly see these arguments as possible. Although it may turn out that Wisdom is right that such direct knowledge is impossible, the quest sets the most exciting goal of neuroscience: to prize open the privacy of consciousness, by reading other brains as all day long we read our own—or rather, as our brain reads itself.

References

Adrian, E. D. (1928). *The basis of sensation: the action of the sense organs.* Cambridge University Press.

Ayer, A. J. (1985). *Wittgenstein.* Weidenfeld and Nicholson, London.

Barlow, H. B. (1977). The languages of the brain. In *Encyclopedia of ignorance,* Vol. II, *Life sciences and earth sciences,* (ed. R. Duncan and M. Weston-Smith), pp. 259–72. Pergamon, Oxford.

Baumgartner, G., von der Heydt, R., and Peterhans, E. (1984). Anomalous contours: a tool in studying the neurology of vision. *Experimental Brain Research,* 413–19.

Gellner, E. (1959). *Words and things: an examination of, and attack on, linguistic philosophy.* Routledge and Kegan Paul, Andover, Hants.

Gregory, R. L. (1972). Cognitive contours. *Nature (Lond.),* **238**, 51–2.

Gregory, R. L. (1986). *Odd perceptions.* Methuen, Andover, Hants.

Gregory, R. L. (1987). In defence of artificial intelligence—A reply to John Searle. In *Mind waves* (ed. C. Blakemore and S. Greenfield), pp. 234–44. Blackwell, Oxford.

Gregory, R. L. and Harris, J. P. (1974). Illusory contours and stereo depth. *Perception and psychophysics,* **15**, (3), 411–16.

Harris, J. P. and Gregory, R. L. (1973). Fusion and rivalry of illusory contours. *Perception,* **2**, 235–47.

Kanizsa, G. (1955). Margini quasi-percettivi in campi con stimolazione omogenea. *Rivista di Psicologia,* **49**, 7–30.

Kanizsa, G. (1979). *Organisation of vision: essays on gestalt perception.* Praeger, New York.

Kenny, A. (1984). *The legacy of Wittgenstein.* Blackwell Scientific Publications, Oxford.

Land, E. H. (1983). Recent advances in retinex theory and some implications for cortical computations: colour vision and natural image. *Proceedings of the National Academy of Science U S A*, **80**, 5163–9.

Mandler, G. (1982). *Mind and body: psychology of emotion and stress.* Norton, New York.

Searle, J. (1980). Minds, brains, and programs. *Behavioral and Brain Sciences*, **3**, 417–57.

Searle, J. (1984). *Minds, brains and science.* B B C Publications, London.

Selby-Bigge, L. A. (ed.) (1978). *David Hume: a treatise of human nature*, (2nd edn.). Oxford University Press.

Shannon, C. E. and Weaver, W. (1949). *The mathematical theory of communication.* University of Illinois Press.

Sillito, A., Gregory, R. L., and Heard, P. F. (1982). Can cognitive contours con cat cortex? Talk presented to the Experimental Psychology meeting, University of St Andrews.

Southall, J. P. C. (ed. and trans.) (1962). *Helmholtz's treatise on physiological optics.* Dover, New York.

Wisdom, J. (1952). *Other minds.* Blackwell Scientific Publications, Oxford.

Wittgenstein, L. (1953). *Philosophical investigations* (trans. G. E. M. Anscombe). Blackwell Scientific Publications, Oxford.

Wittgenstein, L. (1967). *Zettel* (trans. G. E. M. Anscombe). Blackwell Scientific Publications, Oxford.

Woodworth, R. S. (1938). *Experimental psychology.* Holt, New York.

Young, J. Z. (1978) *Programs of the brain.* Oxford University Press.

Zeki, S. (1977). Colour coding in rhesus monkey prestriate cortex. *Brain Research*, **53**, 422–7.

Zeki, S. (1980). The representation of colours in the cerebral cortex. *Nature (Lond.)*, **284**, 412–18.

13

Reduction and the neurobiological basis of consciousness

Patricia Smith Churchland

Introduction

The idea that mental phenomena might be reducible to neurobiological phenomena has seemed to some people to be outrageous, or at least deeply implausible. The reasons for the opposition are varied, the more traditional view being that mental phenomena are properties of a non-physical substance. Though not without support in some circles, substance dualism has come to seem a most remote empirical possibility, given what else we know about the brain, about evolution, and about biology generally.

Nevertheless, philosophers have only rarely been persuaded by the reductionist strategy, not because they are substance dualists, but because, in one way or another, mental categories have seemed necessarily non-biological or anyhow wholly unsuitable to explanation in neurobiological terms. Currently in vogue are two new kinds of dualism: *property* dualism and *theory* dualism. Although neither of the new forms takes the mind to be a non-physical substance, both envision an unbridgeable division between mental events and neurobiological states.

Property dualism essentially says that mental states have special properties that are not explainable in terms of brain states. It is especially appealing to those whose rejection of reductionism is based on a fascination with consciousness and the peculiarly *experiential* aspects of mental states (for example, see Sperry 1980; Nagel 1974; Jackson 1982).

Theory dualism is the favourite of those who focus on the semantic dimension of mental representation, and who see belief–desire (intentional) explanations at the heart of theories in cognitive science (for example, see Fodor 1975, 1981; Pylyshyn 1984). The new breed of dualists are typically materialists, in the sense that they agree that any particular mental state or process is a state or process in the nervous system, but they also believe that psychology is an autonomous level of explanation. On this view, psychological theory will not reduce to neurobiological theory, and hence psychological properties are not explainable in terms of the properties of nervous systems.

The new dualism, both the property and the theory version, draws on several crucial assumptions concerning the nature of reduction, and the conditions to be satisfied in order for a reductive research ideology to be reasonable. In my view, much of the appeal of the new dualism derives from misconceptions concerning what sort of business intertheoretic reduction is. In what follows I shall offer an account of intertheoretic reduction drawn from the work of historians and philosophers of science, followed by a discussion of the possibility of a reductive explanation of consciousness (see also Schaffner 1976; Churchland 1979; Hooker 1981).[1]

Intertheoretic reduction[2]

'Reduction' has come to acquire a whole range of connotations. To some, it reflects the goal that any sound science ought, in the long run, to aim for; to others, it represents the misguided idea that knowledge of fine-grained detail will automatically lead to knowledge of macro-properties, or the equally misguided idea that the study of macro-properties is a waste of time. Before addressing the question of reductionism with respect to neuroscience, it will be useful to have a more precise account of what reduction is. The basis of the analysis I shall offer consists of examples drawn from the history of science, where we can see in examples at arm's length what got reduced to what, and the conditions to be satisfied for reduction to be achieved.

Reduction is first and foremost a relation between *theories*. Simply put, one theory is said to reduce to another theory when the first is explained in terms of the second. Statements about some *phenomenon* reducing to another phenomenon (or set of phenomena) are derivative upon the more basic claim that the *theory* which characterizes the first reduces to the *theory* which characterizes the second.

For example, when it is claimed that light has been reduced to electromagnetic energy, what this means is that (a) the theory of optics has been reduced to the theory of electromagnetic radiation, and (b) the theory of optics is reduced in such a way that it is appropriate to identify light with electromagnetic radiation. Similarly, when we entertain the

---

[1] For a more complete discussion of these issues, see Churchland (1986).
[2] This discussion is based on a more extended treatment in Churchland (1986).

question 'Is light reducible to electromagnetic radiation?' the fundamental question is whether the theory of optics is reducible to the theory of electromagnetic radiation.

In specifying more exactly how theories must be related in order for one to reduce the other, the logical empiricists offered this: the reduced theory must be *logically derived* from the *reducing* theory plus some extra stuff. The extra stuff included boundary conditions, limiting assumptions, approximations, and, most crucially, *bridge principles*. It was the function of bridge principles to connect *properties* comprehended by the reduced theory to properties comprehended by the reducing theory. In the most straightforward case, the bridge principles would *identify* the properties in the reduced theory with properties in the more basic theory. This specification of reduction, essentially involving the *derivation* of the old theory and an *identification* of properties, new and old, was both elegant and appealing, and it came to be regarded as definitive (Nagel 1961).

However, as historians of science examined the dynamics of science, they found it necessary to disensnare themselves from a background myth abetting logical empiricism; namely the myth that science is mainly a smoothly cumulative, orderly accretion of knowledge. Science, it appeared, was a rather more turbulent affair. Sometimes one theory was substantially corrected before a reduction was possible; sometimes one theory was displaced outright rather than smoothly reduced. The problem for the logical empiricist account was that theories invariably had to be corrected and modified to get something derivable from the basic theory. With modifications accomplished, it was of course not the old theory itself which was derived, but a *corrected version* of the old theory (Feyerabend 1963).

Theories range themselves on a spectrum of how much correcting and revamping they require to be rendered suitable for reduction. Some, for example the theory of optics, required relatively little correction in order to reduce to electromagnetic theory; in other cases (thermodynamics to statistical mechanics), the corrections were greater; but in some there was so much correction that almost nothing, save a few low-level, homey generalizations, could be retained. The phlogiston theory of combustion is one such example. Here the correction required was so massive that it seemed appropriate to think of the old ontology as displaced entirely by the new theoretical ontology; that is, we now say that there is no such thing as phlogiston, not that phlogiston reduced to some compound containing oxygen. The spectrum accordingly contains smoothly reduced theories at one end, and theories which were largely eliminated at the other end, and there are intermediate cases in between.

This re-evaluation of the business of reduction turns out to be no mere nicety appended for the sake of historical accuracy. Rather, it has important implications for what we look for in the reductive futures of interrelated sciences. The reason is this: whether *identifications* of properties specified by macro- and micro-theories are forthcoming depends on the extent of the correction and revision required. In the event of considerable revision or a fragmentation of the macro-properties, identifications go by the board. For example, 'impetus' in Aristotlean mechanics cannot be identified with anything in Newtonian mechanics, and, in turn, the 'mass' of Newtonian mechanics cannot be identified with the 'relativistic mass' of relativity theory, though the two are analogues. As it now seems, the genes of early transmission genetics are not identifiable with segments of DNA as specified in modern molecular genetics. As Kitcher (1982) has remarked concerning this latter case, it has turned out that there is no science of the gene, only the science of genetic material.

The general point is this: if capacities, entities, and properties as specified by a given macro-theory fragment as science proceeds, then the identifications which figure so prominently in the traditional account of reduction turn out to have no place in the reductive portrait of that theory. Instead we see reconfigurations, adjustments, and some-times quite revolutionary revisions. Identifications are typical only of the favoured few reductions at one end of the spectrum. They are the badge of the theories lucky enough to reduce smoothly; but, in such cases, the reduced theory must be sufficiently correct in all its essentials that revision and correction is minimal.

This 'brings us to the second major point, a point which was sup-pressed for simplicity in the foregoing discussion. Reduction of one theory to another is usually the final stage in a complicated courtship. Earlier phases involve the co-evolution of the theories, where each provides inspiration and experimental provocation for the cohort theory and where the results of each suggest modifications, revisions, or constraints for the other. As the theories co-evolve, they gradually knit themselves into one another, and points of reductive contact are established and elaborated.

Typically, it is during this gradual co-evolutionary development of theories that the corrections and extensions to both theories are made, and from such theoretical interanimation may ultimately emerge a unified theoretical framework. At this late stage, the reduced theory will have an explanation—of its properties, laws, entities, and so forth—in terms of the reducing theory. However, as Francis Crick has observed, by the the time we get to the point of being able to sit down and effect the derivation of one theory to another, most of the really exciting science is over. By then, the inspired modelling, the wild and woolly imaginative forays, the wall-tumbling experiments, the fitting and revising and revolutionizing, are pretty much behind us.

Admittedly, this is something of an exaggeration, inasmuch as scientific theories are ever incomplete and there is always fun to be had. None the less, what is right about this view is that it emphasizes the importance of the co-evolutionary process in achieving a reductively integrated theory. The logical empiricists, in focusing selectively on the final products of a long history of theoretical co-evolution, overlooked the dynamics of theory revision. This is a serious oversight, since it is frequently in a theory's evolution that the major reductive links are forged and the major revisions—categorial and ontological—are wrought. Since 1953, transmission genetics and molecular genetics have embarked upon such a co-evolutionary development, each correcting and constraining the other, and we can see it continue in the developments of thermodynamics and statistical mechanics, even after the successful reduction of the gas laws. (Hooker 1981).

Reductive achievements sometimes fall short of the complete reduction of one theory to another because the available mathematics are insufficient to the task. Thus quantum mechanics has succeeded in explaining the macro-properties of only the simplest of atoms, and whether more will be forthcoming depends on developments in mathematics. Some people tend to want to make a lot of this (e.g. Popper in Part I of Popper and Eccles 1977), but it has not seemed to me to be a lesson of any great significance anywhere, and not, so far as I can tell, for the reductive future of psychology and neuroscience. In the case of quantum mechanics, the mathematical limitations do not entail that the macro-properties of the more complex atoms are emergent in some spooky sense, but only that mathematical limitations mean we cannot now explain them. Whether the appropriate mathematics will ever be developed is very much an open question, but the important thing in any case is that the general outlines of the reductive story are in place. Moreover, we have no reason at this stage for assuming that a reductive programme in neuroscience will be stopped dead in its tracks because the enabling mathematics peters out.

Consciousness, co-evolution, and reduction

In the domain at issue, namely the mind-brain, what are the theories whose co-evolution might eventually lead them towards reductive integration? In a very loose sense, they are psychological theories concerning macro-phenomena and neurobiological theories concerning micro-phenomena. The sense is very loose for several reasons: (1) both sciences are still in their infancy, and theories of the required kind still have a long co-evolutionary haul ahead of them, and (2) both sciences are finding it increasingly necessary to address intermediate levels of mind-brain organization. For psychology, one major difficulty is that it is still far from clear what the macro-capacities and macro-properties are which need, ultimately, to be explained neurobiologically; for neuro-biology, one major difficulty is that it lacks theories of higher levels of organization—theories which specify the representational and com-putational properties of cell assemblies and, in turn, of collections of cell assemblies. That is, neurobiology needs a theory of what is going on above the level of the single cell.

Much of the discussion concerning the reduction of mental phenom-ena to neurobiological phenomena is conducted on the assumption that a reduction requires the identification of mental phenomena, *as understood within the framework of folk psychology*, with neuro-biological phenomena, as understood within contemporary neuro-science. Given the foregoing discussion concerning reduction in general, we may assume that a reduction of this kind is improbable unless the categories and their embedding framework are fundamen-tally correct.

Should the psychological categories undergo major redrawing and reconfiguration, should our understanding of the psychological capacities be revamped quite radically, then certainly it will not be the original, old-time folk categories which will figure in any reduction to neurobiology. Rather, it will be the newly drawn categories in a newly configured psychological theory. And if I am right in envisaging a co-evolutionary development of the psychological and neurobiological theories, then reductive 'feelers', at the *very* least, are on the cards. Instead of talking in general terms about the reduction of psychology to neuroscience, I want now to focus more narrowly on one domain within the wider psychological framework, namely consciousness.

The question before us is this: can we get a reductive explanation for consciousness? That is, can we understand in neurobiological terms what it is for an organism to be conscious? A straightforward tactic for a materialist is to search for the neurobiological mechanism which re-sults in an organism being conscious, secure in the knowledge that substance dualism is, on empirical grounds, flatly down and out, and hopeful that, in the long run, neuroscientific techniques will reveal the inner secret of what seems terribly mysterious about consciousness. Nor is it difficult to conjure the air of deep mystery about consciousness. Look at a slice of nervous tissue, watch the oscilloscope during recording from a single cell, trace the circuit in the sea hare *Aplysia* that permits the animal to habituate to harmless stimuli; and notice the phenomenological qualities of feeling delight or sympathy, or of seeing blue or hearing a sigh: these seem infinitely beyond the explanatory reach of neurobiology. How on earth can feeling a pain result from ions passing across a membrane?

Nevertheless, our current bafflement does not of itself show that no neurobiological understanding is forthcoming. The nature of light, fire, the heavens, of reproduction and life have also seemed intractably mysterious, yet we now understand quite a lot about them. Pre-scientific intuitions are often the products of a wider framework which is itself a skewed model of reality, and this framework sometimes contributes to bewilderment rather than to clarification. Part of the task of science is to press on even at the risk of shocking our intuitions and revising our pre-theoretic frameworks.

Is it just a blind materialist faith that consciousness is amenable to neurobiological explanation? No, because everything we know about biology, evolution, neuroscience, physics, and chemistry suggests that substance dualism cannot be right, and that mental states are states of the brain. How we see, hear, walk, catch a ball, plan, and problem-solve are operations of the physical brain. For none of these achievements do we have a *complete* explanation, and they remain, one and all, mysterious. Catching an outfield fly is every bit as mysterious, it seems to me, as consciousness. But in all these instances psychology and neuroscience have made progress, and we can begin to see what neurobiological explanations of behaviour will look like. It is, I suppose, possible that we will not after all find satisfactory neurobiological explanations for these things, perhaps because the job is too hard, or even perhaps because they actually *are* the effects of operations in 'spooky stuff'. Logic does not absolutely rule that out, but science does make it highly improbable, and I prefer to go with the probabilities.

The materialist strategy, accordingly, is to seek the neurobiological substrate for consciousness. One way to conceive of the task is on the model of the search for and discovery of the structure of DNA. As is well known, this discovery was profoundly important for genetics and for molecular biology in general, minimally because it allowed us to begin to really explain how traits could be transmitted from parents to offspring. But the implications went much further than that. Since all living things have DNA (except viruses, assuming they are living, which have only RNA), it also yielded a clue to the bond between all living things. In addition, understanding the genetic code provided the key to understanding the microbiological substrate for evolutionary processes. Inspired by this analogy, we may envision the possibility of discovering the neurobiological mechanism of consciousness, and such a discovery might similarly serve to extend explanations, provoke research, and perhaps yield a clue to the bond between human consciousness and that of other animals. To understate the case, such a discovery would be of tremendous importance.

A research vision can have a powerful shaping influence on how people think about a problem and organize their experiments. Thinking about the neurobiological basis for consciousness on the model of looking for the structure of DNA has much to recommend it, principally because it is very direct: consciousness must be a property of the brain, so let us look for the neuronal configurations that produce it. Hard the problem may be, but let us at least find out *how* hard. Even granting these singular virtues, it may nevertheless be profitable to examine this research vision with greater circumspection.

An important question to ask is this: what precisely was the phenomenon or capacity explained by discovering the structure of DNA? What exactly was the problem solved? When we focus this closely, the answer is that first and foremost it solved the copying problem. It explained how genes could be copied, and how, even when a mutant allele was produced by irradiation, it could nevertheless be copied in the gametes. Most basically, it explained how DNA could replicate genetic information. Discovering how genetic information could be replicated was, of course, an important piece in the puzzle concerning the relation between genotype and phenotype, and how phenotypic traits are transmitted. But many additional questions would first have to be answered before the larger puzzle could be solved. For starters, much would have to be learned about the role of enzymes, of transfer RNA and messenger RNA, and so forth. It was only a very small, if crucial, piece of the puzzle of reproduction of organisms. On that topic, much still remains to be learned about cell differentiation and embryological development. Notice, however, that the copying problem itself was quite well-defined; it was a fairly specific and constrained problem. Moreover, research in classical genetics provided a substantial and experimentally integrated framework within which to understand the general phenomenon of genetic transmission and gene expression, and biochemistry also provided a rich theoretical basis within which to conceive of hypotheses and to test them experimentally.

Thus the discovery of the genetic code was made within the context of a substantial macro-level theory (classical genetics), a substantial micro-level theory (biochemistry), and a specific, reasonably well-defined phenomenon which needed to be explained (replication), and in the vicinity of other highly constrained problems concerning the relation between chromosomal material and phenotypic traits.

Given the research vision based on the discovery of the genetic code, the question to be asked is this: Is there comparable infrastructure to support the neurobiological investigation of consciousness, or at least enough to motivate experimental research on its neurobiological basis? Probably not. One problem concerns the background micro-structural theory. Although we know quite a lot at the neurobiological level about how individual neurons work, we know very little about the dynamics of *circuits*—of how *neural networks* achieve their effects. But the more serious problem is this: at the psychological level we do not have anything remotely as well-worked out for consciousness as the principles of classical genetics were worked out for genes and genetic transmission. Consequently, the explanandum—consciousness—is not at all well-defined or circumscribed. In the case of gene replication (the explanadum for DNA structure), scientists had a pretty good handle on what the phenomenon was that needed to be explained; by contrast, in the case of consciousness, even the range and nature of the phenomena to be explained is notoriously unclear and vaguely indicated.

To make matters worse, what we now lump together as 'consciousness' may be not so much a unitary phenomenon admitting of a unitary explanation, but a rag-bag of sundry effects requiring a set of quite different explanations. This want of specificity is very serious, for if we do not know what the phenomena are for which we are looking for an explanation, launching our investigation is rather Pickwickian. We may find ourselves echoing the inimitable words of Raymond Smullyan (1983, p. 19): 'now that we have lost sight of our goal, we must redouble our efforts'. These ideas need considerable discussion to justify, however, and so far I have made claims without justification. That is the focus of the next section.

Consciousness: what is the explanandum?

What are the phenomena to be explained? It seems easy enough to start answering this question, and the favoured place to start is awareness of sensations. Experiencing colours, pains, sounds, smells, tastes, and so forth are typically treated as paradigmatic cases of what it is to be conscious (Dennett 1988, this volume). Consciousness, accordingly, may sometimes be thought of as that general state in which we are aware of sensations: visual, auditory, tactile, olfactory, etc. There are other, more subtle, receptors which yield sensations we may be conscious of. Proprioceptors give us awareness of the position of limbs, the vestibular apparatus gives us awareness of acceleration and balance.

There can also be awareness of visceral circumstance. Sacks' (1985, chapter 3) description of Christina, whose neuropathy had left her without any proprioceptive information, teaches us how important is one's body image and of knowing without looking where the limbs are. Of course when one is conscious, one is aware only of a subset of signals relayed from peripheral receptors. So already there is an interesting question: What are the differences in the state of the conscious brain when it is aware of an input signal and when it is not so aware?

This question provokes us to see that there are other aspects to consciousness, less 'juicy' (immediately salient) than experiencing sensations, but important as we consider the function of consciousness. Planning, deliberating, and deciding appear to be central. For certain tasks, the organism needs to plan, perhaps to imagine the task before actually performing it, to pay attention to certain sensory signals and not others, to prepare, lie in wait, anticipate, figure things out, and to search for sensory signals. In general, sensori-motor control is an absolutely fundamental function of the nervous systems of animals. Animals are essentially movers, and their motor behaviour must fit the circumstances or they will fail to reproduce.

In some role, yet to be determined, short-term memory appears to figure in the story of what it is for an organism to be conscious, at least because reacting to recently occurring events is an essential part of sensori-motor control. We need to take into account events that happened a few moments earlier, or minutes earlier, in order to respond. Even more mysterious, we are also aware of time—of temporal duration, sequences, ordering, of what is long ago and what is not so long ago. And we are aware of space, of where things are in relation to one another, where oneself is—at home, away from home, etc.

We are additionally aware of ourselves as a *self*—as a thing distinct from other objects, as having a certain identity or coherence through time. A person is aware of specific internal states, desires, hopes, motives, intentions, images, imaginings, and day-dreams. Sometimes a person is aware of even more complex internal states, such as that someone looks familiar, or that he has forgotten someone's name, or that he is confused about the effect of reserpine on aminergic cells. Remembering sometimes involves awareness—recollecting the spelling of 'pneumonia', the location of the barn relative to the creek, whether Washington had a beard. A person can be aware of various emotions, such as embarrassment, anxiety, satisfaction, disgust, etc. When we are aware of these things, we can (normally) report them, and this seems to be a central feature of human introspection. In sleep, we are conscious, in the sense that we are arousable and not in coma, but during some stages of sleep, awareness seems entirely absent, while during REM (rapid eye movement) sleep we are aware of dreams but not of the real external world.

This canvas and catalogue of the various aspects of consciousness is done against the background question: What is the explanandum? One cannot help noticing two prominent things: (1) the features considered in the realm of the phenomenon are remarkably diverse, and (2) words such as 'experience', 'awareness', 'introspection' are used to specify the phenomenon, but they are as equally ill-defined as 'consciousness'. Before I consider how not to lose heart in the face of such confusing complexity, I want first to make three philosophical observations.

Defining the words

There is a fatal temptation to try to deal with the problem of the vagueness of 'consciousness' and the related set (awareness, experience, introspection, contemplation, reflection etc.) by giving stipulative definitions, guided perhaps by certain deep and sometimes quirky hunches about the nature and function of consciousness in nervous systems. The difficulty is that, if we are not clear about the phenomena that are meant to be captured under 'consciousness', stipulative definitions will not help significantly. What is needed is a genuine *theory*, one which has real predictive and explanatory power, fits in with other parts of psychology and neurobiology, and is experimentally based. Without such surrounding structure, new definitions tend to look arbitrary. Moreover, as a matter of sociological fact, they are inevitably counter-exampled to death by simple appeals to the core use of the relevant words, to show, in effect, that the new definitions take us beyond the core use with no significant gain in empirical insight or utility.

The scenario, familiar to anyone who has dipped into the literature or attended conferences on consciousness, goes like this:

Proposal: Let us distinguish between two kinds of consciousness, one which involves only (say) discrimination of a physical difference (call this C_1), and C_2, which additionally carries the ability to report.

Reply: But that means thermostats (bit of litmus paper, etc.) are conscious!

Proposal: Well, yes, that *is* a consequence of my definition, but of course thermostats are only C_1—conscious in this narrow sense.

Reply: Fine, but we want to know about the actual phenomenon, about consciousness itself, not about some other phenomenon you have decided to attach the word to.

The general problem is that, unless the new definitions are empirically motivated, no one can see much point in buying into the extensions or re-categorizations. The idea that if only we could get the words correctly defined then we would understand the phenomenon is seductive but misguided. The words will come to have a more precise meaning as they are more deeply embedded within the framework of an empirical theory. This is not to say that no clarification of meaning is useful, but only that, in the absence of additional empirical hand-holds, what we can accomplish through new definition and re-definition alone is actually quite limited. Recognizing that significant meaning change accompanies rather than precedes empirical discovery, some philosophers have called the 'define-the-words-first' strategy *the heartbreak of premature definition*[3]. The point is that if we understood more about the phenomenon we would know what to say 'consciousness' means.

Natural kinds

The list of elements that go into the pot when we are talking about consciousness and awareness is very diverse, and I want now to dwell on that diversity in order to raise the possibility that there may not in fact be a unitary phenomenon at all. That is, there may not be a single type of neuronal configuration which is the substrate of all those cases, unifying the apparent diversity. Intuitively, perhaps, it seems that consciousness is a natural kind, in the sense that it is a unitary sort of phenomenon, yielding to a single, integrated, unified explanation. The paradigm case of a non-natural kind is 'gems' or 'weeds'. Gems turn out to be whatever rare and not very useful stones a given culture chooses to value as precious, and hence will vary as a function of culture, scarcity, etc. Weeds are plants that grow vigourously in a given locale and which a particular gardener person happens not to want in his garden, and what is a weed to one gardener is a prized botanical specimen to another.

By contrast, 'muscle cell', 'gamete', 'protein', 'acid', 'electron' are natural kinds—or, at least, at this stage of science we believe them to be. There are generalizations and counter-factuals that hold true of acids in virtue of acids having the chemical properties and micro-structure they actually have, where these properties are not relativized to the preferences, whims, interests, accidental collections of persons who happen to interact with the acids.

The possibility I want to entertain is that within the very broad class

---

[3] This expression is owed to Daniel C. Dennett.

of states we call *being aware* or *being conscious* there are subclasses which are amenable to different neurobiological explanations (Churchland 1985*a*). That we do in fact use a common expression to cover the diversity is not in itself a decisive consideration, since it is a common feature of folk theories that categorization may be tied to 'appearances', and the transition to scientific theories involves re-categorization, as Nature is carved at her proverbial joints. Thus 'fire' was used to classify not only burning wood, but also the activity on the sun and various stars (actually fusion), lightning (actually electrically induced incandescence), the Northern lights (actually spectral emission), and fire-flies (actually phosphorescence). As we now understand matters, only some of these things involve oxidation, and some processes which do involve oxidation, namely rusting, tarnishing, and metabolism, are not on the 'Fire' list. The classification of these latter processes with the burning of carbon-rich materials was really very surprising, because pre-scientifically the intuition might have been that being very hot was criterial for being the same type of process as wood-burning. It is only when we understand the deeper nature of the phenomenon that we begin to see how the old classification was skewed. Thus the case of fire illustrates both how the intuitive classification can be re-drawn, and how the new classification can pull together superficially diverse phenomena once the underlying theory is available.

Could it be that 'consciousness' in a similar manner is a classification which could be re-drawn as we discover more about the nature of the mind-brain? Perhaps. I have already touched on the diversity in kinds of awareness, but further points should be made. There are some rather obvious differences between, for example, (a) being aware of seeing blue and (b) being aware of ambivalent feelings about hang-gliding, and (c) being aware that one has a tendency to jump to conclusions. The first point is that introspective awareness is internally generated, whereas the visual experience of blue is standardly dependent on receptor stimulation. There are no receptors for introspection, though external stimuli may of course play some role in initiating the process. But my thought that 'there is no axiomatization of arithmetic that is both consistent and complete' is not caused by an external event isomorphic to the thought.

Planning, choosing, and so forth probably involve long-term memory and short-term memory in a manner quite different from experiencing the colour blue. A bird's plan to build a nest may involve use of past history about sites, predators, availability of materials, and so forth. Awareness of movement in one's peripheral vision probably is less connected with memory, but it is not much like a 'hot' sensation such as seeing blue either. Awareness of time presumably has a special relation to memory not shared by tactile experiences or by body-image awareness.

We can note that having a blue visual experience is not (usually) under voluntary control, except as we do a visual search for blue things. Choosing, reflecting, and deliberating do all seem to involve a different degree of voluntariness. It is not that when I want to plan my trip this summer I do an introspective search for a 'trip experience', comparable to searching the external environment for a 'blue experience' or a 'smells-like-a-violet' experience; it is rather, to put it crudely, that I create the plan as I go. Sense of self, on the other hand, does not seem to have a large voluntary component, though of course we do undertake self-improvement and we self-exhort, and so forth. But the self is just there. Awareness of being ill, of being well again, of having changed in character, of feeling energetic, tired, or fed up seem different yet again. Some of them at least appear to require the relevant background concepts (e.g. awareness of being 'fed up'), others seem more minimal in that regard, and some involve language essentially, such as my thought about the Gödel result. Dreaming experiences are in certain respects very unlike waking experiences. As Hobson *et al.* (1987) put it, waking and dreaming seem on the opposite ends of three continua:

(i) stability–instability (of orientation—where one is in space and time);

(ii) congruity–incongruity (of context—what is going on, who belongs, what belongs together);

(iii) confidence–uncertainty (of concept—who the characters are, what the objects are).

Awareness in the dreaming state normally also lacks a self-reflective dimension. Characteristically, one is not surprised or amazed even by very stark incongruity, or by orientational instability. Were such dream events to occur in waking life, we should certainly be aware of and puzzled by their bizarreness. A visit from a long-dead grandfather arriving on a flying fish during a baseball game would, in waking life, be cause for considerable consternation; in a dream, we pretty much just take it as unexceptional (Hobson 1984). Some 'lucidity' can be achieved with practice, so that it is possible for a dreamer to recognize that he is dreaming and that the dream events are not really happening. However, this sort of control and integration of dream data and waking data is not easy to maintain, requires considerable effort, and tends to slip rather quickly. It can occur in the untrained dreamer, but, although the brief lucid period may be memorable, it is in fact a very rare occurrence.

At this stage it is not clear exactly what is implied by this really quite impressive diversity. We simply do not know enough to decide whether the diversity is essentially an articulation of the same basic organizing principles, nor whether we have a genuinely mixed bag, nor, in general, how to conceptualize what is going on. It is, however, important to be alerted both to the amorphous nature of the explanandum and the possibility that consciousness is not a single, unified kind, for it provokes us to focus more closely on the question of research strategies. At bottom, the interesting question for research is this: Where can we get our empirical hooks in and begin to get answers constrained by both psychological and neurophysiological data? I suspect the diversity helps to emphasize the importance of not adopting an *exclusively* top–down research strategy to investigate the nature of consciousness. Important things have been and remain to be discovered at the psychological level, but research conducted exclusively at that level will not enable us to get at the principles of operation, or at the significance and function of the variety of phenomena collected under the rubric of conscious awareness.

A useful result, therefore, of wallowing in the diversity and dwelling on the ill-defined nature of 'consciousness' is that it provokes us to find a more narrowly circumscribed domain—a domain where a phenomenon can be reasonably well-defined and where there are neuroscientific techniques for investigating the neurobiological basis of the phenomenon. (Certainly I do not want to wallow in the mysteries of consciousness merely for the sake of wallowing.) Left open until much later is the question of how the favoured phenomenon does or does not connect with other aspects we pre-theoretically include as part of consciousness. That is, the strategy is to find a domain, relevant to the consciousness, where we have something roughly comparable in specificity and surrounding theoretical structure to the copying problem in genetics. The domain which perhaps comes closest is the cycle of waking–synchronized sleep–REM sleep, and this will be discussed in the next section.

Denormalizing data

If, following Kuhn's (1962) helpful simplifications, we think of 'normal science' as those periods in a science's development where there is a governing paradigm, and hence a relatively stable categorial framework, then 'denormalizing data' are those data which threaten not merely to falsify a local hypothesis, but also to undermine our confidence that the governing framework is basically correct. In the clinical neurology literature, there are studies which collectively have denormalizing implications for the categories of 'consciousness', 'awareness', and related members of the group. These data are denormalizing because they are so profoundly at variance with the conventional wisdom regarding the nature of consciousness that they make us wonder whether the conventional wisdom is itself just shared misconception and thus whether the categories in question fail to do justice to the real nature of the phenomena.

For example, it is generally assumed to be dead obvious that if someone can report on some visual aspect in the environment then he must be consciously aware of it. Yet the blindsight studies (Weiskrantz *et al.* 1974; Weiskrantz 1986, 1988, this volume) show that subjects who are blind in certain parts of their visual field as a result of lesions to their visual cortex may none the less be able to make such reports. That is, they can point with great accuracy to where in the *blind* area of their visual field a light is shining, indicate whether a light is diagonal or horizontal, and so forth. Since there are pathways from the retina to brain areas other than the visual cortex, for example to the superior colliculus, this is not a miracle, but it does present a challenge to our naive assumptions about the connection between reportability and consciousness awareness.

Commissurotomized subjects provide denormalizing data for that 'obvious' assumption according to which the conscious self is a single, unanalysable unity. On this assumption, if the self reports a conscious experience, there is no other part of the self which could be unaware of that experience. It is now well known that data can be presented to one hemisphere of a split-brain patient so that the other hemisphere will be unaware of the data (Gazzaniga and LeDoux 1978; also see Churchland 1986). Multiple personalities, striking and extraordinary as they are, also suggest that there is much to be learned about the so-called unity and coherence of the self.

Denial syndromes are especially puzzling, and perhaps the most remarkable is blindness denial (Anton's syndrome). Sometimes a subject who suffers a sudden onset of blindness as a result of a lesion to the visual cortex will behave as though he is unaware that he cannot see. These patients insist that their vision is fine, though they may hint that the lights are a bit dim, and they confabulate coherently, if erroneously, in answer to questions about what is in their environment. Unlike patients who know they are blind, these patients do not adjust, and persist in bumping into furniture. This syndrome seems almost impossible, because, intuitively, we take it as part of the very concept of consciousness that if one is not having visual experience then one is aware, unequivocally and without inference or reasoning, that one is not having visual experiences.

Hemineglect also has a denormalizing effect on what we conventionally understand about the nature of awareness. It seems obvious that so long as one is capable of having visual experiences in both visual hemifields, one will be aware of both hemifields and of experiences in both hemifields. We naively assume that if you are visually aware, then you are visually aware, and awareness does not just stop at some point in space. Yet that is exactly what happens for patients with hemineglect, whose brain lesion is frequently in the right parietal cortex. They tend not to be aware of left hemispace, and thus do not groom the left side of the body, do not eat food on the left side of the plate, do not notice events in left hemispace, do not complete the left side of drawings and so on (Heilman 1985). Moreover, as Bisiach and Luzzatti (1978) have shown, hemineglect patients also show a deficit in the capacity to form *images* of left hemispace.

Somnambulism challenges the assumption that consciousness and control must go hand in hand. It is part of the conventional wisdom that the conscious self is in control, and what we are in control of we are also conscious of. Somnambulism does not occur during the dreaming phase of sleep, but during synchronized sleep, when we apparently are not consciously aware at all—of the environment or of dreams. There are as well common, everyday occurrences of the decoupling of consciousness and control, such as when a skill is automatized, allowing control either without attention or with intermittent attention.

This list of denormalizing date is by no means complete, but I want to conclude it here with the observation that not all the denormalizing data are pathological. The regular appearance of confabulation in ordinary, everyday explanations of one's behaviour suggests that we do not have anything like unmediated access to our desires, beliefs, decisions, or intentions. As the work of social psychologists (see Nisbett and Ross 1980) demonstrates, in controlled experiments subjects often give explanations for a choice, a preference, a decision when we know that their explanation is in fact incorrect. For example, women shoppers in a shopping mall were asked to choose a pair of panty hose from a table. Unknown to the women, all the hose were identical. When their choices were analysed, a pronounced position effect was evident, in that the women tended to choose hose in the rightmost position. However their self-generated explanations of their choices referred to (non-existent) differences, such as having greater sheerness, a preferred colour, and so forth. None cited the dominant causal factor, namely that they tended to choose objects in the rightmost position. This confabulation seems to be normal, inveterate, and habitual, and does not involve anything like Freudian repression, nor is it done with deliberate or conscious awareness. There are many other studies showing similar results, and they strongly suggest that theorizing about one's own motives, intentions, and other internal states is not anomalous, but commonplace.[4]

As Paul Churchland (1985*b*) puts it, these assorted paradoxical cases in neurology and psychology stand to folk psychology as near-luminous velocities stand to Newtonian mechanics, as very large masses stand to Newtonian gravitational theory, and as very low or very high pressures stand to the classical gas laws. That is, when applied to phenomena outside the comfortable bubble of 'familiar' cases, the categorical and explanatory resources of the old theory prove entirely inadequate. Folk psychology seems to work as well as it does so long as its range of application is narrowly confined—confined to cases where it works.[5]

Waking, sleeping, and dreaming

If there is a domain relevant to consciousness which has sufficient supporting infrastructure and surrounding theory to enable experimental discovery, it is the sleep–dream–awake cycle. There are a number of reasons why this looks like the kind of domain where research will be fruitful. *First*, there are very striking differences at the introspective and phenomenological level. In deep sleep, one is seldom aware at all; in dreaming, one is aware of internally generated images; while being awake is the paradigmatic conscious state, where we are aware of external events and the brain displays complex sensori-motor control. These factors suggest that the cycle is relevant to consciousness, pre-scientifically understood.

Second, there are robust behavioural criteria for identifying the different states, which means that objective measures may be made and animal models can be used. It also means that many invasive neuro-

[4] Gazzaniga and LeDoux (1978) discuss the confabulation of split-brain subjects whose left hemispheres try to explain behaviour initiated by the right when the right hemisphere was given an instruction by the experimenter. Even when the subjects know the experimental situation, and know that as a split-brain subject one hemisphere does not have access to the information of the other, confabulation—natural, facile, coherent confabulation—is typically observed. What does *not* happen is that the subject's left hemisphere will say: 'Well, I don't know why I walked to the door, but I guess my right hemisphere must have got information unavailable to me and went ahead with its plan.' Instead, a likely story will be presented: 'I wanted to get a drink, so I went to walk out.'

[5] Stated thus, the explanatory power of folk psychology seems to be rather crudely trivialized. Admittedly, any theory quite properly suppresses a good deal of empirical evidence as irrelevant and inevitable 'noise'. If the theory is deeply wrong, however, the only way to identify relevant and refuting 'signals' in that 'noise' is by actively exploring alternative theoretical approaches. Moreover, it is deeply suspicious if a theory's bacon is regularly saved by rejecting troublesome data as noise; at some point one has to take seriously the possibility that the theory does not work very well *even in its favoured domain.*

physiological and anatomical techniques can be used to determine correlations between physiological and psychological conditions.

Third, neuroscientific techniques are beginning to reveal the micro-structural properties distinguishing the three states and the neuronal generator for producing shifts in state. Experimental data show that waking, sleeping, and dreaming as described at the psychological level correlate with large-scale functional states involving wide areas of the brain; i.e. with what researchers in the field call 'behavioural states' of the central nervous system (Hobson 1978; Hobson and Steriade 1986; Hobson *et al.* 1986). The general concept of 'state' as it is being developed in sleep and dreaming research will probably be a pivotal conceptual advance as we address other problems such as focal attention, moods, mental set, control, and the voluntary/involuntary distinction.

Rapid eye movement sleep is typical of mammals, with the probable exception of the voluntary breathers such as dolphins and whales (Crick and Mitchison 1986), and, rather curiously, the echidna (Crick and Mitchison 1983). Electroencephalogram (EEG) studies have shown that the brain of the sleeping organism exhibits a rhythmic pattern in shifting from light sleep, to deep sleep, to REM sleep, and around again. As the sleeping period progresses, the REM periods lengthen and the non-REM (NREM) sleep periods shorten (Fig. 13.1). The time constant for states varies as a function of development, and infants spend most of their sleep time (about 80 per cent or eight hours) in REM sleep, which decreases until about adolescence (25 per cent or one-and-a-half hours) when it levels out. There is a small upward swing around age 40 years, declining steadily thereafter. REM-deprived subjects often show REM rebound (longer REM periods when permitted), but some subjects treated for depression with REM-suppressing drugs do not show these effects.

There are five, principal, identifying indicators of REM sleep: muscular atonia, desynchronized EEG, which resembles the waking EEG, irregular muscular twitching, PGO (pontine–geniculo–occipital) waves, and rapid eye movements (Fig. 13.2). The evidence, derived principally by arousing subjects and asking for reports, indicates that subjects dream during REM sleep (summarized in Hobson *et al.* 1986.)

Salient psychological features of REM sleep are familiar: the dreamer is unaware of external events and is aware of internally generated events. It is probably very significant for understanding the neurobiological basis that dream events are virtually never remembered unless the dreamer is awakened. While we may recall rather trivial events in waking life, even highly charged emotional events in dreaming are not recalled unless the dreamer is awakened (Hobson 1984). Also significant

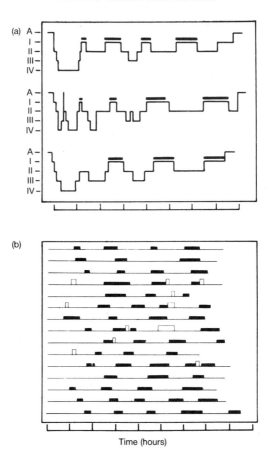

Fig. 13.1. Ultradian sleep cycle of NREM and REM sleep shown in detailed sleep-stage graphs of 3 human subjects (A), and REM sleep periodograms of 15 human subjects (B). In polysomnograms of A, note the typical preponderance of the deepest stages (III and IV) of NREM sleep in the first two or three cycles of the night; REM sleep is correspondingly brief (subjects 1 and 2) or even aborted (subject 3). During the last two cycles of the night, NREM sleep is restricted to the light stage (II), and REM periods occupy proportionally more of the time, with individual episodes often exceeding 60 minutes (all 3 subjects). Same tendency to increase REM sleep duration is seen in B. In these records, all of which begin at sleep onset, not clock time, note the variable latency to onset of first (usually short) REM sleep epoch. Thereafter inter-REM period length is relatively constant. For both A and B time is in hours. [Reproduced from Hobson and Steriade (1986). Neuronal basis of behavioural state control. In *Handbook of Physiology*, (ed. V. Mountcastle) pp. 701–823. American Physiological Society.]

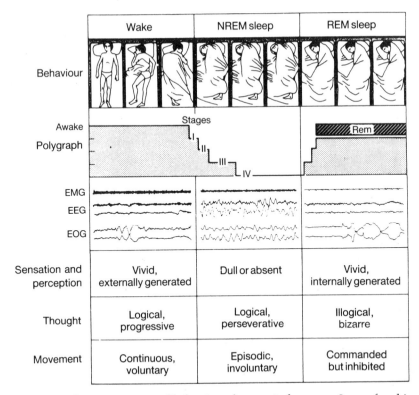

	Wake	NREM sleep	REM sleep
Behaviour			
Polygraph	Awake — Stages I, II, III, IV		Rem
EMG			
EEG			
EOG			
Sensation and perception	Vivid, externally generated	Dull or absent	Vivid, internally generated
Thought	Logical, progressive	Logical, perseverative	Illogical, bizarre
Movement	Continuous, voluntary	Episodic, involuntary	Commanded but inhibited

Fig. 13.2. Characterization of behavioural states in humans. State of waking, NREM sleep, and REM sleep have behavioural, polygraphic, and psychological manifestations. Sample tracings of 3 variables used to distinguish state are also shown: electromyogram (EMG), which is highest in waking, intermediate in NREM sleep, and lowest in REM sleep; and electroencephalogram (EEG) and electro-oculogram (EOG), which are both activated in waking and REM sleep and inactivated in NREM sleep. Each sample record is approximately 20 seconds. The three lower channels describe other subjective and objective state variables. [Reproduced from Hobson and Steriade (1986). Neuronal basis of behavioural state control. In *Handbook of Physiology*, (ed. V. Mountcastle) pp. 701–823. American Physiological Society.]

is the bizarreness of dreams and the lack of insight of the dreamer into his actual state (see earlier discussion, p. 286; Hobson *et al.*, 1987). These are features at the psychological level that a neurobiological account ought to explain. As we shall see, impressive progress has been made in providing just such explanations.

Sensory input is gated during NREM sleep at the thalamo-cortical level via hyperpolarization, but in REM sleep it is gated at the periphery. Motor output is blocked during NREM sleep by disfacilitation of brain-stem neurons, but, by contrast, during REM sleep there is active inhibition of motor neurons by neurons in the pons. This produces a kind of paralysis, which can be abolished by making a small lesion near the locus coeruleus, whereupon the animal will move about during REM sleep, apparently in accordance with dream events. There are ten recorded patients (all male) who have REM without atonia. They behave in accordance with the motor demands of their dream narrative, and consequently crash into walls and furniture. This behaviour contrasts quite markedly with somnambulists who typically manoeuvre quite well, open doors, and so on.

Sensory input is evidently not completely blocked or completely unprocessed, however, since adult sleepers show considerable navigational and orientation facility (they do not fall out of bed, they tend not to crash into each other, etc.) and sleepers can be aroused from synchronized sleep (also called deep sleep or S-sleep) or REM sleep by specific auditory stimuli, such as a whimper of a baby, even though they may be deaf to louder sounds the brain knows it can safely ignore. Arousal from S-sleep is normally easier than arousal from REM sleep.

The neurophysiological level has been investigated using mainly three types of technique: (1) recordings for correlation studies, with the aim of determining the behaviour of populations of neurons under different behavioural states, (2) lesions, to determine the effect of removing a section of tissue, and (3) pharmacological intervention, most recently, by micro-injections of specific chemicals to mimic, enhance, or disrupt the function of relevant endogenous neurochemicals. Neuroanatomical techniques have been used to determine projection paths of various neuronal groups, and neuropharmacological techniques are used to determine the kinds of neurochemicals involved in the crucial structures and the role of those chemicals in the sleep–dream cycle.

So far the neurobiological story is incomplete, and I shall present only the barest outline of the theory[6], but even that much yields a picture that underscores the fruitfulness of experimental investigation of a seemingly unapproachable mystery concerning consciousness (Hobson and Steriade 1986). The main anatomical structure involved is the brain stem and, even more specifically, the reticular formation in the pons (Fig. 13.3). It now appears that there are three principal elements underpinning behavioural states and their cycles. The first is the aminergic neurons in the locus coeruleus (in the pons). They are very active during the waking state, then they decrease their activity with

[6]The best review articles are Hobson (1984), Hobson and Steriade (1986), and Hobson *et al.* (1986).

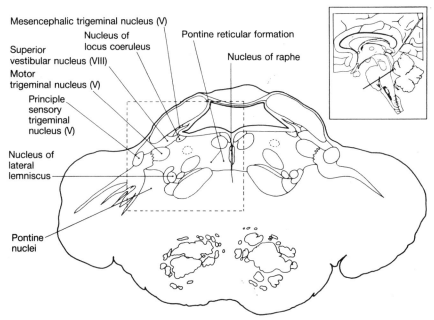

Mesencephalic trigeminal nucleus (V)

Nucleus of
locus coeruleus

Pontine reticular formation

Superior
vestibular nucleus (VIII)

Nucleus of raphe

Motor
trigeminal nucleus (V)

Principle
sensory
trigeminal
nucleus (V)

Nucleus of
lateral
lemniscus

Pontine
nuclei

Fig. 13.3.   Cross section of the brain stem showing the location of the locus coeruleus and the nucleus of raphe. Magnified × 20. Figure at top right shows the level of the section in the nervous system. [Reproduced from DeArmond *et al.* (1976) *Structure of the Human Brain: A Photographic Atlas*, (2nd edn), by kind permission of Oxford University Press.]

the onset of S-sleep, and they have a very low level of activity during REM sleep. There are also aminergic neurons in the dorsal raphe nucleus which may have a modulatory role in the states. Second, there is a special group of neurons with huge cell bodies in the tegmental field (hereafter referred to as 'FTG neurons') which are very active during REM sleep. These cells are believed to release acetylcholine as a neurotransmitter and have receptor sites for acetylcholine. Injecting the centre of this area with carbachol, an acetylcholine agonist, produces REM sleep in the animal (Steriade and Hobson 1976). The third element is that in general the activity of cholinergic neurons in the brain stem is high both in waking and in REM sleep, but low in S-sleep (Figs 13.4 and 13.5).

The inactivity of the aminergic neurons during REM sleep begins to make sense when we see that they project diffusely all over the brain (Fig. 13.6), including the thalamus, the hypothalamus, and all over the cortex. A decrease in their activity suggests that their dynamics is an important element in the account of the REM state and how it differs

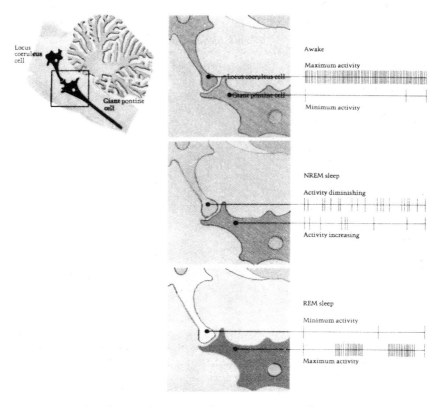

Fig. 13.4. The electrical activity of giant pontine cells, as measured by the frequency of action potentials, is lowest during waking and highest during dreaming sleep. Exactly the opposite occurs in the nearby locus coeruleus, or LC cells. In other words, the increase in the rate of giant cell activity at the onset of dreaming sleep is mirrored by a *decrease* in the activity of LC neurons, suggesting that the two cell groups are reciprocally mediated. [Reproduced from Hobson, *et al.* (1977) *An Experimental Portrait of the Sleeping Brain*, for the Carpenter Center for the Visual Arts, Harvard University. © J. Allan Hobson and Hoffman-La Roche Inc.]

from the awake state. In the waking state, the locus coeruleus is crucially involved in alerting and orienting the brain to particular sensory inputs.

Does the reduction of the aminergic output from the pons mean that in sleep and dreaming the brain just loafs and is generally inactive? Not at all, and this is what is especially intriguing about the diminution of aminergic activity. Neuronal activity in REM sleep and in waking is *comparable*, and activity during NREM sleep is down by only about 20

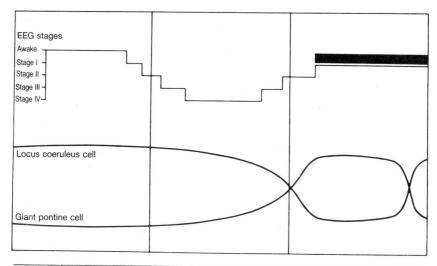

Perception	vivid, externally-generated	dull or absent	vivid, internally-generated
Thought	logical, rational	repetitious or absent	illogical, delusional
Movement	voluntary	infrequent but possible	impossible

Fig. 13.5. Schematic diagram showing the relation between sleep-wake stages, as measured by the EEG, and activity of selective brain stem cells. REM sleep is indicated by a solid black line and appears in the third panel. The locus coeruleus cells begin to decrease in activity during deep sleep and show least activity during REM sleep. The giant pontine cells show the reverse pattern of behaviour. [Reproduced from Hobson *et al.* (1977). *An Experimental Portrait of the Sleeping Brain*, for the Carpenter Center for the Visual Arts, Harvard University. © J. Allan Hobson and Hoffman-La Roche Inc.]

per cent. This includes such ostensibly unlikely places as the cerebellum, which busily hums away during REM sleep. It is the ratio of aminergic activity to activity involving other neurochemicals in those regions receiving aminergeric projections that distinguishes REM sleep from the waking state. The acetylcholine/monoamine[7] ratio is substantially higher during REM sleep, and this is a rather subtle effect which does not show up on gross measures of neuronal activity.

The functional significance of this change in ratio is not yet understood, but because aminergic neurochemicals have an important role in learning in the systems so far studied, a leading hypothesis is that the major effect of decreased aminergic activity is a reduction in plasticity (Flicker *et al.* 1981). As Hobson (1984, p. 250) suggests, 'it is the

[7]The monoamines are norepinephrine (NE) from neurons in the locus coeruleus, and serotonin (5-HT) from neurons in the dorsal raphe nucleus.

Patricia Smith Churchland

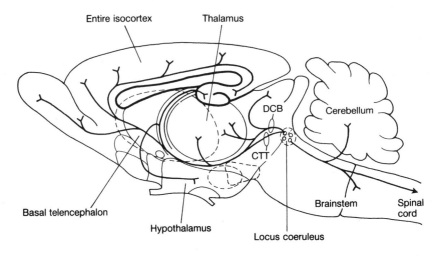

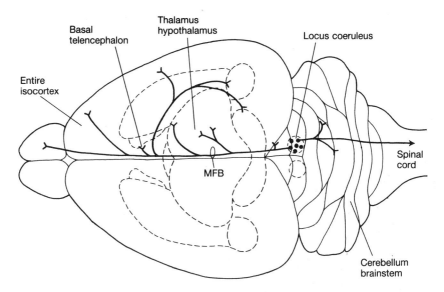

Fig. 13.6. Schematic diagram showing the very diffuse projection paths of neurons in the locus coeruleus of a rat brain. Upper figure is a view from the side of a section made through the middle of the brain (sagittal section); lower figure is top view of the brain (horizontal section). DCB = dorsal catecholamine bundle. CTT = central tegmental tract. MFB = medial forebrain bundle. [Reproduced from Angevine and Cotman (1981) *Principles of Neuroanatomy*, by kind permission of Oxford University Press.]

postsynaptic instruction about the messages—rather than the messages themselves—that differentiates the waking from the dreaming sleep state'. Under conditions of decreased aminergic activity, cortical neurons may process as usual whatever information they receive, but the network does not learn and remember. Since the alerting and orienting mechanism of the locus coeruleus is damped, attentional and reflective operations are also largely eliminated.

During REM sleep, increased activity by cholinergic cells in the pons activate the thalamus and the sensory cortex, and this activity shows up in polygraph recordings as the PGO (pontine-geniculate-occipital) waves. They also activate oculo-motor and vestibular nuclei in the brain stem. The cortex does not know whether the inputs it gets are generated internally or not, it just processes the data it gets (Hobson 1984), trying, as one might say, to make sense of whatever it gets. In the absence of external data to constrain the information that is received by the cortex, bizarre and haphazard images are bound to occur. Copies of the signals sent to the oculo-motor muscles from brain-stem structures (what is called 'efference copy') probably reach cortical areas, and may be a significant component in the information the cortical networks try to incorporate into a coherent picture (Fig. 13.7).

The brain-stem regions are very complex, and their architecture does not readily suggest functional hypotheses about what the networks are doing. Certainly many questions about the sleep–wake cycle remain. For example, is there a general utility resulting from the increased activity of cholinergic FTG cells during REM sleep? Why are there not just two states—S-sleep and being awake? What is the role of the other neurochemicals in the reticular formation, such as the peptides? Most pressingly, perhaps, precisely how does decreased activity in aminergic cells of the locus coeruleus and the raphe affect the processing and network function of cells to which they project? Although we do not yet have answers to these questions, several points are worth emphasizing: (1) the questions are far more specific and well-defined than they could have been even 30 years ago, and (2) the result is that testable hypotheses can be generated.

Accordingly, I think there is much that is satisfying and inspiring in the research on the sleep–wake cycle. At the very least—and this is no mean thing—we have got our empirical hooks into a phenomenon that appears to be a global state (behavioural state) of the nervous system. If we conceive of the behavioural state in terms of a state space, the dimensions of which will include activity of aminergic cells, the neuro-modulatory role of peptides such as histamine and substance P, the activity of cholinergic cells, the state of motor neurons, sensory gating activity, and so forth, then we can begin to get a feel for what it is to be

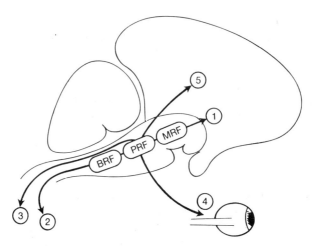

Processes accounted for:

(1) Activation of forebrain;

(2) Blockade of exteroceptive input;

(3) Blockade of motor output;

(4) Oculomotor activation;

(5) Provision of forebrain with internally generated information.

Fig. 13.7. Schematic diagram of a sagittal section of the cat brain showing the main physiological elements in dream state generation. (BRF = bulbar reticular formation; PRF = pontine reticular formation; MRF = midbrain reticular formation.) [Reproduced from Hobson and McCarley (1977), The brain as a dream state generator: an activation-synthesis hypothesis of the dream process. *American Journal of Psychiatry.*]

awake or in REM sleep in terms of the shape of a volume of the relevant state space. Naturally it is disappointing that we do not yet have physiological techniques for tapping into events at the network or circuit level, for we need to know not only the cellular but also the network effects of decreased aminergic activity. Nevertheless, computer modelling may be of considerable help in this regard, and new techniques may become available.

The general approach to global states also raises the question of how we might modify connectionist models (Sejnowski and Hinton 1985; Kienker *et al.* 1986; Lehky and Sejnowski 1989) to accommodate such features as global state control and the modulation of behavioural states. Although existing network models generally do not include such features, the network devised by Kienker *et al.* (1986) to solve figure – ground problems does have the kind of attentional mechanism which serves to bias the network. This enables the network to switch figure and ground, as in the familiar face–vase diagram.

If we are expecting to find in the neurobiology of sleep and dreaming a mechanism that turns the light off and turns it on again, then the odds are that we shall be frustrated. The data available so far indicate that the neurobiological explanation is not going to look remotely like that, and consequently that being awake and being asleep should not be understood on that model. Instead, the theory will be a network-to-network account, involving the role of sensory input in constraining the sequence of internally generated images, of the functioning of short-term memory when sensory input is admitted, of emotions, of appropriate motor assembling, short-term memory, long-term memory, and goodness knows what else. If it is even roughly right, this theory indicates that we will have to re-think our pre-scientific intuitions concerning what it is to be conscious.

Concluding remarks

Where then does this leave the question of the neurobiological reduction of consciousness? It is possible, for all I have said, that consciousness will reduce smoothly, in the classical sense, to neurobiological phenomena; that is, in the way that light reduced to electromagnetic radiation (theory of optics to theory of electromagnetic radiation), or temperature in a gas to mean molecular kinetic energy. But smooth reductions are atypical in the history of science, and more commonly old theories are corrected, revised, and modified in various degree as points of reductive contact are made. For a theory to reduce smoothly, its categorial framework must be pretty much correct. While the ways for a categorial framework to get it wrong are legion, getting it right is a much smaller target. As points of reductive contact are made between psychology and neurobiology, and as the two fields co-evolve, we can expect modification and corrections to the received wisdom at all levels of description. In such fashion, the various sciences of the mind-brain will likely converge upon unified explanations. Perhaps not of 'consciousness', for in the evolved framework that may have gone the way of 'caloric fluid' or 'vital spirit' (see pp. 274–277 above).

The prospect that consciousness might not yield to a classical reduction should not, of course, give any comfort to a dualist. For the classical reduction may be missed, not because there is spooky stuff or spooky properties, but because the folk psychological categories lack sufficient integrity to stick. In that case, the reduction is either a revisionary reduction—of revised psychological categories to neurobiological categories—or, more radically, an outright replacement of the old folk notion of consciousness with new and better large-scale concepts.

Acknowledgements

I am particularly indebted to Allan Hobson for teaching me about the importance of behavioural states, and to Francis Crick for provoking me to think about the appropriate domain for neurobiological investigation. Others whose conversation helped me in various ways are Edoardo Bisiach, Paul Churchland, Daniel Dennett, Warren Dow, Patricia Kitcher, Philip Kitcher, Tony Marcel, Terry Sejnowski, Steve Stich, and David Zipser.

References

Angevine, J. B. Jr and Cotman, C. W. (1981). *Principles of neuroanatomy*. Oxford University Press.

Bisiach, E. and Luzzatti, C. (1978). Unilateral neglect of representational space. *Cortex*, **14**, 129–33.

Churchland, P. M. (1979). *Scientific realism and the plasticity of mind*. Cambridge University Press.

Churchland, P. M. (1985a). Reduction, qualia, and the direct introspection of brain states. *Journal of Philosophy*, **82**, 8–28.

Churchland, P. M. (1985b). On the speculative nature of our self-conception: a reply to some critics. In *New essays in the philosophy of mind*, (ed. J. MacIntosh and D. Copp), supplemental volume of the *Canadian Journal of Philosophy*, pp. 157–73.

Churchland, P. S. (1986). *Neurophilosophy: toward a unified science of the mind-brain*. MIT Press, Cambridge, MA.

Crick, F. H. C. and Mitchison, G. (1983). The function of dream sleep. *Nature*, **304**, 111–14.

Crick, F. H. C. and Mitchison, G. (1986). REM sleep and neural nets. *Journal of Mind and Behavior*, **7**, 229–50.

DeArmond, S. J., Fusco, M. M. and Dewey, M. M. (1976). *Structure of the human brain*, (2nd edn). Oxford University Press.

Dennett, D. C. (1988). Quining qualia. In *Consciousness in contemporary science*, (ed. A. J. Marcel and E. Bisiach), p. 42. Oxford University Press.

Feyerabend, P. (1963). How to be a good empiricist. In *Challenges to empiricism*, (ed. H. Morick), pp. 164–93. Wadsworth, Belmont, CA.

Flicker, C., Robert W., McCarley, R. W., and Hobson, J. A. (1981). Aminergic neurons: state control and plasticity in three model systems. *Cellular and Molecular Neurobiology*, **1**, 123–66.

Fodor, J. (1975). *The language of thought.* Thomas J. Crowell, New York.

Fodor, J. (1981). *Representations.* MIT Press, Cambridge, MA.

Gazzaniga, M. S. and LeDoux, J. E. (1978). *The integrated mind.* Plenum, New York.

Heilman, K. M. (1985). Neglect and related disorders. In *Clinical neuropsychology* (2nd edn), (ed. K. Heilman and E. Valenstein), pp. 268–307. Oxford University Press.

Hobson, J. A. (1978). What is a behavioral state? In *Neuroscience Symposia, Vol. 3, Aspects of behavioral neurobiology,* (ed. J. A. Ferrendelli), pp. 1–15. Society for Neuroscience, Bethesda, MD.

Hobson, J. A. (1984). How does the cortex know when to do what? A neurobiological theory of state control. In *Dynamic aspects of neocortical function,* (ed. G. M. Edelman, W. E. Gall, and M. W. Cowan), pp. 219–57. Neurosciences Institute/Wiley, New York.

Hobson, J. A. and McCarley, R. W. (1977). The brain as a dreamstate generator: an activation–synthesis hypothesis of the dream process. *American Journal of Psychiatry,* **134**, 1335–48.

Hobson, J. A. and Steriade, M. (1986). Neuronal basis of behavioral state control. In *Handbook of physiology, section I, The nervous system,* (ed. V. Mountcastle), pp. 701–823. American Physiological Society, Bethesda, MD.

Hobson, J. A., Spagna, T., and Earls, P. (1977). *Dreamstage: an experimental portrait of the sleeping brain.* Hoffman–LaRoche.

Hobson, J. A., Lydic, R. and Baghdoyan, H. A. (1986). Evolving concepts of sleep cycle generation: from brain centers to neuronal populations. *The Behavioral and Brain Sciences,* **9**, 371–448.

Hobson, J. A., Hoffman, S. A., Helfand, R., and Kostner, D. (1987). Dream bizarreness and the activation synthesis hypothesis. *Human neurobiology.*

Hooker, C. A. (1981). Toward a general theory of reduction. Parts 1–III. *Dialogue,* **20**, 38–59, 201–36, 496–529.

Jackson, F. (1982). Epiphenomenal qualia. *Philosophical Quarterly,* **32**, 127–36.

Kienker, P. K., Sejnowski, T. J., Hinton, G. E., and Schumacher, L. E. (1986). Separating figure from ground with a parallel network. *Perception,* **15**, 197–216.

Kitcher, P. (1982). Genes. *British Journal for the Philosophy of Science,* **33**, 337–59.

Kuhn, T. S. (1970). *The structure of scientific revolutions,* (2nd edn). University of Chicago Press.

Lehky, S. and Sejnowski, T. J. (1989). Neural network models of visual processing. In *Computational neuroscience,* (ed. E. Schwartz). MIT Press, Cambridge, MA.

Nagel, E. (1961). *The structure of science.* Harcourt, Brace and World, New York.

Nagel, T. (1974). What is it like to be a bat? *The Philosophical Review.* **83**, 435–50.

Nisbett, R. and Ross, L. (1980). *Human inference: strategies and shortcomings of social judgment.* Prentice-Hall, Englewood Cliffs, NJ.

Popper, K. R. and Eccles, J. C. (1977). *The self and its brain.* Springer, Berlin.

Pylyshyn, Z. (1984). *Computation and cognition.* MIT Press, Cambridge, MA.

Sacks, O. (1985). *The man who mistook his wife for a hat.* Summit Books, New York.

Schaffner, K. F. (1976). Reductionism in biology: prospects and problems. In *PSA Proceedings 1974,* (ed. R. S. Cohen, C. A. Hooker, A. C. Michalos, and J. W. van Evra), pp. 131–56. Reidel, Dordrecht.

Sejnowski, T. J. and Hinton, G. E. (1987). Separating figure from ground with a Boltzmann Machine. In *Vision, brain, and co-operative computation,* (ed. M. Arbib and A. R. Hanson), pp. 703–24. MIT Press, Cambridge, MA.

Smullyan, R. (1983). *5000 B.C. and other philosophical fantasies,* St Martin's Press, New York.

Sperry, R. W. (1980). Mind–brain interaction: Mentalism, yes; dualism, no. *Neuroscience,* **5,** 195–206.

Steriade, M. and Hobson, J. A. (1976). Neuronal activity during the sleep–waking cycle. *Progress in Neurobiology,* **6,** 155–376.

Weiskrantz, L. (1986). *Blindsight: a case study and its implications.* Clarendon Press, Oxford.

Weiskrantz, L. (1988). Some contributions of neuropsychology of vision and memory to the problem of consciousness. In *Consciousness in contemporary science,* (ed. A. J. Marcel and E. Bisiach), p. 183. Oxford University Press.

Weiskrantz, L., Warrington, E. K., Sanders, M. D., and Marshall, J. (1974). Visual capacity in the hemianopic field following a restricted occipital ablation. *Brain,* **97,** 709–28.

14

Information-processing models of consciousness: possibilities and problems

Tim Shallice

Consciousness and the methods of cognitive psychology and neuropsychology

At face value the existence of consciousness is one of the greatest, if not the greatest, of the unsolved problems of science. Yet throughout this century many have dismissed the mystery of the relation between awareness and mechanism as a pseudo-problem stemming from an attempt to incorporate into science archaic philosophical notions. The 'delusion', however, does remain very difficult to eradicate. Moreover the development of cognitive psychology and neuropsychology has produced severe internal problems for any advocate of the dismissal position.

In cognitive psychology and neuropsychology our explanations of particular findings may at least on the surface be as mechanistic as ones in say molecular biology or sensory physiology. However, our methods of study have one characteristic very different from those applied, say, to a cell or a retina. The subject of our studies is aware.

This is much more critical than would appear from many written papers. So a cognitive neuropsychologist attempts to be sensitive to the experience of the patient both in everyday life and in particular test situations. Many if not most cognitive psychologists prefer, as far as possible, to try out new experimental situations so as to experience them as a subject would and also to ask subjects themselves how they experience them. Most of the time such information is used informally. It sensitises one to particular artefacts of design or procedure and suggests possible explanations of results.

In many parts of cognitive psychology and neuropsychology it is in fact the norm for accounts of personal experience to be used only informally. Such information does not become part of the phenomena publicly described in an empirical paper. So in, say, a choice reaction-time experiment the results reported can appear to be no different in type from those produced in other areas of science. No novel epistemological situation seems to arise in the use of such data.

Even in such areas of cognitive psychology much more clear-cut public use of experiential findings can occur. An excellent example comes from the study by Sperling and Speelman (1970), which was concerned with modelling short-term storage (STS) processes mathematically. Their procedure was to estimate STS capacity by measuring performance in memory-span tasks under a variety of factorially manipulated conditions, including two different vocabularies ('acoustically similar' letters which rhyme vs. 'acoustically different' letters which do not), three different retrieval conditions, three different exposure durations, and two different modalities of presentation (auditory, visual). In the visual conditions twelve letters in an array are presented to the subject for a total of 0.2 sec., 2.0 sec., or 12 sec. The authors showed that the results of all but one of the different conditions used fitted fairly well with a linear relationship between two variables known to be important in short-term memory research. One is span—the maximum number of items that can be repeated without error. The other was the similarity decrement: how much smaller the span is when all the items are 'acoustically' similar (e.g. they rhyme) than when they are dissimilar (see Baddeley 1976).

When questioned after the experiment all subjects reported that in the visual conditions they used a rote, verbal, sequential rehearsal strategy at the two shorter exposures—they repeated the letters continually to themselves—but only five subjects (set R) reported doing this for the longer 12-sec exposure, the one condition that did not fit the linear model. Six (set C) reported that in this condition they looked for patterns among the letters, formed associations between them, or used some other non-rote type of coding. When the results of the 12-sec-exposure visual condition were considered separately for these two groups of subjects, those of the R group obeyed the general linear relationship well, but those of the C group were quite discrepant. For these subjects there was no similarity decrement at all in this condition. Clearly for theoretical modelling of the mechanisms involved it is inappropriate to average across the two subgroups who used different strategies. In the mathematical model of auditory-verbal short-term storage developed by the authors, goodness-of-fit was therefore assessed leaving out the visual 12-sec-exposure results except for the set-R subjects. In practice this seems an appropriate procedure. However it is critically dependent upon the use of subjective reports.

A second reason why the awareness of the subject is important for our science is that there are areas of psychology where accounts of phenomenal experience form the bulk of the experimental findings (see

Marcel 1988, this volume). So our knowledge of perception, for example, is almost entirely based on what Weiskrantz (1977) called commentary procedures. One might argue that the perceptual processes under study and commentary procedures themselves are carried out by what are effectively quite distinct systems. Thus, as far as our knowledge of perceptual processes is concerned, the mechanisms by which the commentary is produced can be considered a 'black box', whose internal workings are basically irrelevant. This does not apply, however, to the use of protocols in the study of thinking (e.g. see Newell and Simon 1972; Ericsson and Simon 1985). Here the processes reported on and those involved in producing the protocol seem much less disjoint. Even clearer examples of findings where phenomenal accounts cannot be ignored are those in which their interest as pre-theoretical generalizations requires them to be couched at least in part in terms of the experience of the subject. Examples include neuropsychological research on blindsight (e.g. Weiskrantz *et al.* 1974; Weiskrantz 1986) and on certain aspects of amnesia (e.g. see Weiskrantz 1977; Weiskrantz and Warrington 1979; Mandler 1989, and, in normal subjects, the work of Marcel (1983) and others on pattern masking (but also see Holender 1986).

A final reason is that our experiments nearly always depend upon subjects understanding and applying instructions which use terms like 'see', 'hear', 'judge', 'remember', and so on. We are, in other words, treating our subjects as responsible conscious agents. Sometimes these instructions can have considerable phenomenal complexity. The work on mental imagery by Kosslyn and his colleagues contains many examples. Consider the following procedure used by Kosslyn *et al.* (1983 p. 295):

On hearing the name of a scene, the subject was to form a mental image of that scene and respond by pressing the button labelled *True* as soon as the image was clear. Subjects were instructed to make all images as clear, detailed, bright, and vivid as possible. After a pause, the subject heard the probe phrase. Subjects were to mentally 'focus' on the first object named and then to scan directly across the image until reaching the second object named. On 'seeing' the second object clearly, the subject was to respond by pressing one of two buttons.

If one accepts that cognitive psychology and neuropsychology are making progress by using a computational/information-processing theoretical framework, then the paradox exists for the 'ghost in the machine' position of advances in 'mechanistic' theorizing being based at least in part on the use of such instructions and on accounts of experience. Clearly the fields must remain intellectually incomplete unless they can explain how experiential terms like 'see', 'remember', 'guess', 'dream', 'image', and so on are understood and used by the entity whose behaviour is being explained through mechanistic models.

Possible lines of attack on the problem

It might be possible to attempt to maintain a neo-behaviourist position and hold that our present explanations of our actions in experiential terms may in future seem equivalent to Aristotelian accounts of physical phenomena (see Churchland 1988, this volume). So current descriptions of cognition using such concepts would have no simple equivalent in a well-developed scientific account of people. Effective use of phenomenological reports in present-day science would be analogous to some limited domain where in the early history of physics Aristotelian concepts played a progressive role. This response echoes one of Feyerabend's (1975) beliefs—that it is unnecessary to justify the scientific procedures we use; all that is required is that they work. This, though, has the problem that applies generally to Feyerabend's approach to science; it is essentially anti-rational.

There is a more general problem for this approach. If there is no need to understand how experiential concepts can be effectively applied within present-day cognitive psychology, because they will be replaced by more adequate ones derived from future computational science and neuro-science, current explanations of behaviour in terms of experiential concepts are downgraded by contrast with those using current computational science and neuro-science. For it is implicitly assumed that our future scientific understanding of people will derive more from the latter than the former. In our current social climate this could well have the effect of legitimizing reductionist explanations of action in the social and human sciences at the expense of ones couched at least in part in experiential concepts. The idea of self-awareness crucial for both therapeutic and social understanding of people loses its conceptual foundation. The negative effect on the social and human sciences could be analogous to that which occurred in psychology after the rise of behaviourism as a paradigm for 'hard-core' areas of the field (for discussions of related issues see Burnham 1968; Chomsky 1972; Chorover 1979).

The alternative to the neo-behaviourist dismissal of the problem is to treat accounts of phenomenal experience as one valid and partially independent procedure for understanding human thought and action. From this perspective, the obvious approach is to attempt to obtain some link between mechanistic explanations and experiential accounts of the same events. There is, though, a major difficulty. When one obtains a link between two sets of scientific concepts a range of analogies can be used as vehicles for the link and there will be properties in common between the concepts. Genes and DNA, for instance, each have the property of being a unit containing two elements which have closely related characteristics. The representations underlying thought about mechanistic systems are, however, as 'distant' from those concerned with describing phenomenal experience as any two sets of representations can be within the cognitive system.

Only one level of scientific theorizing appears to offer any purchase at all for producing a useful link with accounts of conscious experience. This is the information-processing/computational one. The 'grain' of this set of concepts appears compatible with mental terms. For instance, if consciousness could be related to such system-level concepts one would obtain obvious explanations for certain of the seemingly paradoxical findings of modern neuropsychology—dual awareness in split-brain patients (see Gazzaniga 1988, this volume) and the critical areas of lack of awareness in blindsight patients (see Weiskrantz 1988, this volume). Thus, in the split-brain case, the lesion does not basically alter the *local* mode of functioning of individual neurons in a different fashion from any other lesion which reduces their input but which does *not* lead to two streams of awareness. Yet, in the split-brain case, the lesion results in a *system* which, though previously working in an integrated fashion, now operates in many respects as two non-communicating subsystems. By contrast, non-functionalist theorists have been forced to resort to bizarre positions such as that there are two unitary consciousnesses that co-exist in the normal person unbeknown to each other—one in the left hemisphere and the other in the right (Puccetti 1973). In these respects psychology contrasts with all other sciences, which may produce paradoxes about consciousness (e.g. Puccetti and Dykes 1978; Libet *et al.* 1983) but which do not appear to have concepts of the appropriate 'grain' to produce any positive advance.

A programme for producing links between experiential concepts and information-processing ones could begin from a number of starting points. Conceptually the simplest approach, and also the one most frequently adopted, is to presuppose that there is one fundamental phenomenological concept, namely 'consciousness', and to attempt to obtain a correspondence for that. An information-processing/computational theory is developed in which a particular type of information flow, structure, or process has properties isomorphic with those of consciousness (for instance, see Shallice 1972, 1978; Mandler 1975, 1985; Baars 1983; Johnson-Laird 1983; and, for a more physiological version, O'Keefe 1985).

It is far from clear that such a programme will work. The functionalist approach undoubtedly has major philosophical problems (e.g. Block 1980). Moreover the mapping relations required can hardly be as straightforward as those obtaining between, say, molecular biology and genetics. On the computational approach it is natural to treat experiential terms as signifying precise inner states of the organism. Yet, as the Wittgenstein/Ryle tradition teaches us, experiential concepts are not clearly circumscribed scientific concepts. Mental processes are not transparent to description, as the history of introspectionist psychology illustrates in practice (Humphrey 1951).

Before, however, considering weaker forms of the functionalist research programme, such as attempting to differentiate conscious experiences into a number of forms and modelling each separately (see Oatley 1988, this volume), or attempting to produce quite distinct explanations for individual groups of mental terms such as verbs of perception and verbs of thinking, it is appropriate to examine the simple strong form. This is that if 'conscious' is used in what Wisdom (1963) calls 'its fundamental sense', namely implying either 'feels' or 'is aware', then the conscious/non-conscious distinction does map onto some system-level property of cognition. The aim of this chapter is to assess this possibility.

The properties of consciousness

To obtain links between two sets of concepts it is obviously necessary to be able to spell out the properties of each set. Cognitive scientists concerned about mapping experience onto mechanism tend to produce lists of the properties of 'consciousness' used as the noun derived from the fundamental sense of the adjective 'conscious'. Thus Mandler (1975) has a list, Dennett (1978) another, and Johnson-Laird (1988, this volume) yet another. In general the properties are held to be self-evident and are not defended. The effort goes into developing the computational theory. Take one of my earlier papers as an example (Shallice 1972). I blandly listed a number of properties of consciousness. Consciousness was held to have the property of intentionality, to have a variety of different types of content such as percepts, intentions, images, memories, and to have only one of these at a time. The generalizations were neither defended nor considered provisional.

To construct a more adequate set of properties requires consideration of various research traditions. They include phenomenology, Anglo–American philosophy of mind, pre-behaviourist psychology—in particular the work of James (1890) and of the Würzburg school—and current cognitive psychology. It is impossible to produce a properly supported set of properties in an article of this length. I will therefore produce a list of properties, very much of a rough and ready sort, but which is at least somewhat more extensive than those normally used as starting points for computational theories of consciousness. I will not attempt, however, to defend the generalizations in detail, on the assumption that even if individual ones are clumsy or inadequate the safest policy lies in increasing their number. The only general principle I will use is that the concepts in which the generalizations are expressed will be not too far removed from everyday mental terms. I will assume that the more immediately understandable and illustratable the generalization the better. Thus a concept like 'qualia', which is itself subject to extensive philosophical debate (see Dennett 1988, this volume), will be ignored. I will also duck the thorny issue of what grounds one could have for accepting the generalizations.

The properties of experience I want to consider may be divided into two types—those concerning structure and those concerning content. This division is a loose one—the structure and the contents of experience cannot strictly be varied orthogonally. The structural ones are best approached in a step-wise way.

'Consciousness' is always changing but from waking to sleeping is sensibly continuous (property 1a). In temporal cross-section it has a foreground and a background; the degree of contrast between foreground and background varies (property 1b)

As far as the first property is concerned, James's (1890) succinct argument is sufficient, i.e. that consciousness does not appear to itself chopped up in bits. The second property is standardly accepted as a useful analogy (e.g. Hamilton 1859; James 1890). Yet some argue that it does not apply to all mental states. Take the standard phenomenological account of attention, again that given by James (1890 pp. 403–4):

'Everyone knows what attention is. It is the taking possession by the mind, in clear and vivid form, of one out of what seem several simultaneous possible objects or trains of thought. Focalisation, concentration, of consciousness are of its essence. It implies withdrawal from some things in order to deal effectively with others.'

However, he continues:

... [it] is a condition which has a real opposite in the confused, dazed, scatter-brained state which in France is called *distraction*, and *Zerstreutheit* in German. ... The eyes are fixed on vacancy, the sounds of the world melt into confused unity, the attention is dispersed so that the whole body is felt, as it

e, at once, and the foreground of consciousness is filled, if by anything, by a
t of solemn sense of surrender to the empty passing of time (p. 404).

Kinsbourne (1988, this volume) gives another example where the 'foreground' is relatively less important, namely the experience of looking at a view. Breadth of conscious experience rather than focalization is of its essence and this is probably true quite frequently in aesthetic experience. Therefore I will assume that the relative importance of foreground and background aspects of consciousness may vary across mental states (for a conflicting view, see Evans 1970).

The primary structuring feature of consciousness—attention—has a number of subvarieties (property 2)

James (1890) differentiated between passive, reflex, non-voluntary, or effortless attention, on the one hand, and active or voluntary attention, on the other. Evans (1970) differentiates two voluntary forms—executive and interrogative. Whether it is appropriate to categorize attention phenomenologically into specifically three types remains unclear. However, there are clearly qualitatively different subvarieties.

One special form of attention is reflexive consciousness

In executive attention one is doing something; in interrogative attention one is inspecting or engaged in preparatory waiting. The two can be combined and given a specific object in the act of attending to one's own thought process—reflection. Some, e.g. Johnson-Laird (1983), have elevated this distinction into a key one, and have even claimed that awareness is, in principle, indefinitely recursive. This, though, appears to be an unnecessarily abstract extrapolation from the more direct lower-level description of experience. As Mandler (1985) has pointed out, experience is limited to only one or two cycles of reflection at a given time. So I am *now* reflecting on reflecting. I am possibly *now* reflecting on reflecting on reflecting. Yet this state is so fragile I can hardly hope to freeze it sufficiently firmly for another cycle to be possible. Therefore, in line with the policy of limiting properties of awareness to *relatively* direct descriptions of experience, I will view this facility as following from an elaboration of property 2. The facility is special in what it allows, namely the possibility of categorizing and describing experience. The possibility for categorizing and describing should not though be identified with what it is that can be categorized and described.

Consciousness—in the sense of the contents of the foreground of attention—is very limited in capacity (property 3)

This is normally treated as following fairly self-evidently from the first property. However, it is spelled out in rather different and possibly contradictory ways. Thus there is the view put forward by Hamilton (1859) that consciousness is limited but can be extended to a small number of items—according to him, six. The other view, following Ribot (1890) and James (1890), is that the foreground of consciousness is relatively unitary in that its contents are at any one time structured around a single 'master idea' or form a single 'complex object'.

Hamilton's view stemmed from considering the type of situation we now know as subitizing (e.g. Kaufman *et al.* 1949), where subjects appear to estimate the number of elements in a visual array in a 'single mental act' if their number is small (Jevons 1871). In fact performance in such situations appears to be based on recognizing individual canonical patterns or to use mental counting (Mandler and Shebo 1982). Indirectly, then, the Ribot/James view is supported.

An obvious challenge to the Ribot/James view occurs in the phenomenology of dual-task experiments such as in reading one piece and listening to another. Thus Spelke *et al.* (1976) argue that such studies did not always support the view that consciousness is unitary. They contrasted the account of Solomons and Stein (1895), where subjects claimed to cease to be aware of the secondary task, with that of Downey and Anderson (1915), who reported no full loss of consciousness of either task when subjects read and wrote together. There do seem to be experiential differences between studies, but it is not clear that any subjects report roughly equal awareness between the two tasks. Recent dual-task experiments have unfortunately not tended to pay as much attention to phenomenology as earlier ones. My own view, having been involved in a variety of such experiments over the last three years or so (see Shallice *et al.* 1985), is that in dual task experiments where the tasks are conceptually quite distinct one cannot at any one time equalize one's attention between the tasks. One task is carried out in a considerably more automatic mode than the other.

Ribot's thesis can therefore be defended as a generalization only if one incorporates a number of provisos into it. One possibility—a position I have previously argued (Shallice 1978)—is that a major divergence from unitariness occurs only under highly specific conditions, namely where a pair of unrelated tasks are practised in combination. A more plausible possibility is that the unitariness of awareness at any one time is not absolute; this position in fact merely extends the qualification of the foreground/background idea as expressed in property 1b. However, in most cases there seems to be a fairly sharp change in focus between the most attended 'master idea' and the rest of the contents of consciousness at any time. In this sense consciousness is unitary (property 3). While Ribot's seems to me to be the generally preferable position, one could not claim that it is self-evidently true.

There are a variety of types of experience (properties 4–14)

The flow of experience can be 'frozen' for a short period of time and a sparse and not very reliable account given of it. In Woodworth's (1938) terms 'a series of such statements does describe the *general course* of a thinking process—just as naming the towns through which you have driven maps the route you have taken'. The most obvious characteristic of such states is their great variety. Language has a plethora of words for describing experience. Almost all experiences have contents—what one sees, hears, remembers, feels: the so-called property of intentionality (property 4a). The contents are the objects of mental operations—one hears A, guesses B, remembers (as an act) C, imagines D (property 4b). These include markers to future action—one intends X, gets ready for Y. To make a listing of these contents, operations, and markers, to decide how far they are distinguishable one from another and under what conditions they occur, is much more difficult. I am not aware of any attempt to provide a fairly exhaustive account of the different types of experience that we have. A very preliminary listing is given below in terms of abstract but commonplace characterizations of the varieties of experience as different properties. What would be much harder is to map the types of experience onto the types of cognitive tasks being carried out when the experiences occur.

The psychological operations which are best investigated from a phenomenological perspective are perceptual ones. Our knowledge of the phenomenology of perception is a long-established part of psychology. Objects are experienced as having colour, shape, position, size, and so on, and as forming scenes (property 5). This property should also include the way that we are not aware and in general cannot be of many sorts of information used in the perceptual process—take binocular disparity, for example.

Other types of mental operation are much less well explored. Take, for instance, action. Its phenomenology is also complex. Thus actions are experienced as being in one of three qualitatively distinct classes —those of which one is not aware, ideo-motor ones, and willed ones (see Norman and Shallice 1986) (property 6). Some actions can occur without one even being aware of eliciting them or carrying them out. Take, as an example, the seemingly meaningless side-products of concentration. Thus the last action I have just discovered myself doing is the strange one of pushing the end of my pen into the flesh on my chin. It was certainly not accompanied by any awareness prior to my decision to look to see what actions I was in fact carrying out and, but for that, would have remained at a pre-conscious level.

A qualitatively distinct type appears to be what James (1890) called 'ideo-motor' acts, where 'movement follows unhesitatingly and immediately the notion of it in the mind'. 'Whilst talking I became conscious of a pin on the floor, or of some dust on my sleeve. Without interrupting the conversation I brush away the dust or pick up the pin'. Ideo-motor acts of this sort fit with Searle's (1983) characterization of intention-in-action. They can take place, however, with what Ach (1905) called 'awareness of determination' (property 6a); this arises because one had a general intention to act in a certain way when a particular type of signal occurred, say in a choice reaction-time task, even though no specific conscious intention had been present on that trial. Third, there are consciously willed acts. In this case we are all familiar with the phenomenological gap between the occurrence of an intention to make a 'basic action' and its execution (property 6b).

This type of subdivision does not exhaust the complexity of the phenomenology of action. Consider, for instance, playing a shot from the baseline in tennis. I, at least, would seem to be vividly aware—and probably must be to have any sort of stab at hitting the ball—of the ball coming from my opponent's racket towards me and of where I want to send it. By contrast, in, say, archery, the action—and its goal—does not have to be selected, it just has to be executed as effectively as possible. It is sometimes claimed that this is best done in some form of semi-trance condition (Herrigel 1972).

Similar complexities occur with respect to thought processes. Superficially it seems that higher level thought processes require awareness, but lower level ones might not. Thus awareness seems to be involved in the exercising of choices and judgments and what Mandler (1975) calls trouble-shooting—dealing with an unexpected failure to reach a goal (property 7). As A. D. Leslie (personal communication) points out, non-routine 'decoupled' operations like pretending are only carried out on objects of which one is conscious. Yet even here there are problems. Thus judgments may be conscious and yet have no conscious contents —as in the Wurzburg school's 'Bewusstseinslagen' (see Humphrey 1951). More problematic still, the routine/non-routine contrast does not always work. Thus a routine cognitive operation like carrying out mental arithmetic requires awareness (property 7a). By contrast, the thought operations that most immediately precede the 'coming-to-mind' of a new idea do not. If, for instance, in answering the question of how long is the spine one thinks of the size of a shirt, the processes by which shirt 'comes to mind' are not ones of which we are aware (property 7b). Again, then, criteria do not seem to be available for when consciousness need or need not be involved in thought processes.

For language and memory the situation *may* be more straightforward. In comprehending an utterance one is in general aware of the meaning of what the person says (property 8a). The same seems to apply to producing an utterance, although, as James (1890) pointed out, it is difficult to characterize just what one is aware of in this situation. As Merleau-Ponty (1962) pointed out, speech does not translate ready-made thought, but accomplishes it. One is frequently not aware of the thought in any clear sense before the utterance occurs (property 8b). Certain sorts of memories too seem to be associated with a special form of awareness when they are retrieved, as has been much discussed in the psychological literature in the last ten years (e.g. Tulving 1972, 1983; Weiskrantz 1977). The recalling or recognizing of events that have occurred to one, people one has met, places one has visited, all seem to occur with an experiential quality in common, which is quite different from that of, say, retrieving the meaning of a word or a *type* of object which one may not have encountered for an equivalent period of time (property 9a). Moreover, what one remembers is typically—although possibly not always (see Erdelyi 1988, this volume)—something of which one was conscious (Fisk and Schneider 1984) (property 9b).

There are, of course, a variety of other types of experience. Thus there are emotions and feelings (property 10), intentions—an especially slippery set with a complex relation with awareness (see Anscombe 1957; Boden 1981; Baars and Mattson 1981) (property 11), ones which serve to direct and comment on the thought processes themselves (property 12)—like the state of knowing (in say tip-of-the-tongue situations) or of doubt in decision-making situations. Then one has images (property 13), dreams (property 14), and a large range of states, excellently reviewed by Reed (1972) under the title 'anomalous experiences'. In addition there are the types of phenomena discovered in cognitive psychology and neuropsychology which have a critical aspect which can only be described in experiential terms. Two which seem particularly relevant are blindsight (e.g. Weiskrantz *et al.* 1974; Weiskrantz 1986) and perception under pattern-masking conditions (e.g. Marcel 1983).

Correspondences with information-processing theories

The purpose of listing this limited and rough-and-ready set of the properties of conscious experience is to provide—provisionally—one side of the mapping relation with information-processing/computational theory. If one turns to the complementary side, any suggested information-processing correspondence for conscious experience can in turn be assessed in two ways—in terms of its adequacy as an information-processing/computational theory per se and in terms of its adequacy for providing a correspondence. 'Mechanistic' theories can be developed specifically to provide a correspondence for consciousness (e.g. Dennett 1978). However, for psychology to be productive and not merely produce an intellectual echo, the mechanistic theory has to be advocated or at least supported for reasons internal to cognitive psychology. There seems little point in discussing information-processing theories which are inadequate in their own mechanistic terms. For instance, it is not useful to consider Atkinson and Shiffrin's (1971) view that consciousness can be identified with the contents of a general purpose short-term memory system which contains both representations of just preceding input, the results of rehearsal and of current processes, as their theory is generally accepted as having been superseded (e.g. see Baddeley and Hitch 1974; Crowder 1982).

There are a wide variety of theories providing potential correspondences. Rather than assessing them in turn—far too daunting a task—I want instead to introduce a framework which I believe can be defended in standard information-processing terms, independent of consciousness. It is an extension of one developed by Norman and myself (Norman and Shallice 1980; see also Norman 1986). Our aim was to make certain of the structural aspects of experience in action more transparent (property 4 above) by mapping the conscious/non-conscious distinction into a processing one after the fashion of Posner and Snyder (1975) and Shiffrin and Schneider (1977). As an *information-processing model* the theory has a family resemblance with many others (e.g. Reason 1979; Mandler 1985). The general points I wish to make are, however, somewhat independent of the particular theory, but it provides a convenient basis for discussing them. I will distinguish four levels of processing (see Fig. 14.1).

I. A very large set of special-purpose processing systems—analogous to special-purpose microcomputers—resembling Fodor's (1983) 'modules' but with three modifications (see Shallice 1984):

(i) The variety of types of local functional architecture—i.e. the way subsystems interact—is greater than Fodor envisages.
(ii) The overall functional architecture is not totally innately determined. Subsystems can be acquired in the learning of skills, as, for instance, in learning to read.

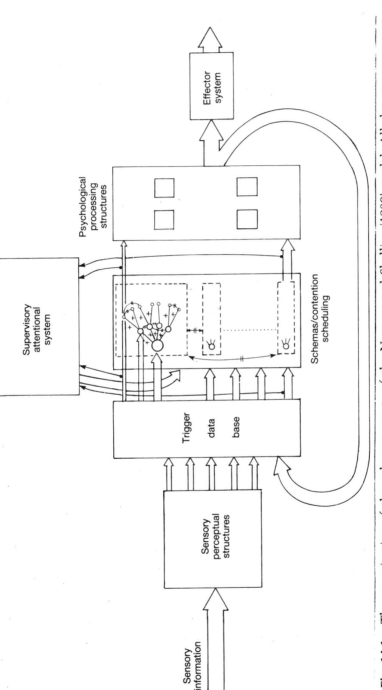

Fig. 14.1. The organization of the subcomponents of the Norman and Shallice (1980) model. All the structures not represented by the Supervisory Attentional System and Schemas/Contention Scheduling can be considered as approximately modular. Within the Contention Scheduling system excitatory and inhibitory connections are shown differently. In the rest of the figure solid lines denote flow of activation and information used for variable-binding operations—namely the inputs that determine the specific targets of the operation on that particular application; narrow lines denote information transmission used for any purpose.

Effector system

Psychological processing structures

Supervisory attentional system

Schemas/contention scheduling

Trigger data base

Sensory perceptual structures

Sensory information

(iii) 'Modularity'—in this weaker sense—does not just apply to input systems. It can be assumed that, in addition to processors, relevant subsystems include special-purpose stores and special-purpose systems for channelling information-flow through particular subsets of processing modules (mediating lower-level attentional functions).

II. A very large set of action and thought schemas [in the sense of schema of Rumelhart and Ortony (1977)]—well-learned, *highly* specific but complex programs for operations such as eating soup with a spoon, carrying out long division, travelling to work, and so on which can call other such programs that 'run' on the processing modules. These schemas are held to be activated from a variety of sources in parallel. Potential competition and interference between them is resolved—at a routine level—by a special-purpose system, 'contention scheduling', that selects the schema or schemas to be carried out on the basis of activation level and lateral inhibition (see Norman and Shallice 1980).

III. A Supervisory System whose function is to modulate the operation of contention scheduling by activating or inhibiting particular schemas (Luria 1966, Norman and Shallice 1980). It has access to representations of operations on the previous level, of the organism's goals, and of the environment—a facility in common with Johnson-Laird's (1983) operating system. It comes into play when the operation of routinely selected schemas does not directly satisfy goals; for instance, when one is in a novel situation or when an error has occurred.

IV. A language system which can operate in two modes. In comprehension mode, it can control the operation of the rest of the cognitive system by activating a specific schema or schemas and setting any variables necessary for its operation on that particular occasion or by providing input to the Supervisory System. In 'slave' or production mode it produces a representation of the state of the Supervisory System or of the schemas selected in contention scheduling in an utterance.

The framework was designed to reflect the phenomenological distinction between willed and ideo-motor action in the operation of contention scheduling with and without modulation by the supervisory system.

The general approach of cognitive psychologists taking a functionalist perspective is to take a processing model and associate some aspect of the processing of some component (or of information flow to and from it) with consciousness. The problems latent in the procedure can be illustrated with the present model. What should be that critical component? In principle it offers various alternatives for a correspondence for consciousness which in fact relate to existing theories:

1. It (consciousness) could correspond to inputs to the language system. This would be rather similar to Dennett's (1969) earlier position.

2. It could correspond to the processing carried out by the Supervisory System or to information-flow to and from it. This would have similarities to positions advocated with varying degrees of detail by MacKay (1966), Marshall and Morton (1978), and Mandler (1985).

3. It could be identified with the selection of schemas (and their bound variables) in contention scheduling including the setting of the variables which determine their goal or target on that operation (their bound variables). Thus, in a previous paper (Shallice 1972), I argued that consciousness corresponds to the selector input, which leads to an action system (or schema in the present terminology) becoming 'dominant' because it controls action, is stored in (episodic) memory, and can be spoken about. That version of the model contained no Supervisory System. Models which identify consciousness with the execution of a high-level, fairly serial thought/action operation in similar fashion (e.g. Posner and Klein 1973; Johnson-Laird 1983) conflate this level with the previous one.

4. It could be identified with the operation of some specific part of the modular processing system. Two main suggestions of this type have been made. Consciousness has frequently been identified with the contents of short-term stores (e.g. Atkinson and Shiffrin 1971). In addition, Marcel (1983) has identified perceptual awareness with the product of a selective and synthesizing system, which might be characterizable as the operation of a high-level perceptual module.

5. If one were to add an 'episodic memory' system (see Tulving 1983) as an isolable component, or more plausibly as a distinct way of utilizing knowledge systems, which has its own specific input and output, then consciousness could correspond to this input and output. This proposal about functional organization is admittedly highly controversial even from a purely information-processing standpoint (see the discussion in Tulving *et al.* 1984).

Do any of these five possible levels of processing provide a correspondence for all or almost all of the properties of conscious experience? This is what a standard functionalist theory of consciousness would seem to require.

It might seem a simple matter to decide this using the present framework. However it is not. There are three reasons for this. First, the properties of consciousness are too vaguely specified. Second, large-scale information-processing theories are very loosely characterized; adding a connection, a constraint, or another subsystem to a model is unfortunately only too easy. Probably theories will become tighter, if only gradually, and phenomenological generalizations may well be capable of improvement.

There is a third reason, which is, however, a severe objective constraint. The human information-processing system is so constructed that in most mental operations which have a conscious correspondence many subsystems are involved. There is no way to determine for any particular type of operation treated in isolation which of the subsystems or processes are the critical ones responsible for awareness.

Consider, as a simple example, focal perceptual awareness. On a type-4 theory of the correspondence, awareness arises because a synthesized set of verified perceptual hypotheses has been achieved or because information has been entered into a short-term store. On a type-3 theory it is because an action or thought schema triggered by a perceptual input has been selected in contention scheduling. On type-1, type-2, or type-5 theory it is because the language system, Supervisory System, or episodic memory system, respectively, received a perceptual input. All the accounts are equally satisfactory and equally vague; they *might* be held to support a general functionalist position but not any specific one. A similar argument in reverse could probably be given for the relation between awareness and trouble-shooting. Trouble-shooting, it can be argued, necessarily requires system *X*, where *X* could be chosen to fit the theory being advocated.

Or take the foreground/background relation. For each hypothetical level of correspondence one could point to more parallel-operating 'lower' levels as corresponding to the background. On the 'synthesized percepts' view the background could be unsynthesized percepts. On the contention scheduling position the contents of short-term stores and the output of on-line perceptual modules act as triggers to activate action and thought schemas in an analogous fashion to the contents of short-term memory in production-system simulations of cognition, such as those of Newell and Simon (1972). In general these outputs are not sufficiently effective as triggers for their schemas to become strongly activated and so selected. However, given the appropriate combination of other triggers converging on the same schema they can become (partial) causes of schema selection. The background of consciousness would therefore correspond to triggering-inputs to contention scheduling potentially capable of assisting in schema selection (for a related view, see Shallice 1978). On the Supervisory System position this argument can be extended further. Schemas selected without any intervention from the Supervisory System can be included in the background of consciousness.

Again, the accounts are equally 'satisfactory', if one views them from the perspective of trying to support the claim of a particular correspondence, but equally vague if one contrasts the three explanations. The argument could probably be extended to the issue of the 'unitariness' of consciousness. The same applies for explanations of, say, blindsight or of the dual awareness in split-brain patients or of the effects of pattern masking. In the blindsight case the positions differ only with respect to the overall system the relevant information is assumed to fail to reach. Individual properties of consciousness, then, do not on the whole differentiate between rival possibilities for the correspondence.

There are some experiential generalizations which are difficult for some theories. Take the contrast between the types of attention suggested by Evans (1970)—spontaneous and executive/interrogative— or in James's (1890) terms between ideo-motor and willed thought. All would appear to require perceptual synthesis. The fourth type of correspondence would therefore not seem to give any account of this experiential distinction. The contention scheduling theory, by contrast, explains it simply according to the source of the triggering activation for the selected schema—as not involving the Supervisory System or as involving it (see Norman and Shallice 1980). The same type of explanation could be turned on its head for the Supervisory System correspondence. An ideo-motor action is one which the Supervisory System 'knows about' only as input; a willed one is 'known about' because it is initiated by the system itself. The other two theories can both only provide *ad hoc* explanations.

If one takes all the properties and makes a rough assessment of how easily the potential correspondences being considered can provide an explanation for each property, the theories fall into three groups. There are many properties which are difficult for any theory that consciousness corresponds to an aspect of the operation of a processing module. There are only a few properties which are difficult for the language system and episodic-memory system correspondences to explain, but there are not all that many correspondences which seem 'natural' or fairly tight; many seem rather *ad hoc*. The Supervisory System and contention scheduling alternatives do give 'natural' explanations for over half the properties.

However, even for these two positions there are some phenomenological generalizations that provide difficulty. Take the contention scheduling position (the type-3 theory). Rubbing one's cheek while talking, changing gear to negotiate a simple bend while driving a familiar road; actions such as these on the Norman and Shallice (1980) theory all require schema selection in contention scheduling. Yet one of the standard presuppositions about awareness is that in some sense highly routine actions of this sort are made 'automatically'. In the original version of the theory an action or thought schema is selected and effected if its activation level reaches its threshold. A schema controlling a very routine action might, however, be held to have a very low threshold, so it is very easily selected; then awareness can be made to correspond to the selected schemas which are strongly activated in contention scheduling and to their 'bound variables' on that particular operation. Indeed, this makes functional sense. Low-level activation will not lead to inhibition of other schemas which do not compete directly. Yet, looked at more broadly, if a schema on being selected does not interfere with the simultaneous selection of other schemas, if its occurrence could have been so well predicted that its selection does not need to be stored, if it does not engage language or reflection processes in any way, it remains an autonomous process which does not impinge on the current or future operation of the rest of the cognitive system. How could the system—as a whole—be other than 'unconscious' of this subprocess? Therefore a necessary consequence of the contention scheduling position would be that consciousness corresponds to an operation of the subsystem when it necessarily affects many other subsystems. This provokes a further question: Why should there be just one subsystem of this type?

A somewhat complementary argument can be presented for the idea of a correspondence with some state of the Supervisory System. States of mind such as being in doubt or in the tip-of-the-tongue state or intending to do something in the sense of having a conscious purpose (see Baars and Mattson 1981) certainly lend themselves to the position that a state of the Supervisory System could be a correspondence for consciousness. Yet take instead an action that is triggered in contention scheduling fairly rapidly—like, in certain sports, selecting and modulating an appropriate stroke within a second or so after the necessary predisposing perceptual conditions occur—as in the tennis example given earlier under property 6. Since the Supervisory System operates slowly and with considerable variance, when rapid precise behaviour is required it can hardly play a major role. Or take carrying out routine mental arithmetic where the Supervisory System is held not to be required; one is aware of the intermediate steps in obtaining the solution (property 7a). Is one's awareness a mere epiphenomenon in these cases as in the case of the reflex withdrawal when touching something hot? It certainly seems to be more. Is one's awareness to correspond, then, to some state of the

Supervisory System which oversees but in fact takes no action? This position seems to attribute to the Supervisory System properties which are dangerously homonculus-like. Or take reverie, where possibly the Supervisory System may not be functioning. Or frontal lobe patients in whom, following Luria (1966), I have argued that it is malfunctioning (Shallice 1982). Consciousness does not disappear in these cases. Moreover, patients with frontal lobe lesions have difficulty carrying out certain high-level operations of which one is not generally aware (those covered by property 7b) (e.g. Shallice and Evans 1978).

Could it be that property-12 examples, such as to be in doubt or to intend to do something, correspond merely to states of the Supervisory System *as they affect other subsystems* which they directly influence; e.g. schemas selected by contention scheduling or by the language system? After all as Baars and Mattson (1981) point out, there are layers of 'intentions' superordinate to any particular conscious purpose of which we are not aware when carrying out the purpose, and yet that must in some sense be represented in the Supervisory System. They are, however, not directly involved in the activation of schemas in contention scheduling.

Consider one final potential paradox. One simulation that has been thought to provide a useful analogy with high-level human supervisory self-programming functions is Sussman's (1975) program HACKER, which learns from its own errors (see Boden 1977). Within the program there is a clear split between the 'doing' part and the 'self-programming' part, which comes into play when the program is in the so-called CAREFUL mode; this mirrors the distinction between schema operation under contention scheduling alone and when modulated by the operation of the Supervisory System. To learn from its mistakes the self-programming part of HACKER requires an episodic memory known as CHRONTEXT, in which, when CAREFUL mode is operative, is stored a detailed chronological record of the state of its world and of the program's actions. If human episodic memory had a similar evolutionary function then the basic input to episodic memory would need to be the schemas selected in contention scheduling together with their bound variables [and maybe, in addition, some state-of-the-world information, possibly related to the automatic memory of Hasher and Zachs (1979)]. When retrieved at some later time the information would need, however, to be transmitted not to contention scheduling but to the Supervisory System. Thus if what you remember is what you were conscious of at some earlier time (property 9b), and the activity of remembering is itself necessarily conscious (property 9a), then 'consciousness' does not correspond to the operation of the same systems on the two occasions.

Conclusions

The discussion in the previous section has been very limited. Only a sub-set of possible phenomenological generalizations have been discussed. Moreover, only a single information-processing theory has been considered. However, it seems possible to draw two initial conclusions. First, the conceptual systems of phenomenology and information-processing/computational psychology are sufficiently loose and remote from each other that one may be able to produce apparently plausible correspondences to 'consciousness' from a variety of loci within information-processing theories. One should therefore be suspicious of mechanistic correspondences for 'consciousness' which are based on broad information-processing theories and a limited range of phenomenological generalizations.

The other general conclusion is that it may not be possible to produce a satisfactory correspondence for all the phenomenological general-izations that can be advanced with the operation of a single component within an information-processing model. Yet this would appear to be the explicit or implicit aim of most previous attempts to explain 'consciousness' from an information-processing or functionalist approach.

There seem to be a number of ways one could respond to this problem. One could, for instance, argue that in concentrating on the functional difference—within the total system—of the operations performed by contention scheduling and the Supervisory System that a similarity between their functions may be being overlooked. It may be possible to view the operation of the Supervisory System as equivalent to the activation of high-level general purpose schemas which in turn can 'call' or activate lower-level specific-purpose schemas, with greater interference existing both within the set of higher level schemas and within the set of lower level ones than between the two sets. The conscious contents of focal attention could then correspond to the strongly acti-vated schemas and their bound variables, a related position to that held in my 1972 paper (Shallice 1972). In its overall computational function the subsystems to which consciousness relates would be similar to Johnson-Laird's (1983) operating-system, and from a neuroscience perspective there would be some links with the position taken by Kinsbourne (1988, this volume).

Even if such a model were appropriate in information-processing terms, the correspondence suggested for consciousness may still be inadequate. Some of the times when our awareness corresponds—in some sense—to high-level control functions the process seems to be one which can be quite sharply defined in time. When one has a feeling of knowing something, or realizes that one half-recognizes a face or that one has made an error, the experience does seem as specific in time as when one, say, sees a light come on. Yet when one is 'in doubt' or is 'planning to do something' or 'miserable' it seems inappropriate to view the experience as corresponding to the state of the cognitive system over a few hundred milliseconds. Yet that is what would be entailed in simply extending the general correspondence approach to experiences related to states of the Supervisory System. An experience has a referent which has—implicitly—a certain time span. If the time spans differ between experiences then will not, too, the processes or states being 'known' and maybe, too, the processes doing the 'knowing'?

A second possibility to which this argument appears to lead is to divide the concept of consciousness into a number of subvarieties, as Oatley (1988, this volume) has suggested, and model each of them through the operation of separate mechanisms. Thus Oatley's Helmholtzian consciousness could correspond to inputs activating schemas in contention scheduling, his Woolfian consciousness to schemas *selected* in contention scheduling and their bound variables, and the Vygotsky/Mead types to the operation of the Supervisory System. Yet is it an accident or an artefact that the same term 'consciousness' is applied to them all?

A third approach would also accept that the attempt to construct a correspondence between consciousness and the operation of *one* information-processing subsystem is misguided. However, simply subdividing the concept into different subvarieties which could be separately modelled would also be rejected. Instead it would build on the way that any mental operation which gives rise to awareness necessarily involves a number of different subsystems. Assume that the human information-processing system is so constructed that four different types of general purpose control subsystem exist. These would be two main ones: contention scheduling and the Supervisory System, and two subsidiary ones: the language system and the episodic memory process (see Fig. 14.2). Quite often, as with the selection of a very overlearned schema in contention scheduling, an operation in one control system would have no counterparts in the others. However, very frequently when one of these subsystems is operating its operation would be complementary to the operation of one or more of the others; they would be operating on representations which relate to the same event or mental operation and contain information from the same schemas. In these situations, there would be a coherent pattern of control over all *other* subsystems which is shared between those control subsystems that are active. Might not this coherent shared control be the basis for

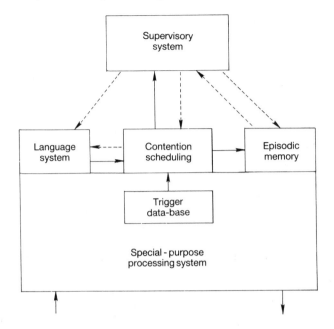

Fig. 14.2. Suggested flow of information between control systems. Where 'boxes' are contiguous they represent separate processes, but do not necessarily involve physically distinct structures. For instance, the 'episodic memory' unit need not be totally physically distinct from the special-purpose processing system. It represents the particular processes by which event-specific traces (engrams) are laid down and retrieved. Information transfer is represented by a solid line if it is obligatory and by a dotted one when optional. The flow from episodic memory to the Supervisory System is optional when retrieval is voluntary, obligatory when it is spontaneous.

'consciousness'? Its 'contents', in so far as they could be precisely characterized, would correspond to the information flow between the control subsystems and, in turn, to the flow of information and control between them and all the rest of the cognitive system.

It would not be meaningful to specify for any cycle of control subsystem operations which part of the cycle corresponded to consciousness. Its contents would correspond to the information that flowed through them, maybe undergoing transformation in the process. It would not be possible to say that in some sense one was or one was not aware of the processes (or their results) occurring in control subsystem 1 *before* those occurring in control subsystem 2. So property 8b would follow: one is frequently not aware of the thought in any clear sense before the utterance occurs. On the rare occasions when the operation of the four control systems was not complementary, again it should be difficult to apply the concept 'consciousness' clearly. The difficulty in describing what one experiences in dual-task situations (see the discussion of property 3) would be an example. Another would be an action lapse of the 'capture error' variety (e.g. see Reason 1979; Norman 1981). It can be argued that this type of lapse is the consequence of 'capture' in contention scheduling by a triggering input incompatible with the requirements of the Supervisory System (see Norman and Shallice 1980). To ask whether the person in the middle of an action lapse is conscious of what they are doing leads to a paradoxical conclusion. In some sense they are, but in another they are not.

Why might the content of consciousness correspond to the information flow through one cycle of control subsystem operations? A system of the complexity of the one underlying human cognition needs not only to have access to certain of its own internal states but also the capacity to identify and even to characterize them abstractly. Assume, for instance, that a system with these capacities has a functional architecture consisting of a number of high-level processors which directly or indirectly have access to the output of a large set of other processors, or can modulate their operation. The low-level processors, however, each communicates with only a limited set of their fellows. Assume, further, that there is considerable overlap between the low-level operations (and their results) to which each of the set of high-level processors can have access at any particular time, and also considerable overlap between those operations to which each does not have access, but that the overlap relationship is not one of complete consistency across central processors. There are, let us assume, situations where communication is occurring between a lower level process and some, but not all, of the central processors; in addition, no specific central processor is specifically privileged in this respect. There would then be a strong family resemblance between the lower level operations and their results to which higher level 'access' was available at any particular time, and also the processors which have access would come from a small subset. Then it would be economic for the system when characterizing itself to have a concept which captured these two family resemblances, and this concept would correspond to something real about the system. There should then be a family resemblance between the different senses of the system's 'everyday language' use of the concept—call it 'conscious*'—as indeed is widely accepted about the concept 'conscious' in our use of the term (e.g. Wisdom 1963). Further, the operations which corresponded to 'conscious*' in information-processing terms would also need to be related by means of a family resemblance. There would be no single processing correspondence for all the generalizations the system itself could derive about 'conscious*'. Moreover, if it tried to resolve the paradox of why a single processing correspondence did not exist for 'conscious*', then different apparently conflicting ideas for overcoming the problem might well be different ways of viewing a more complex whole. Perhaps 'conscious' is in fact 'conscious*'.

Acknowledgements

I should like to thank Daniel Dennett, Nick Maxwell, and the editors for helpful comments on an earlier draft of this paper.

References

Ach, N. (1905). *Ueber die Willenstätigheit und das Denken*. Vardenhoek, Göttingen.

Anscombe, G. E. M. (1957). *Intention*. Blackwell Scientific Publications, Oxford.

Atkinson, R. C. and Shiffrin, R. M. (1971). The control of short-term memory. *Scientific American*, **224**, 82–90.

Baars, B. J. (1983). Conscious contents provide the nervous system with coherent global information. In *Consciousness and self regulation*, vol. 3, (ed. R. J. Davidson, G. E. Schwartz, and D. Shapiro), Plenum, New York.

Baars, B. J. and Mattson, M. E. (1981). Consciousness and intention: a framework and some evidence. *Cognition and Brain Theory*, **4**, 247–63.

Baddeley, A. D. (1976). *The psychology of memory*. Harper and Row, New York.

Baddeley, A. D. and Hitch, G. (1974). Working memory. In *The psychology of learning and motivation*, vol. 8, (ed. G. H. Bower), pp. 47–89, Academic Press, New York.

Block, N. (1980). Troubles with functionalism. In *Readings in the philosophy of psychology*, vol. 1, (ed. N. Block) Harvard University Press.

Boden, M. A. (1977). *Artificial intelligence and natural man*. Basic Books, New York.

Boden, M. A. (1981), *Minds and mechanisms: philosophical psychology and computational models*. Harvester Press, Brighton, Sussex.

Burnham, J. C. (1968). The new psychology: from narcissism to social control. In *Change and continuity in twentieth century America: the 1920s*, (ed. J. Braeman, R. Bremner, and D. Brody), Ohio University Press.

Chomsky, N. (1972). Psychology and ideology. *Cognition*, **1**, 11–46.

Chorover, S. L. (1979). *From Genesis to genocide*. MIT Press, Cambridge, MA.

Churchland, P. S. (1988). Reduction and the neurobiological basis of consciousness. In *Consciousness in contemporary science*, (ed. A. J. Marcel and E. Bisiach), p. 273. Oxford University Press.

Crowder, R. G. (1982). The demise of short-term memory. *Acta Psychologica*, **50**, 291–323.

Dennett, D. C. (1969). *Content and consciousness*. Routledge and Kegan Paul, Andover, Hants.

Dennett, D. C. (1978). *Brainstorms*. MIT Press, Cambridge, MA.

Dennett, D. C. (1988). Quining qualia. In *Consciousness in contemporary science*, (ed. A. J. Marcel and E. Bisiach), p. 42. Oxford University Press.

Downey, J. E. and Anderson, J. E. (1915). Automatic writing. *American Journal of Psychology*, **26**, 161–95.

Erdelyi, M. H. (1988). Hypermnesia and insight. In *Consciousness in contemporary science*, (ed. A. J. Marcel and E. Bisiach), p. 200. Oxford University Press.

Ericsson, K. A. and Simon, H. A. (1985). *Protocol analysis: verbal reports as data*. MIT Press, Cambridge, MA.

Evans, C. O. (1970). *The subject of consciousness*. George Allen and Unwin, Hemel Hempstead, Herts.

Feyerabend, P. (1975). *Against method: outlines of an anarchistic theory of knowledge*. New Left Books, London.

Fisk, A. R. and Schneider, W. (1984). Memory as a function of attention, level of processing and automatization. *Journal of Experimental Psychology: Learning, Memory and Cognition*, **10**, 181–97.

Fodor, J. A. (1983). *The modularity of mind*. MIT Press, Cambridge, MA.

Gazzaniga, M. S. (1988). Brain modularity: towards a philosophy of conscious experience. In *Consciousness in contemporary science*, (ed. A. J. Marcel and E. Bisiach), p. 218. Oxford University Press.

Hamilton, Sir W. (1859). *Lectures on metaphysics and logic*, vol. 1. Blackwood, Edinburgh.

Hasher, L. and Zachs, R. T. (1979). Automatic and effortful processes in memory. *Journal of Experimental Psychology*, **108**, 356–88.

Herrigel, E. (1972). *Zen in the art of archery*. Routledge and Kegan Paul, Andover, Hants.

Holender, D. (1986). Semantic activation without conscious identification. *Behavioral and Brain Sciences*, **9**, 1–66.

Humphrey, G. (1951). *Thinking: an introduction to its experimental psychology*. Methuen, Andover, Hants.

James, W. (1890). *The principles of psychology*. Holt, Rinehart and Winston, New York.

Jevons, W. S. (1871). The power of numerical discrimination. *Nature*, **3**, 281–2.

Johnson-Laird, P. N. (1988). A computational analysis of consciousness. In *Consciousness in contemporary science*, (ed. A. J. Marcel and E. Bisiach), p. 357. Oxford University Press.

Kaufman, E. L., Lord, M. W., Reese, T. W., and Volkmann, J. (1949). The discrimination of visual number. *American Journal of Psychology*, **62**, 498–525.

Kinsbourne, M. (1988). Integrated field theory of consciousness. In *Consciousness in contemporary science*, (ed. A. J. Marcel and E. Bisiach), p. 239. Oxford University, Press.

Kosslyn, S. M., Reiser, B. J., Farah, M. J., and Fliegel, S. L. (1983). Generating visual images: units and relations. *Journal Experimental Psychology (General)*, **112**, 278–303.

Libet, B. G., Leason, C. A., Wright, E. W., and Pearl, D. K. (1983). Time of conscious intention to act in relation to onset of cerebral activity (readiness-potential). *Brain*, **106**, 623–42.

Luria, A. R. (1966). *Higher cortical functions in man*. Tavistock Publications, Andover, Hants.

Mackay, D. M. (1966). Cerebral organisation and the conscious control of action. In *Brain and conscious experience*, (ed. J. C. Eccles), Springer, New York.

Mandler, G. (1975), *Mind and emotion*. John Wiley, New York.

Mandler, G. (1985). *Cognitive psychology: an essay in cognitive science*. Erlbaum, Hillsdale, NJ.

Mandler, G. (1989). Memory: conscious and unconscious. In *Memory—an interdisciplinary approach*, (ed. P. R. Solomon, G. R. Goethals, C. M. Kelley, and B. R. Stephens), Springer, New York.

Mandler, G. and Shebo, B. J. (1982). Subitizing: an analysis of its component processes. *Journal of Experimental Psychology*, **111**, 1–22.

Marcel, A. J. (1983). Conscious and unconscious perception: an approach to the relations between phenomenal experience and perceptual processes. *Cognitive Psychology*, **15**, 238–300.

Marshall, J. C. and Morton, J. (1978). On the mechanics of Emma. In *The child's conception of language*, (ed. A. Sinclair, R. J. Jarvella, and W. J. M. Levelt), pp. 225–39, Springer, Berlin.

Merleau-Ponty, M. (1962). *Phenomenology of perception*. Humanities Press, New York.

Newell, A. and Simon, H. A. (1972). *Human problem solving*. Prentice-Hall, Englewood Cliffs, NJ.

Norman, D. A. (1981). Categorisation of action slips. *Psychological Review*, **88**, 1–15.

Norman, D. A. (1986). Reflections on cognition and parallel distributed processing. In *Parallel distributed processing*, vol. 2, (ed. J. L. McClelland and D. E. Rumelhart), pp. 531–52, MIT Press, Cambridge, MA.

Norman, D. A. and Shallice, T. (1980). Attention to action: willed and automatic control of behaviour. *Center for Human Information Processing Technical Report* No. 99. Reprinted in revised form in *Consciousness and self regulation*, vol. 4, (ed. R. J. Davidson, G. E. Schwartz, and D. Shapiro), pp. 1–18, Plenum, New York, 1986.

Oatley, K. (1988). On changing one's mind: a possible function of consciousness. In *Consciousness in contemporary science*, (ed. A. J. Marcel and E. Bisiach), p. 369. Oxford University Press.

O'Keefe, J. (1985). Is consciousness the gateway to the hippocampal cognitive map? A speculative essay on the neural basis of mind. In *Brain and mind*, (ed. D. Oakley), Methuen, Andover, Hants.

Posner, M. I. and Klein, R. M. (1973). On the functions of consciousness. In *Attention and performance IV*, (ed. S. Kornblum), pp. 21–35. Academic Press, New York.

Posner, M. I. and Snyder, C. (1975). Facilitation and inhibition in the processing of signals. In *Attention and performance V*, (ed. P. M. A. Rabbitt and S. Dornic), pp. 669–82, Academic Press, London.

Puccetti, R. (1973). Brain bisection and personal identity. *British Journal of the Philosophy of Science*, **24**, 339–55.

Puccetti, R. and Dykes, R. W. (1978). Sensory cortex and the mind-brain problem. *Behavioral and Brain Sciences*, **3**, 337–75.

Reason, J. T. (1979). Actions not as planned. In *Aspects of consciousness, Vol. 1.*, (ed. G. Underwood and R. Stevens), pp. 67–89, Academic Press, London.

Reed, G. (1972). *The psychology of anomalous experience*. Hutchinson, London.

Ribot, T. H. (1890). *The psychology of attention*. Longmans & Co., London.

Rumelhart, D. E. and Ortony, A. (1977). The representation of knowledge in memory. In *Schooling and the acquisition of knowledge*, (ed. R. C. Anderson, R. J. Spiro, and W. E. Montague), Erlbaum, Hillsdale, NJ.

Searle, J. R. (1983). *Intentionality: an essay in the philosophy of mind.* Cambridge University Press.

Shallice, T. (1972). Dual functions of consciousness. *Psychological Review*, **79**, 383–93.

Shallice, T. (1978). The dominant action system: an information-processing approach to consciousness. In *The stream of consciousness: scientific investigations into the flow of human experience*, (ed. K. S. Pope and J. L. Singer), pp. 117–57, Plenum, New York.

Shallice, T. (1982). Specific impairments of planning. *Philosophical Transactions of the Royal Society of London*, **B**, **298**, 199–209.

Shallice, T. (1984). More functionally isolable subsystems but fewer 'modules'? *Cognition*, **17**, 243–52.

Shallice, T. and Evans, M. E. (1978). The involvement of the frontal lobes in cognitive estimation. *Cortex*, **14**, 294–303.

Shallice, T., Mcleod, P. D., and Lewis, K. (1985). Isolating cognitive modules with the dual-task paradigm: are speech perception and production separate processes? *Quarterly Journal of Experimental Psychology*, **37A**, 507–32.

Shiffrin, R. M. and Schneider, W. (1977). Controlled and automatic human information processing II. Perceptual learning, automatic attending, and a general theory. *Psychological Review*, **84**, 127–90.

Solomons, L. and Stein, G. (1895). Normal motor automatism. *Psychological Review*, **3**, 492–512.

Spelke, E., Hirst, W., and Neisser, U. (1976). Skills of divided attention. *Cognition*, **4**, 215–30.

Sperling, G. and Speelman, R. G. (1970). Acoustic similarity and auditory short-term memory. In *Models of human memory*, (ed. D. A. Norman), Academic Press, New York.

Sussman, G. J. (1975). *A computational model of skill acquisition.* American Elsevier, New York.

Tulving, E. (1972). Episodic and semantic memory. In *Organisation of memory*, (ed. E. Tulving and W. Donaldson), pp. 381–403, Academic Press, New York.

Tulving, E. (1983). *Elements of episodic memory.* Oxford University Press.

Tulving, E., *et al.* (1984). Multiple book review of *Elements of episodic memory, Behavioral and Brain Sciences*, **7**, 223–68.

Weiskrantz, L. (1977). Trying to bridge some neuropsychological gaps between monkey and man. *British Journal of Psychology*, **68**, 431–45.

Weiskrantz, L. (1986), *Blindsight*. Clarendon Press, Oxford.

Weiskrantz, L. (1988). Some contributions of neuropsychology of vision and memory to the problem of consciousness. In *Consciousness in contemporary science*, (ed. A. J. Marcel and E. Bisiach), p. 183. Oxford University Press.

Weiskrantz, L. and Warrington, E. K. (1979). Conditioning in amnesic patients. *Neuropsychologia*, **17**, 187–94.

Weiskrantz, L., Warrington, E. K., Sanders, M. D., Marshall, J. (1974). Visual capacity of the hemianopic field following a restricted occipital ablation. *Brain*, **97**, 709–28.

Wisdom, J. (1963). *Problems of mind and matter.* Cambridge University Press.

Woodworth, R. S. (1938). *Experimental psychology.* Holt, Rinehart & Winston, New York.

15

The control operations of consciousness

Carlo Umiltà

Introduction

As made clear in the title, this chapter is about *control*. Control is clearly not the only aspect of consciousness (for other aspects, see, besides the other chapters contained in this volume, Humphrey 1983; Klatzky 1984; Mandler 1984; Rychlak 1986; Yates 1985). But the reason for stressing it when dealing with the concept of consciousness in scientific theories is that, in my opinion, this is the aspect of consciousness which is most likely to have a causal role. Therefore it ought to be included in any scientific account of human behaviour. Whereas other aspects of consciousness may be epiphenomenal, the specific conscious experience that accompanies control is not epiphenomenal and has very important effects on human information processing. It is true that one can think of an organism, or even a machine for that matter, which exhibits control without being conscious, but that does not seem to be the way control is achieved in human beings [for a criticism of the causal role of consciousness, however, see Bisiach (1988, this volume); also see Gregory (1988, this volume) for an interesting discussion of the issue].

The role played by consciousness is that of allowing some voluntary organizational and regulatory influence over the myriad of unconscious operations that go on in our mind at any given moment. More specifically, the point I want to make is that voluntary control over cognitive processes *depends on* the phenomenal experience of being conscious. It is the very fact that we are aware, have phenomenal experience, of some cognitive processes that can bring them under voluntary control.

A somewhat similar position has recently been expressed by Libet (1985; also see Sperry 1986) for the conscious control of movement. Besides not being limited to the control of movement, the thesis advocated in this chapter differs from that espoused by Libet in other respects. First, apparently Libet believes that there are no cerebral processes for conscious control. Although I will not discuss this point here, I am inclined to believe that there must exist a cerebral basis for the mechanisms that produce voluntary control. Second, I do not intend to confine the role of consciousness to that of exerting a sort of 'stop-or-go' control over operations that begin unconsciously. I do agree that one of the most important functions (perhaps, *the* most important function) of consciousness is that of blocking ongoing cognitive processes. However, it seems to me that, at this stage, it would be mistaken to rule out the possibility that consciousness can also play a role in the voluntary initiation of some cognitive processes.

In the following pages, I will try to buttress the notion that control depends on consciousness in three ways. First, it will be shown that this notion is implicit in a common-sense account of consciousness. Then it will be shown that the literature supports the notion of a central processor (or control system, or operating system: these terms will be used interchangeably from now on), which is in charge of the control operations and whose activities are conscious. Finally, I will attempt to provide empirical evidence in favour of the thesis by showing that conscious control has an influence on information processing.

There are a few points that, for sake of clarity, need to be stated at this early stage, even though some of them will be discussed at some length later. The first concerns the distinction between the phenomenological and information-processing concepts of consciousness (for a comprehensive discussion, see Johnson-Laird 1988; Marcel 1988; and Shallice 1988; all in this volume). From an information-processing point of view consciousness is a mechanism (namely, the central processor), whereas from a phenomenological point of view consciousness means the experiential equivalents of the contents and operations of that mechanism. Since this distinction is relevant for the intents of this chapter, the two meanings of consciousness will be kept separated, with the exception of the next section, in which to be too rigorous would be premature. Also, the term attention can be employed either in the sense of an information-processing mechanism or in the sense of the experiential equivalents that accompany its activities. Since, however, this distinction is less important here, the two senses of the term will often be conflated. Although attention is no doubt strictly related to awareness, consciousness and attention are not identical. In fact, I wish to argue that one of the most important control operations of consciousness is that of allocating attention to some cognitive processes at the expense of others. That is, the idea is that consciousness *uses* attention to exert its control functions.

The second point concerns the fact that attention can have two meanings, from a purely information-processing point of view. In a sense to attend is to select for a special type of processing, which perhaps takes place in a memory system called 'working memory' (e.g. see

Klatzky 1984). However, several students of attention (e.g. Kahneman 1973; Wickens 1984) postulate that to attend is to commit a certain amount of limited resources to some cognitive processes (but, for a criticism of the concept of resources, see Navon 1984, 1985). Here the terms attention and attentional resources will be used interchangeably, because both concepts are compatible with the notion that consciousness can privilege some cognitive processes, either by selecting them for access to working memory or by allocating to them resources that are limited.

The third point is the distinction between cognitive processes and their products, or, put in another way, between mental representations (products) and the mental operations (processes) that operate on them. Both mental representations and processes may or may not be conscious, and it can well be that, by being conscious, mental representations become qualitatively different (Marcel 1983*b*, 1988, this volume). There can be little doubt, however, that consciousness is crucial in the case of processes, because conscious mental processes can be submitted to direct control, whereas unconscious ones cannot.

The fourth point concerns the relationships between consciousness and self-awareness. I believe that these two concepts should not be confused. In fact, a distinction can be made, at least in principle, between the mental representations or processes of which one is conscious (i.e. consciousness) and the state of being conscious of some mental representations or processes (i.e. self-awareness). In a certain sense, consciousness is a phenomenal experience that refers to the state of an object (i.e. an internal representation), whereas self-awareness refers to the state of an agent.

A common-sense account of consciousness illustrated through an apologue

According to a legend reported in the writings of Cicero, Livy, and Plutarch, during the siege of Syracuse (212 B C) a Roman soldier, angry at not getting a reply, killed Archimedes who was busy solving a problem of geometry. This sad story can be taken as a starting point for discussing some issues concerning consciousness.

First, one wonders why there was no reply to the soldier's question. Presumably, it was because Archimedes was paying attention to the problem, of whose features he was aware, and not to the soldier, of whose presence he was not aware. If this is the reason why Archimedes failed to answer that fateful question, one can argue that consciousness is accessed through attention. It would also seem that attention is able to handle a limited amount of information only, or that consciousness has a limited capacity, or both. At any rate, we can be aware of only a few events at a time.

A second interesting issue is how Archimedes's life could have been saved. It is likely that, if the soldier had spoken very loudly or if he had uttered the scientist's name, there would have been a reply, even though Archimedes was apparently deeply involved in solving the problem and was paying attention to nothing else. It would thus seem that consciousness can also be accessed automatically, without the deployment of attention. Notice that in this case the product of the perceptual process becomes conscious, whereas the process itself is not conscious and need not be so in order to be fully effective.

Archimedes, however, could have taken a more active part in getting out of trouble. Under the circumstances (the Roman army was in the process of conquering Syracuse after a long siege) being approached by bad-tempered soldiers was only to be expected. This being the case, a sensible thing to do on Archimedes's part would have been to look out of the corner of his eye, thus devoting some attentional resources to detecting soldiers and some to solving problems. Had he succeeded, it is likely that the soldier would not have gone undetected, even if it remains to be seen whether Archimedes could have dealt effectively with the problem. In this case we have a conscious intention (i.e. being ready to detect soldiers) which makes it more probable that the product of perceptual processing (i.e. the soldier) will access consciousness. In the example above (when the soldier is supposed to have called out the scientist's name) only the product of the perceptual process would have been conscious.

Archimedes could have had a still more active attitude by consciously deciding to stop poring over his drawings and to have a look around once in a while. In other words, he could have allocated his attentional resources sometimes to the problem and sometimes to the outside world. By an appropriate timing of such shifts of attention, he could have made sure of discovering any soldier. In this case the state of affairs in Archimedes's head becomes fairly complex. First of all, the above strategy requires a mental state that can be termed 'self-awareness'. That is, one has to be aware that those cognitive processes that are presently active occur in one's mind and are volitional, in the sense that one can have some influence on them. Second, there needs to be a control system, whose contents and operations are conscious, which allows the voluntary control of the goals and time courses of at least some of the cognitive processes. It is important to realize that prominent among the functions of the control system is that of bringing cognitive processes to a halt. If it lacked such capability, and all cognitive processes were ballistic (that is, they could not be stopped once started), there would be no basis for a strategy like the one suggested above. In our story, Archimedes could not have intentionally stopped thinking of the problem to look around. Assuming that he had adopted the attention-shifting strategy, we can be reasonably sure that the product of his perceptual process, assisted by attention, would have reached consciousness and the soldier's question would have got a polite answer. In this example the happy end requires the intervention of consciousness at three levels: self-awareness activating the control system; the control system bringing about a conscious strategy to regulate cognitive processes; the strategy causing attention to be assigned in timely fashion to the perceptual process, whose product would have thus a ready access to consciousness.

At this point it might be useful to summarize the main features of the three scenarios capable of saving Archimedes's life. Notice that in every one of them the important outcome is the same: Archimedes becomes aware of being questioned by an angry soldier and the soldier's question gets access to Archimedes's consciousness. In the first scenario (when the scientist's name is called aloud) consciousness is accessed automatically because no conscious cognitive process concerning the soldier takes place before the stage of becoming aware of being questioned. Archimedes's behaviour follows directly from stimulus presentation.

In the second scenario (when he tries to maintain two independent goals for his cognitive processes) his behaviour is still purely stimulus-driven. The access of the relevant information to consciousness is, however, facilitated by the foreknowledge (a conscious mental state) that a Roman soldier is likely to come into sight. Here we have a stimulus-driven response that is rendered more probable because attentional resources have been intentionally allocated to the relevant cognitive process. In some sense one might say that here the behaviour is not volitional, but it is not unintentional either because it is guided by a conscious goal.

In the third scenario (when Archimedes adopts an attention-shifting strategy) his behaviour becomes fully volitional. The relevant information gets a ready access to consciousness since he is self-aware and because of this attains conscious strategic control over those cognitive processes that are aimed at the intended goal.

From the foregoing apologue the following propositions have emerged that deserve closer examination.

A. Consciousness can be identified with the phenomenal experience of the contents and operations of a limited capacity system, which, to some extent, controls cognitive processes.

B. The control system exercises its functions either directly by consciously initiating and stopping cognitive processes (that is, by affecting their time courses) or indirectly by setting specific conscious goals to them.

C. The control system works by selectively allocating attentional resources to some mental representations and cognitive processes, which become conscious, at the expense of others, which remain unconscious.

D. Even if the control system can regulate the inflow of information, some events have automatic and privileged access to it and, therefore, can become conscious without the mediation of attention.

E. Self-awareness is a necessary condition for the control system to become fully operational.

Supporting the common-sense account of consciousness through a review of the literature

Having established a framework for a discussion of the control aspect of consciousness, I will consider briefly in this section the pertinent literature. In the next section I will present some empirical evidence from our laboratory, which seems to be relevant to some of the issues discussed here.

Proposition A

Several authors have proposed that it is the activities of a 'central processor' which give rise to consciousness (see Mandler 1975; Posner and Boies 1971; Shallice 1972; but for dissenting opinions see Allport 1980a, 1988, this volume; Bisiach 1988, this volume). Actually, Shallice (1972; also see 1978) did not speak of a specialized central processor whose activities are conscious. Instead he first suggested that consciousness is the result of an action system becoming 'dominant', and, subsequently, maintained that it originated from an action or thought schema being selected through a special purpose mechanism, called 'contention scheduling', whose operations are in turn modulated by a supervisory system (Shallice 1988, this volume; Norman and Shallice 1986). The central processor has also been called the operating system (Johnson-Laird 1983), the executive system (Logan and Cowan 1984), and, as already said, the supervisory system (Norman and Shallice 1986), to name just a few of the terms used. At times special emphasis has been given to some of its features by pointing to its limited capacity (Posner 1978; Posner and Boies 1971), to its role as a device for determining goals (Shallice 1972, 1978), or to its serial mode of information processing (Johnson-Laird 1983; Klatzky 1984; Mandler 1975; Norman and Shallice 1986). In any event, it is assumed that more cognitive processing goes on in our mind than we can be aware of and that consciousness is limited to the contents of the central processor, that is, to the strategical control operations and the products of some cognitive processes (Bagnara 1984; Carr 1979; Johnson-Laird 1983, 1988, this volume; Klatzky 1984; Mandler 1984; Posner 1978; Underwood 1982; Weiskrantz 1988, this volume; but for various kinds of correspondences between consciousness and structures or processes see, in this volume, Bisiach 1988; Marcel 1988; Shallice 1988). The vast majority of cognitive processes instead take place in parallel, outside the limited capacity central processor, and do not give rise to the phenomenon of consciousness (however, for a critical view of the distinction between automatic unconscious processing and controlled conscious processing, see Kahneman and Triesman 1984; White 1982).

There are two reasons why the central processor cannot monitor all mental operations. One is that lower order mental operations occur in parallel and are simply too many to be handled by the central processor, which operates in series through the use of limited capacity attentional mechanisms (Johnson-Laird 1983; Posner 1978). The other is that the central processor is slow acting and it takes time for it to summon the attentional mechanisms; thus the very fast lower order mental operations escape its control (Neely 1977; Posner 1980a; Underwood 1982).

In brief, the notion that seems to emerge is that of consciousness as the experiential equivalent of a central processor, which governs the multitude of unconscious lower order mental operations, without, however, having complete control over them. This is of course reminiscent of the 'homunculus', namely the little man in the head, which is in charge of every operation we are unable to explain in a more mechanistic way. The central processor is no doubt a version of the homunculus. The central processor perhaps becomes preferable to the previous versions of the homunculus if we can be specific about the processes that do not require the intervention of the homunculus, thus leaving for it a limited number of well-defined functions (Attneave 1960; Bagnara 1984).

Proposition B

The human mind comprises two qualitatively different operating modes, usually referred to as automatic and controlled processing

[Jonides *et al.* 1985; Hasher and Zacks 1979; Logan 1978; Posner 1978; Posner and Snyder 1975; Schneider *et al.* 1984; Shiffrin and Schneider 1977; Underwood 1982; but see again Kahneman and Treisman (1984) for a somewhat contrary opinion, and White (1982) for a cautionary note concerning the association of control with consciousness]. Automatic processing is fast, parallel, relatively immune to interference, and fairly effortless, and it is not limited by short-term memory capacity and perhaps cannot be stopped once it begins. Controlled processing is slow, serial, effortful, capacity-limited, subject to interference, and can be stopped.

By definition, controlled processing is under direct conscious control of the central processor, which forms intentions and issues commands to realize those intentions. The role of the central processor can be viewed as that of yielding a strategical organization of cognitive processes by a series of acts of control (Carr 1979; Ceci and Howe 1982; La Berge *et al.* 1977; Logan 1978, 1979, 1983; Logan and Cowan 1984; Logan *et al.* 1983). That is not to say that in the case of controlled processing the central processor has detailed access to the inner workings of mental operations, but that it exercises a strategical organizing influence by initiating and stopping sequences of such operations (Logan 1978, 1979, 1985a, 1985b; Logan and Cowan 1984; Osman *et al.* 1986; Zbrodoff and Logan 1986). In this regard it should be kept in mind that, even though inhibition is not likely to be the only act of control in the central processor's repertoire, it certainly is a most important one; after all, cognitive processes can be put into an organized sequence by stopping or slowing down some of them.

The role of the central processor is less apparent in the case of automatic processing, which, by definition, should not be under direct conscious control. If it were true that automatic processing is purely ballistic (Logan 1983), then any direct control by the central processor would be denied. However, Logan and Cowan (1984) have proposed a distinction between an early stage of motor response execution that can be inhibited and a subsequent ballistic stage that cannot be stopped once it has been started. If this distinction can be extended to other cognitive processes (Logan 1985b; Osman *et al.* 1986; Zbrodoff and Logan 1986), that would indicate a way for the central processor to exercise some direct control over automatic processing.

Even if the volitional organization of mental operations is largely limited to controlled processing, the central processor can exercise strategical, though indirect, control over automatic processing by the conscious knowledge of the particular goal that is to be attained. In other words, as suggested by Allport (1980b; also see Keele and Neil 1978), when a goal-state becomes active in the central processor (that is, phenomenologically, when a goal-state becomes conscious), an auto-matic cognitive process is triggered whenever the appropriate environ-mental conditions arise. This is similar to Shallice's (1972, 1978) notion according to which indirect control is exercised through conscious activation and inhibition of goal-states (action systems in his termin-ology), which, in turn, modulate the likelihood of automatic processes being initiated (also see Norman and Shallice 1986).

In summary, the central processor can influence both controlled and automatic processing, but does so in different ways. In the case of controlled processing, it affects directly the temporal sequence of men-tal operations; in the case of automatic processing, control is the indirect consequence of a conscious goal-state. The central processor, in addition, might play a more direct role in assembling and sequencing automatic processes if it is true that their early stages (i.e. those occurring before the processing becomes ballistic) can be consciously brought to a halt.

Propositions C and D

Attention is closely associated with consciousness, and some authors even seem to consider the two terms to be synonymous (e.g. Klatzky 1984; Logan and Cowan 1984; Schneider *et al.* 1984). No doubt, the tendency to conflate the two terms originates from the observation that the central processor, whose activities are thought to be conscious, and attention (in the sense of either a structure or a process) have many features in common. Both are capacity limited and slow acting, the contents of both are active in working memory, both show a serial processing mode and require general mental resources to operate, both are instrumental for monitoring and issuing commands to lower order independent processors, both intervene in planning and decision-making. However, there are good reasons for keeping the two concepts separated.

First, even if attention plays an important role in allowing infor-mation to enter the central processor (that is, phenomenologically, consciousness) it is not always necessary and there are cases in which the central processor is accessed as bypassing attentional mechanisms (see the reviews in Ceci and Howe 1982; Posner 1978, 1980*a*; Schneider *et al.* 1984; Underwood 1982). Such 'interrupts' are not to be confounded with those instances or events which can influence behaviour without attention being directed to them and without them becoming conscious [see the reviews in Dixon 1971, 1981; Marcel 1983*a*, 1983*b*; Underwood 1976, 1978, 1982; but also see Holender (1986) for a dissenting opinion and related commentaries for a comprehensive discussion]. In these latter cases we infer that subjects can, for example, recognize un-attended words (i.e. the unattended words activate the corresponding memory traces) because of their effects upon the responses to other, attended, words. The unattended words, however, do not become conscious. In the former cases (i.e. when an interrupt occurs) the unattended word (for example, the listener's name) instead gains immediate access to the central processor and becomes conscious. Attention seems to be a privileged route for events to enter the central processor and become conscious, but not the only one available. According to Norman and Shallice (1986) it is even possible to be aware of performing an action without paying attention to it. This happens, for example, in the case of well-learned routine actions.

Second, and more importantly, there is clear evidence that we have the power of controlling attention, of deploying it for the purpose of enhancing (or inhibiting) the processing of certain information. For example, we can intentionally direct attention to certain locations in the external world or in our long-term memory (e.g. see the reviews in Ceci and Howe 1982; Posner 1978, 1980a, 1980b; Underwood 1982). The central processor must therefore be in charge of strategically allocating attentional mechanisms to some events and not to others.

The distinction between attention and intention to attend (see Ceci and Howe 1982) is crucial in understanding why attention and con-sciousness must not be confounded. When attention is automatically directed to an event without intention, the product of attending only has access to the central processor and becomes conscious. When attention is intentionally directed to an event, it is the central processor itself that controls the process of attending, and thus the intention to attend is conscious along with the product of attending.

In brief, the central processor, besides rendering conscious those events that, with or without attention, have access to it, exercises strategical control over cognitive processes by selectively allocating attention to them. That means, phenomenologically, that conscious-ness and attention are not the same thing: the former uses the latter to attain strategical control over lower order cognitive processes.

Proposition E

Self-awareness is not just a special form of consciousness. In fact, it is the capacity of self-awareness that underlies the unique subjective experience of being in control of the operations of our mind; that is, that the mind is in control of the mind (Johnson-Laird 1983; White 1982). This is crucial among the activities of the central processor since the awareness of being capable of control is a prerequisite for exercising such control. The feeling that someone else is in control, or that nobody is in control, or even no feeling of control at all, would render any act of control over cognitive processing utterly impossible.

No doubt there are cases in which self-awareness and feeling of control exist independently. In dreams we experience thoughts that we know belong to us, but over which we cannot exert any conscious control. That is true also of pathological thoughts of depressive patients. Similarly, paranoid schizophrenics feel that someone else is in control of their own thoughts. These, however, are examples of self-awareness, in the sense of the term used here, without an accompanying feeling of control, whereas my point is that there cannot be *intentional* control without self-awareness. To have an intention means to have decided to achieve a particular goal, and for that to happen an organism must have two types of awareness: a conscious representation of the goal and the feeling that it *itself* can do something for achieving that goal (Humphrey 1983; Johnson-Laird 1983, 1988, this volume). This point was well argued by Johnson-Laird (1983) in describing an automaton which is able to cope with its environment (also see Broadbent 1977).

The first step toward such an automaton would be a machine that comprises a number of parallel and independent information-processing devices, very similar to the modules proposed by Fodor (1983; also see Shallice 1988, this volume). This machine, however, would not be able to show any degree of organized behaviour. The second step would be to add a central processor that is not on a par with the other modules because it is the only one that can access the response devices and run them according to a wired-in set of motor programmes. Now the machine would display an organized, though stereotyped, behaviour. The situation changes radically if the central processor is rendered capable of control over the activities of the parallel, lower-order modules by means of attentional subroutines. The automaton would show a high degree of flexibility in its behaviour, which would, however, remain essentially stimulus-driven. The feature of consciousness could be fitted into the automaton as a mere epiphenomenon of the activities of the central processor. As Johnson-Laird (1983) has aptly pointed out, in order for the automaton to have intentions (i.e. strategical control), it must be endowed with the crucial component of self-awareness.

Even though ordinary common sense and introspection seem to be in accordance with the idea that self-awareness is distinct from consciousness, the risk inherent in accepting lightheartedly such a notion is evident. A second homunculus, whose functions promise to be even more difficult to specify than those of the central processor, is likely to emerge within the first homunculus (i.e. the central processor itself). What is worse, there is no guarantee that this second homunculus will be the last [this is, of course, the trap of an infinite regress; see Gregory (1988) and Weiskrantz (1988), both in this volume].

Finding empirical evidence in favour of the common-sense account of consciousness

From the foregoing discussion, it appears that the central processor exercises strategic control over the lower-order mental operations in essentially two ways. One is indirect and entails a state of conscious knowledge [a goal-state in Allport's (1980b) terminology, or a dominant action system in Shallice's (1978)], brought about by the activation of the central processor, which in turn modulates the activities of the lower-order processors through attentional facilitation or inhibition. The other is more direct and consists in the construction of conscious strategies by the central processor, which issues commands to stop processing before it becomes ballistic. In our laboratory we conducted a series of experiments that might shed some light on the operations of the central processor.

The two experimental tasks were a *same–different* pattern discrimination and a *right–left* locational discrimination. The stimuli were segments that formed either alphabetical letters or nonsense figures. These patterns were shown, two at a time, to the right or left of a central fixation mark on a cathode ray tube screen driven by a computer. It is important to keep in mind that these same stimuli had shown a very consistent 50–70 msec advantage for *same* responses in a number of previous experiments (Bagnara *et al.* 1982; Simion *et al.* 1988; Taylor 1976).

In experiment 1 a pair of letters was shown for 150 msec and the subjects were instructed to depress as fast as possible the right-hand key on the computer keyboard if the letters appeared on the right side and to depress the left-hand key if they appeared on the left side. No mention whatsoever was made of the fact that the two letters could be the same or different. Not surprisingly, reaction time (RT) to *same* or *different* pairs was nearly identical (289 msec vs. 295 msec, a nonsignificant difference; also see Table 15.1 in which the results of all seven experiments are shown). In experiment 2 the task for which response latencies were measured was again *right–left* discrimination, but this time the subjects (a new group) were instructed also to report verbally, after the manual response, whether the letters were the same or different. The instructions stressed the importance of being fast in pressing the correct key and also put some emphasis on the accuracy of the *same–different* discrimination. However, there was no particular time pressure for

Table 15.1 Mean R Ts (in msec) for the *right–left* locational discrimination (primary task) to pairs of patterns (letters in experiments 1, 2, 5, 6, and 7, nonsense figures in experiments 3 and 4), which could be the same or different. Percentage of errors is shown within parentheses: The first figure refers to the manual response and the second to the verbal response. The stimuli were shown for 150 msec in experiments 1–4 and for 2000 msec in the other experiments. In experiments 2, 4 and 6 the subjects had to report verbally (secondary task) whether the patterns were the same or different. In experiment 7 they had to report verbally whether an arrow (shown 1800 msec after stimulus onset) pointed to the upper or the lower letter. In the other experiments there was no verbal report.

	Response		
Experiment	Same	Different	Mean
1	289 (0.9)	295 (1.0)	292 (1.0)
2	392 (0.9–12.5)	400 (0.3–3.7)	396 (0.6–8.1)
3	310 (0.6)	315 (1.0)	312 (0.8)
4	384 (0.4–11.5)	388 (0.3–0.3)	386 (0.4–6.0)
5	330 (0.3)	337 (0.3)	334 (0.3)
6	453 (0.7–1.5)	451 (0.0–1.7)	452 (0.3–1.6)
7	339 (0.9–1.4)	353 (0.5–1.2)	346 (0.7–1.3)

making the verbal response since the interstimulus interval was 2 sec. Overall response latency showed a significant 104 msec increase in comparison with experiment 1 (392 msec for *same* pairs and 400 msec for *different* pairs; the difference between the two types of response did not reach significance here nor in any of the following experiments).

Experiments 3 and 4 replicated these findings by employing, with two groups of new subjects, nonsense patterns instead of letters. Mean R T was 310 msec for *same* pairs and 315 msec for *different* pairs without verbal report, vs. 384 and 388 msec, respectively, with verbal report (a significant lengthening of 74 msec).

The instructions of experiments 2 and 4 explicitly suggested the strategy of first making the manual response concerning the *right–left* discrimination and then comparing the two patterns for the *same–different* discrimination. It is possible, however, that the subjects did not adopt that strategy due to the rather short stimulus exposure (150 msec). Therefore experiments 5 and 6 exactly replicated experiments 1 and 2 with the exception that the stimuli were shown for 2000 msec (two groups of new subjects participated). Even if now there was plenty of time to adopt, in experiment 6, the strategy of emitting the manual response before initiating the *same–different* comparison, R T was 330 msec for *same* pairs and 337 msec for *different* pairs without verbal report, and increased to 453 msec and 451 msec, respectively, with verbal report. Again, the verbal report brought about a significant increase in response latency of 118 msec.

At this point we wondered whether it could have been the mere fact of having to perform two tasks that had yielded the lengthening of RT. There are some studies that render this a likely interpretation. For example, Mcleod (1977) showed that performing simultaneously two tasks that apparently did not interfere still produced a worsening in speed of response. Even with a totally automatic secondary task, the mere existence of the intention to perform it caused a decrement in the other, primary task; and this was also true when the two tasks did not overlap in time (Mcleod and Posner 1984). In a similar vein, a study by Noble *et al.* (1981) showed that just knowing that one has to do something else later interferes with performance of a first, very simple RT task. To test this possibility, in experiment 7 a new group of subjects executed the *right–left* discrimination to pairs of letters shown for 2000 msec and then had to say whether an arrow, which appeared 1800 msec after stimulus onset, pointed to the upper or lower letter in the pair. In this case the subject knew, as in the previous experiments, that there were two tasks to execute, but nothing could be done about the second until the arrow was shown. RT was 339 msec for *same* pairs and 353 msec for *different* pairs; that is significantly faster than in experiment 6 and not different from experiment 5. Therefore it can be safely concluded that it was not the mere fact of having to perform two tasks that produced the slowing down of responses in experiments 2, 4, and 6. It remains to be seen whether the interpretation I will propose for our results can be extended also to the results of the studies cited above.

In discussing the foregoing results, the first thing to consider is that the central processor could not monitor the elementary mental operations involved in the execution of the two tasks. This is because it can consciously monitor and control the general goals and, at least to a certain extent, the overall organization of the lower-order independent processors, but cannot inspect their internal operations, since only their products have access to consciousness.

That the central processor exercised some sort of general control is attested by the delay brought about by the presence of the goal-state 'compare patterns to decide whether they are same or different'. When this goal-state was active (experiments 2, 4, and 6) response latency was on the average about 100 msec slower than when it was not active (experiments 1, 3, and 5), even though the stimuli were identical and always required a *right–left* discrimination. Stimulus presentation is not a sufficient condition for the process of comparison to be initiated. It is the central processor that triggers the comparison process by the appropriate goal-state. Of course, the goal-state is not a sufficient condition either, as shown by the fact that in the absence of the relevant information the process corresponding to the active goal-state does not begin (see experiment 7, in which the arrow was presented 1800 msec after the onset of the two letters).

Granted that the slowing down of the *right–left* discrimination observed in experiments 2, 4, and 6 was due to the concomitant *same–different* discrimination, it remains to be explained how the interference originated. One possibility is that the *same–different* discrimination was ballistic and thus could not be stopped once it had begun as a consequence of the goal-state. It had to run on to completion before any manual response could be emitted and thus delayed the *right–left* discriminative response. That this was not the case is shown by the absence of a *same–different* effect. If the responses concerning the locational discrimination were emitted after the completion of the comparison process, then the *right–left* responses to *same* pairs should have been faster than those to *different* pairs. This is because it has been shown, by using identical stimuli, that the outcome of the comparison is faster for *same* than for *different* pairs (Bagnara *et al.* 1982; Simion *et al.* 1988; Taylor 1976).

Another possible explanation is that the two processes (the *right–left* and the *same–different* discriminations) began simultaneously and went on to completion in parallel. If they competed for general (i.e. attentional) processing resources, or specific processing resources, or even specific processing mechanisms (e.g. see Kahneman 1973; Wickens 1984), an interference effect would have occurred with a consequent slowing down in RT. No *same–different* effect was found because the *same-different* comparison was still in progress when the faster locational discrimination terminated and the manual response was emitted.

Even though this explanation is plausible, there is indirect evidence against it. As for a capacity interference, it is difficult to figure out why there should be a competition for general resources between two tasks that both possess all the features of automatic processing, according to Schneider *et al.*'s (1984) criteria. No doubt the two tasks were not under step-by-step attentional control and only their products (that is, the two patterns were on the right or left, were same or different) entered consciousness. Of course, a capacity interference would have been very likely if the two decisional stages overlapped in time (e.g. see Keele and Neil 1978). This, however, was clearly not the case because the *same–different* decision could be delayed for about 2 sec. A structural interference at the output stage or a competition for specific motor resources seem to be ruled out because the outputs were different (manual in one case and verbal in the other) and, more importantly, the two responses took place about 2 sec apart. There is still the possibility of a competition for some visual mechanisms or for specific visual resources. I am not aware, however, of any evidence of dual task interference for processing shape and location of a pair of rather simple visual patterns. On the contrary, Duncan (1984) has shown that all the features of a visual pattern (shape and position included) can be processed simultaneously without giving rise to mutual interference.

In my opinion, the most likely explanation is one that makes reference to the role of the central processor in constructing conscious strategies. The delay caused by the verbal report is attributable to the strategy consciously adopted by the subject with the intent of achieving the best organization of the cognitive processes. From this point of view experiment 6 is the crucial one. In it both the instructions and ordinary good sense suggested a precise strategy for enhancing the speed of the *right–left* discrimination; namely, that of first depressing the key corresponding to the side of the stimulus and then processing the letters to decide whether they were the same or different: The subjects knew very well that the stimuli stayed on for at least 1.5 sec after the manual response.

I propose that RT in experiment 6 was about 120 msec slower than in experiment 5 because in the former, but not in the latter, the central processor was involved in organizing a processing sequence aimed at maximizing the speed of response for the locational discrimination. In experiment 5 no such conscious strategy was needed because there was only one task to perform.

The activity of the central processor in the dual-task condition is likely to concern an attempt to bring under control the *same–different* discrimination by inhibiting, stopping, or delaying it. In fact, if both the *right–left* and *same–different* discriminations ran on to completion automatically, no conscious strategy would be possible. A prerequisite of such strategical control by the central processor is the ability to affect the time course of cognitive processes (Logan and Cowan 1984).

Assuming that in experiment 6 (and, perhaps, also in experiments 2 and 4) a stop command was issued by the central processor to inhibit the *same–different* discrimination (which, otherwise, would have occurred, due to the presence of the goal-state and the availability of the relevant information), we have three ways of explaining the lengthening of RT. The first is that the delay is due to the stop-command response latency; that is, the time that elapses between the moment the central processor issues the stop command and when the cognitive process actually stops (Logan and Cowan 1984). This explanation, however, implies that the *right–left* discrimination is somehow inhibited and does not begin until the *same–different* one comes to a halt. If not, the locational discrimination should take place unaffected and with no delay (remember that there is no reason to expect a dual-task interference effect).

The second explanation is that the process of stopping the *same–different* discrimination and that of performing the primary task (i.e. the locational discrimination) are not independent and compete for general or specific processing resources. If this were the case, then there would be fewer resources available to the locational discrimination, hence the slower RT. However, this interpretation is not convincing because it has been demonstrated (Logan and Cowan 1984) that the stop process and the primary task process are in fact independent and there is no interference between them. Of course, it is possible that, whereas stopping the pattern discrimination *per se* did not interfere with the locational discrimination, the interference was produced by the activation of the procedure necessary for bringing the pattern discrimination under control. However, this is very likely to be an aspect of the cognitive organization whose establishment, as argued below, is responsible for the delay.

The slowing down of response latency caused by the verbal report can best be explained if one assumes that the mental operations for organizing cognitive processes require general and/or specific processing resources, as suggested by Logan *et al.* (1983). In other words, there would be substantial costs involved in the construction of a strategy by the central processor. It is the resources expended in strategy construction, not those expended in executing the stop command, which is the product of the strategy, that reduce the resources available to the primary task (i.e. discriminating right from left). This does not contradict the notion that the *right–left* discrimination is automatic and thus does not require processing resources. In fact, there is some evidence that strategy construction mostly requires specific motor resources, besides some general resources as well (Logan *et al.* 1983). It is likely, therefore, that the observed interference took place between the process of constructing a strategy and that of deciding to emit the motor response, both of which occurred at a conscious level and occupied the central processor [see Keele and Neil (1978), and also Libet (1985), who maintains that the consummation of a motor action can depend on a conscious control function].

Martin (1984) has postulated that the execution of two or more concurrent tasks can be achieved through a temporally fine-grained decision process, which in turn necessitates the operation of a fine-timing mechanism. When this fine-timing mechanism is also involved in performing one of the tasks, interference occurs. Martin's notion no doubt bears some resemblance to the explanation I offered for the interference effect discussed here. However, she explicitly states that the competition for the fine-timing mechanism arises only when one of the tasks requires the detection of targets of very short duration (50 msec or less). This was clearly not the case in the present experiments, in which the stimuli stayed on for 2000 msec.

In conclusion, the results of this series of experiments can be interpreted by assuming that a conscious strategy was constructed and its construction had a measurable cost in terms of time. However, this is not the only point under discussion here. Two, perhaps more important, questions have been asked: (a) Is the notion of a central processor necessary for explaining the construction of the strategy? and (b) Is it necessary that the operations of the central processor be conscious? Of course, the thesis advocated in this chapter demands affirmative answers to both questions: the central processor *is* necessary and its operations *must* be conscious.

It seems fair to say that the most convincing way to account for the construction of the strategy is to invoke the activities of a central processor, which exerts an organizing influence over the lower-order cognitive processes. The central processor plans and realizes a specific sequencing of the various processes; namely, it plans and realizes a strategy. If its activities must be conscious in order to be effective, one would expect to find also a precise sequencing of the experiential equivalents of such activities; that is, the subjects should have been aware, first of stopping the *same–different* discrimination (or, more generally, of favouring the *right–left* dicrimination), then of performing the *right–left* discrimination, and, finally, of performing the *same–different* discrimination. However, when questioned, they did not report being clearly aware of the first, crucial stage, even though their phenomenal experience included the decision to perform the locational discrimination first or the pattern discrimination second. In other words, what was clear in their reports was the awareness of having decided *not* to execute the pattern discrimination first. Even if the matter no doubt demands further examination (the subjects were questioned after a certain interval had elapsed after the experiment, and not all of them could be retraced), such subjective reports seem to be compatible with the notion that consciousness accompanied the establishment of at least the general outline of the strategy. Of course, that does not demonstrate that consciousness was *necessary*. However, it is not easy to figure out how the strategy could have been established in the absence of conscious awareness of the intended sequencing of the two tasks. In general, it seems that, for an effective control to occur, at least the following aspects must be conscious: the intention to attain a

Carlo Umiltà

goal, the goal, a rough representation of the environmental conditions, and an outline of the strategy planned to attain the goal [see Marcel (1988, this volume) for a very similar view].

Acknowledgements

Preparation of this chapter and execution of the experiments reported in it were supported by grants from the Consiglio Nazionale delle Ricerche and the Ministero della Pubblica Istruzione. The chapter was written in part while the author was fellow of the Institute for Advanced Studies of the Hebrew University of Jerusalem. The author wishes to thank Sebastiano Bagnara, Roberto Nicoletti, and Francesca Simion for running the experiments and for discussing many of the issues raised in the chapter. He is also indebted to Edoardo Bisiach, Anthony Marcel, and Tim Shallice for helpful comments on a first draft.

References

Allport, D. A. (1980*a*). Patterns and actions: cognitive mechanisms are content specific. In *Cognitive psychology: new directions*, (ed. G. Claxton), pp. 26–64. Routledge and Kegan Paul, Andover, Hants.

Allport, D. A. (1980*b*). Attention and performance. In *Cognitive psychology: new directions* (ed. G. Claxton), pp. 112–53. Routledge & Kegan Paul, Andover, Hants.

Allport, D. A. (1988). What concept of consciousness? In *Consciousness in contemporary science*, (ed. A. J. Marcel and E. Bisiach), p. 159. Oxford University Press.

Attneave, F. (1960). In defense of homunculi. In *Sensory communication*, (ed. W. Rosenblith), pp. 777–82. MIT Press, Cambridge, MA.

Bagnara, S. (1984). *L'attenzione*. Il Mulino, Bologna.

Bagnara, S., Boles, D. B., Simion, F., and Umiltà, C. (1982). Can an analytic/holistic dichotomy explain hemispheric asymmetry? *Cortex*, **18**, 67–78.

Bisiach, E. (1988). The (haunted) brain and consciousness. In *Consciousness in contemporary science*, (ed. A. J. Marcel and E. Bisiach), p. 101. Oxford University Press.

Broadbent, D. E. (1977). Levels, hierarchies, and the locus of control. *Quarterly Journal of Experimental Psychology*, **29**, 181–201.

Carr, T. H. (1979). Consciousness in models of human information processing: primary memory, executive control and input regulation. In *Aspects of consciousness. Volume 1. Psychological issues*, (ed. G. Underwood and R. Stevens), pp. 123–53. Academic Press, London.

Ceci, S. J. and Howe, M. J. A. (1982). Metamemory and effects of intending, attending, and intending to attend. In *Aspects of consciousness. Volume 3.*

Awareness and self-awareness, (ed. G. Underwood), pp. 147–64. Academic Press, London.

Dixon, N. F. (1971). *Subliminal perception: the nature of a controversy.* McGraw-Hill, London.

Dixon, N. F. (1981). *Preconscious processing.* John Wiley, New York.

Duncan, J. (1984). Selective attention and the organization of visual information. *Journal of Experimental Psychology: General,* **113**, 501–17.

Fodor, J. A. (1983). *The modularity of mind.* MIT Press, Cambridge, MA.

Gregory, R. L. (1988). Consciousness in science and philosophy: conscience and con-science. In *Consciousness in contemporary science,* (ed. A. J. Marcel and E. Bisiach), p. 257, Oxford University Press.

Hasher, L. and Zacks, R. T. (1979). Automatic and effortful processes in memory. *Journal of Experimental Psychology: General:* **108**, 356–88.

Holender, D. (1986). Semantic activation without conscious identification in dichotic listening, parafoveal vision, and visual masking: a survey and appraisal. *Behavioral and Brain Sciences,* **9**, 1–66.

Humphrey, N. (1983). *Consciousness regained.* Oxford University Press.

Johnson-Laird, P. N. (1983). *Mental models.* Cambridge University Press.

Johnson-Laird, P. N. (1988). A computational analysis of consciousness. In *Consciousness in contemporary science,* (ed. A. J. Marcel and E. Bisiach), p. 357. Oxford University Press.

Jonides, J., Naveh-Benjamin, M., and Palmer, J. (1985). Assessing automaticity. *Acta Psychologica,* **60**, 157–71.

Kahneman, D. (1973). *Attention and effort.* Prentice-Hall, Englewood Cliffs, NJ.

Kahneman, D. and Treisman, A. M. (1984). Changing views of attention and automaticity. In *Varieties of attention,* (ed. R. Parasuraman and D. R. Davies), pp. 29–61. Academic Press, New York.

Keele, S. W. and Neil, W. T. (1978). Mechanisms of attention. In *Handbook of perception. Volume 9,* (ed. E. L. Carterette and M. P. Friedman), pp. 3–47. Academic Press, New York.

Klatzky, R. L. (1984). *Memory and awareness.* Freeman, New York.

LaBerge, D., Petersen, R. J., and Norden, M. H. (1977). Exploring the limits of cueing. In *Attention and performance VI,* (ed. S. Dornic), pp. 285–306. Erlbaum, Hillsdale, NJ.

Libet, B. (1985). Unconscious cerebral initiative and the role of conscious will in voluntary action. *Behavioral and Brain Sciences,* **8**, 529–66.

Logan, G. D. (1978). Attention in character classification tasks: evidence for the automaticity of component stages. *Journal of Experimental Psychology: General,* **107**, 32–63.

Logan, G. D. (1979). On the use of concurrent memory load to measure attention and automaticity. *Journal of Experimental Psychology: Human Perception and Performance,* **5**, 189–207.

Logan, G. D. (1983). On the ability to inhibit simple thoughts and actions: I. Stop-signal studies of decision and memory. *Journal of Experimental Psychology: Learning, Memory and Cognition,* **9**, 585–606.

Logan, G. D. (1985a). Executive control of thought and action. *Acta Psychologica,* **60**, 193–210.

Logan, G. D. (1985*b*). On the ability to inhibit simple thoughts and actions: II. Stop-signal studies of repetition priming. *Journal of Experimental Psychology: Learning, Memory, and Cognition*, **11**, 675–91.

Logan, G. D. and Cowan, W. B. (1984). On the ability to inhibit thought and action: a theory of an act of control. *Psychological Review*, **91**, 295–327.

Logan, G. D., Zbrodoff, N. J., and Fostey, A. R. W. (1983). Costs and benefits of strategy construction in a speeded discrimination task. *Memory and Cognition*, **11**, 485–93.

Mandler, G. (1975). *Mind and emotion*. John Wiley, New York.

Mandler, G. (1984). *Mind and body*. W. W. Norton, New York.

Marcel, A. J. (1983*a*). Conscious and unconscious perception. Experiments on visual masking and word recognition. *Cognitive Psychology*, **15**, 197–237.

Marcel, A. J. (1983*b*). Conscious and unconscious perception. An approach to the relations between phenomenal experience and perceptual processes. *Cognitive Psychology*, **15**, 238–300.

Marcel, A. J. (1988). Phenomenal experience and functionalism. In *Consciousness in contemporary science*, (ed. A. J. Marcel and E. Bisiach), p. 121. Oxford University Press.

Martin, M. (1984). Attentional constraints in the detection of brief auditory targets: temporally fine-grained processing in speech perception. *Journal of Experimental Psychology: Human Perception and Performance*, **10**, 794–811.

Mcleod, P. (1977). A dual-task response modality effect: support for multi-processor models of attention. *Quarterly Journal of Experimental Psychology*, **29**, 651–67.

Mcleod, P. and Posner, M. I. (1984). Privileged loops from percept to act. In *Attention and performance X*, (ed. H. Bouma and D. G. Bouwhuis), pp. 55–66. Erlbaum, Hillsdale, NJ.

Navon, D. (1984). Resources—a theoretical soup stone? *Psychological Review*, **91**, 216–34.

Navon, D. (1985). Attention division or attention sharing? In *Attention and performance XI*, (ed. M. I. Posner and O. S. M. Marin), pp. 133–46. Erlbaum, Hillsdale, NJ.

Neely, J. H. (1977). Semantic priming and retrieval from lexical memory: roles of inhibitionless spreading activation and limited-capacity attention. *Journal of Experimental Psychology: General*, **106**, 226–54.

Noble, M. E., Sanders, A. F., and Trumbo, D. A. (1981). Concurrence costs in double stimulation tasks. *Acta Psychologica*, **49**, 141–58.

Norman, D. A. and Shallice, T. (1986). Attention to action: willed and automatic control of behavior. In *Consciousness and self regulation: advances in research. Volume 4*, (ed. R. J. Davidson, G. E. Schwartz, and D. Shapiro), pp. 1–18. Plenum Press, New York.

Osman, A., Kornblum, S., and Meyer, D. E. (1986). The point of no return in choice reaction time: controlled and ballistic stages of response preparation. *Journal of Experimental Psychology: Human Perception and Performance*, **12**, 243–58.

Posner, M. I. (1978). *Chronometric explorations of mind.* Erlbaum, Hillsdale, NJ.

Posner, M. I. (1980*a*). Mental chronometry and the problem of consciousness. In *Structure of thought: essays in honour of D. O. Hebb,* (ed. R. Klein and R. Jusczeck), pp. 95–113. Erlbaum, Hillsdale, NJ.

Posner, M. I. (1980*b*). Orienting of attention. *Quarterly Journal of Experimental Psychology,* **32,** 3–25.

Posner, M. I. and Boies, S. J. (1971). Components of attention. *Psychological Review,* **78,** 391–408.

Posner, M. I. and Snyder, C. R. R. (1975). Facilitation and inhibition in processing signals. In *Attention and performance V,* (ed. P. M. A. Rabbitt and S. Dornic), pp. 669–82. Academic Press, New York.

Rychlak, J. F. (1986). The logic of consciousness. *British Journal of Psychology,* **77,** 257–67.

Schneider, W., Dumais, S. T., and Shiffrin, R. M. (1984). Automatic and control processing and attention. In *Varieties of attention,* (ed. R. Parasuraman and D. R. Davies), pp. 1–27. Academic Press, New York.

Shallice, T. (1972). Dual functions of consciousness. *Psychological Review,* **79,** 383–93.

Shallice, T. (1978). The dominant action system: an information processing approach to consciousness. In *The stream of consciousness,* (ed. K. S. Pope and J. L. Singer), pp. 117–57. Plenum Press, New York.

Shallice, T. (1988). Information-processing models of consciousness: possibilities and problems. In *Consciousness in contemporary science,* (ed. A. J. Marcel and E. Bisiach), p. 305. Oxford University Press.

Shiffrin, R. M. and Schneider, W. (1977). Controlled and automatic human information processing: II. Perceptual learning, automatic attending, and a general theory. *Psychological Review,* **84,** 127–90.

Simion, F., Bagnara, S., and Umiltà, C. (1988). Immaginare e vedere: le caratteristiche dei processi di generazione e di confronto delle immagini visive. *Giornale Italiano di Psicologia,* **15,** 491–513.

Sperry, R. W. (1986). The new mentalist paradigm and ultimate concern. *Perspectives in Biology and Medicine,* **29,** 413–22.

Taylor, D. A. (1976). Holistic and analytic processes in the comparison of letters. *Perception and Psychophysics,* **20,** 187–90.

Underwood, G. (1976). Semantic interference from unattended printed words. *British Journal of Psychology,* **67,** 327–38.

Underwood, G. (1978). Attentional selectivity and behavioural control. In *Strategies of information processing,* (ed. G. Underwood), pp. 235–66. Academic Press, New York.

Underwood, G. (1982). Attention and awareness in cognitive and motor skills. In *Aspects of consciousness. Volume 3. Awareness and self-awareness,* (ed. G. Underwood), pp. 111–45. Academic Press, New York.

Weiskrantz, L. (1988). Some contributions of neuropsychology of vision and memory to the problem of consciousness. In *Consciousness in contemporary science,* (ed. A. J. Marcel and E. Bisiach), p. 183. Oxford University Press.

White, P. (1982). Beliefs about conscious experience. In *Aspects of conscious-*

ness. Volume 3. Awareness and self-awareness, (ed. G. Underwood), pp. 111–25. Academic Press, New York.

Wickens, C. D. (1984). Processing resources in attention. In *Varieties of attention,* (ed. R. Parasuraman and D. R. Davies), pp. 63–102. Academic Press, New York.

Yates, J. (1985). The content of awareness is a model of the world. *Psychological Review,* **92**, 249–84.

Zbrodoff, N. J. and Logan, G. D. (1986). On the autonomy of mental processes: a case study of arithmetic. *Journal of Experimental Psychology: General,* **115**, 118–30.

16

A computational analysis of consciousness

P. N. Johnson-Laird

This chapter argues that the problems of consciousness are only likely to be solved by adopting a computational approach and by setting a number of tractable goals for the theory. Four such phenomena need to be explained: the division between conscious states and unconscious mental processes; the relative lack of conscious control over many emotions and behaviours; the unique subjective experience of self-awareness; and that aspect of intentions that is missing from goal-directed computer programs and automata. The chapter outlines a theory of consciousness based on three main assumptions: the computational architecture of the mind consists in a hierarchy of parallel processors; the processor at the top of the hierarchy is the source of conscious experience; this processor—the operating system—has access to a model of itself, and the ability to embed models within models recursively. If the thesis of the chapter is correct, the four problems of consciousness can be solved once we understand what it means for a computer program to have a high-level model of its own operations.

The problems of consciousness

What should a theory of consciousness explain? This is perhaps the first puzzle about consciousness, because unlike, say, the mechanism of inheritance, it is not clear what needs to be accounted for. One might suppose that a theory should explain what consciousness is and how it can have particular objects; that is, the theory should account for the qualitative aspects of the phenomenal experience of awareness. The trouble is that there are no obvious criteria by which to assess such a theory. Indeed, this lack of criteria lent respectability to the behaviouristic doctrine that consciousness is not amenable to scientific investigation; i.e. that it is a myth that the proper study of nerve, muscle, and behaviour will ultimately dispel. A prudent strategy is therefore both to take a different approach to consciousness and to suggest a more tractable set of problems for the theory to solve. My approach here will be to assume that consciousness is a computational matter that depends on how the brain carries out certain computations, not on its physical constitution. I will outline four principal problems that a theory of consciousness should solve. I will then propose a theory of mental architecture—a conjecture about how the mind is organized —and, finally, I will show that it could provide answers to these problems.

1. *The problem of awareness.* When someone speaks to you, you can be aware of the words they utter, you can be aware of the meaning of their remarks, and you can be aware of understanding what they are saying. And if none of this information is available to you, then you can be aware of that, too. Yet there is much that is permanently unavailable to you. You cannot be aware of how you understand the speaker's remarks or of the form in which their meaning is represented in your mind. In fact, you can never be completely conscious of how you exercise any mental skill. A theory of consciousness must explain how it is possible for there to be this division between conscious states and unconscious processes.

2. *The problem of control.* You cannot consciously control all of your feelings. You can feign happiness and sadness, but it is difficult, if not impossible, for you to evoke these emotions merely by a conscious decision: your best hope is to use a strategy, such as thinking of a particular situation that evoked the relevant emotion, but even this method may fail. Conversely, a particular feeling may overwhelm you despite all your efforts to resist it. This lack of control extends, of course, to behaviour. You may, for instance, intend to give up smoking but be unable to stick to your intention. Some individuals can exert a tight control on themselves and on their expressions of emotion; others, as Oscar Wilde said of himself, can resist everything except temptation. There do indeed seem to be differences in will-power from one individual to another, though the topic is almost taboo in cognitive psychology.

3. *The problem of self-awareness.* You can be aware of the task that you are carrying out and be so absorbed in it as to forget yourself. Alternatively, you can be self-aware and be conscious of what you yourself are currently doing. Sometimes, perhaps, you can even be aware that you are aware that you are aware ... and so on. Self-awareness is, of course, essential for a sense of one's integrity, continuity, and individuality.

4. *The problem of intentions.* To act intentionally is at the very least to decide to do something (for some reason or to achieve some goal) and then in consequence to do it. There are many computer programs, however, that are governed by internally generated goals; for example, the programming language PLANNER enables programs to be written that set up goals, and that then seek to achieve those goals by simulating the action of a non-deterministic automaton (see Hewitt 1972). These programs may provide a reasonable account of unconscious intentions —goals that influence your behaviour without your conscious realization, but the programs do not have your capacity to formulate conscious intentions, because they have no awareness of what they are doing (see Marcel 1988, this volume). A theory of consciousness should elucidate the component of intentional behaviour that is missing from such programs. Metacognition is similarly missing from programs: you can, for example, reason about how you carry out some task, such as reasoning itself, and then use the results of such cogitations to modify your performance. Programs can carry out many tasks but they do not yet reason about their own performance. As we shall see, self-awareness, intentions, and metacognition all appear to call for the same computational mechanism.

Psychologists and others have, of course, proposed theories of consciousness. They have tried to account for it in terms of the evolution of more complex brains (e.g. John 1976), or of more complex behaviours culminating in linguistic communication and social relations (e.g. Mead 1934). But consciousness is unlikely to be merely a consequence of more neurones with more connections between them; it almost certainly depends on how the neurons are connected and the nature of the computations they carry out. Likewise, if language and society could have evolved without consciousness, why should they need it, and how would they be able to awaken our slumbering minds? Psychologists have also identified consciousness with the contents of a limited capacity processing mechanism (Posner and Boies 1971), with a device that determines what actions to take and what goals to seek (Shallice 1972), and with a particular mode of information processing that affects the mental structures governing actions (Mandler 1975). These claims are plausible, but they were not addressed to the four problems above, and they do not solve them. They account for some of the characteristics of consciousness, but they might well apply to a device such as a computer running a PLANNER-like program. My aim is to sketch a computational approach to consciousness that may lead to solutions to the unsolved problems of awareness, control, self-awareness, and intentional behaviour.

Hierarchical parallel processing

From the simple nerve networks of coelenterates to the intricacies of the human brain, there appears to be a uniform computational principle: asynchronous parallel processing. That mental processes occur in parallel is also borne out by the fact that, for example, language is organized at different levels—speech sounds, morphemes, sentences, and discourse—and processed at these levels contemporaneously. There are good reasons to suppose that one processor in the parallel system cannot directly modify the internal instructions of another, because such interactions—even if they were physically possible— would produce highly unstable and unpredictable consequences. A more plausible form of interaction relies solely on the communication of messages between the processors. These messages may take the form of predictions, constraints on processing, the results of computations, emergency signals, and other such interrupts.

The most general conception of a system of parallel processing is of a set of finite-state automata that have channels between them for communicating data, and that operate serially and according to the asynchronous principle that each processor starts to compute as soon as it receives the data that it needs. Other parallel systems are special cases of this design; for example, 'connectionist' systems in which only information about level of activation is passed from one processor to another (e.g. see Anderson and Hinton 1981; Rumelhart and McClelland 1986), systems in which all the processors are synchronized by reference to an internal clock (see Kung 1980), and vector machines in which all the processors carry out the same procedure (see Kozdrowicki and Theis 1980). It is important to keep in mind the distinction between a function and an algorithm for computing that function, because there are infinitely many different algorithms for computing any computable function (e.g. see Rogers 1967). Moreover, although any function that can be computed in parallel can be computed by a serial device, there are many algorithms that run on parallel computers that cannot run on serial ones. Hence if consciousness depends on the computations of the nervous system, then it is likely to be a property of the algorithms that are used to carry out those computations rather than a property of their results (Johnson-Laird 1983a): it ain't what you do, it's the way that you do it! One unfortunate consequence of this consideration is the lack of any characteristic hallmark in behaviour indicating that an organism is conscious. Some psychologists might at this point abandon the study of consciousness on the grounds that accounts of it are unlikely to have testable consequences in behaviour. Such empirical pedantry may be premature: the development of computational accounts may lead to other forms of testing.

There are some problems—the parsing of certain abstract languages, for example—that can be shown to be solvable in principle but not in practice: they are to computation what Malthus's doctrine of population growth is to civilization. A problem is inherently intractable when any algorithm for it takes a time that grows exponentially with the size of the input (e.g. see Hopcroft and Ullman 1979). For an input of n, where n is small, an algorithm may be feasible, but even if the time it takes is proportional, say, to 2^n, then, because such exponentials increase at so great a rate with an increase of n, a computer the size of the universe operating at the speed of light would take billions of years to compute an output for a relatively modest input. Parallel processing is of no avail for rendering such problems tractable. What it does is to speed up the execution of algorithms that take only a time proportional to a polynomial of the size of the input. If many processors compute in parallel, they can divide up the task between them whenever there are no dependencies between the computations. Such a division of labour not only speeds up performance, but it also allows several processors to perform the same sub-task so that should one of them fail the effects will not be disastrous, and it enables separate groups of processors to specialize in different sub-tasks. The resulting speed, reliability, and specialization have obvious evolutionary advantages.

But parallel computation has its dangers too. One processor may be waiting for data from a second processor before it can begin to compute, but the second processor may itself be waiting for data from the first. The two processors will thus be locked in a 'deadly embrace' from which neither can escape. Any simple nervous system with a 'wired in' program that behaved in this way would soon be eliminated by natural selection. Higher organisms, however, can develop new programs—they can learn—and therefore there must be mechanisms, other than those of direct selective pressure, to deal with pathological configurations that may arise between the processors. A sensible design is to promote one processor to monitor the operations of others and to override them in the event of deadlocks and other pathological states of affairs. If this design feature is replicated on a large scale, the resulting architecture is an hierarchical system of parallel processors: a high-level processor that monitors lower level processors, which in turn monitor the processors at a still lower level, and so on down to the lowest level of processors governing sensory and motor interactions with the world. A hierarchical organization of the nervous system has indeed been urged by neuroscientists from Hughlings Jackson to H. J. Jerison (for the history of this idea see Oatley 1978), and Simon (1969) has argued independently that it is an essential feature of intelligent organisms.

The operating system

In a simple hierarchical computational device, the highest level of processing could consist of an operating system. The operating system of a digital computer is a suite of programs that allows a human operator to control the computer. There are instructions that enable the operator to recover a program stored on a magnetic disk, to compile it, to run it, to print out its source code, and so on. When the computer is switched on, its resident monitor is arranged to load the operating system either automatically or as a result of some simple instructions. The notion that the mind has an operating system verges, as we shall see, on the paradoxical, but it has some relatively straightforward consequences. The operating system must have considerable autonomy, though it must also be responsive to demands from other processors. It must be switched on and off by the mechanisms controlling sleep, though other processors continue to function. It must depend on a second level of processors for passing down more detailed control instructions to still lower levels, and for passing up interpreted sensory information. Doubtless, there are interactions between processors at the same or different levels, and facilities that allow priority messages from a lower level to interrupt computations at a higher level. However, the operating system does not have complete control over the performance of the hierarchy. In particular, as Oatley and I have argued (Oatley and Johnson-Laird 1987), emotional signals constitute a separate channel of communication in the hierarchy and set processors into characteristic modes that predispose the organism towards particular classes of behaviour. Conflicts within the hierarchy between different emotional modes are resolved, not by the operating system alone, but by some general architectural principle such as lateral inhibition.

The hierarchy of communicating parallel processors imposes one great virtue on the operating system: it can be relatively simple, because it does not need to be concerned with the detailed implementation of the instructions that it sends to lower level processors. It specifies what they have to do (e.g. to walk, to think, to talk) but not how they are to carry out the computations that underlie these tasks. It receives information from the lower processors about the results of computations, but not about how they were obtained. Thus vision makes explicit to the operating system *what* is *where* in the scene before us: we have little or no access to the sequence of representations that vision must depend on (Marr 1982). The visual world is presented in a way that is as real as the stone that Dr Johnson kicked in order to refute idealism. This phenomenal reality—our direct awareness of things—is a triumph of the adaptive nature of the mind. If we were aware that the visual world is a representation, then we would be more likely to doubt its veridicality and to treat it as something to be pondered over—a potentially fatal debility in the event of danger. Psychology is difficult just because there is an evolutionary advantage in a seemingly direct contact with the world and in hiding the cognitive machinery from consciousness.

Some empirical evidence

Let us take stock of the theory so far. The brain is a parallel computer that is organized hierarchically. Its operating system corresponds to consciousness and it receives only the results of the computations of the rest of the system. Such a system can begin to account for the division between conscious and unconscious processes (the problem of awareness) and it can also allow the lower level processors a degree of autonomy (the problem of control). There are at least three clinical syndromes that corroborate this division. First, there is the phenomenon of 'blindsight' described by Weiskrantz *et al.* (1974). After damage to the visual cortex, certain patients report that they are blind in parts of the visual field, and their blindness is apparently confirmed by clinical tests. Yet, more subtle testing shows that the patients are able to use information from the 'blind' part of the field. It seems that their sight in the affected regions has continued to function but no longer yields an output to the operating system: they see without being conscious of what they see. Second, there are the 'automatisms' that occur after epileptic attacks. In this state, patients seem to function completely without consciousness and without the ability to make high-level decisions. They may be capable of driving a car, for example, but unable to respond correctly to traffic lights (Penfield 1975). Evidently, the attack leads to a dissociation between the operating system and the multiple processors. Third, there are the well-attested cases of hysterical paralysis. Prolonged stress may lead to paralyses that have, unbeknownst to the patient, no neurological explanation. They can often be cured, as the late Lord Adrian showed during World War I, by similarly duping the patient into believing that electrical stimuli will produce a cure (see Adrian and Yealland 1917). Since these patients are not malingering, they provide us with clear examples of a reaction that is outside the knowledge and control of the operating system.

Why is it that the contents of the operating system—more precisely, its working memory—are conscious, but all else in the hierarchy of processors is unconscious? In other words, what is it about the operating system that gives rise to the subjective experience of awareness? Of course, it is logically possible that each processor is fully aware but cannot communicate its awareness to others—an analogous view was defended by William James in order to explain the phenomena of hysteria without having to postulate an unconscious mind. The view that I shall defend, however, is that the operating system's potential capacity for self-awareness is what gives rise to consciousness. Whenever any computational device is able to assess how it itself stands in relation to some state of affairs, it is—according to my hypothesis —conscious of that state of affairs. What needs to be elucidated is therefore how a computational device could assess its own relation to some state of affairs. It is to this problem that I now turn.

Self-awareness and the embedding of models

Reflection on the human capacity for self-reflection leads inevitably to the following observation: you can be aware of yourself. You also understand yourself to some extent, and you understand that you understand yourself, and so on. . . . The idea is central to the subjective experience of consciousness, yet it seems as paradoxical as the conundrum of an inclusive map. [If a large map of England were traced out in accurate detail in the middle of Salisbury Plain, then it should contain a representation of itself within the portion of the map depicting Salisbury Plain (which in turn should contain a representation of itself [which in turn should contain a representation of itself (and so on *ad infinitum*)])]. Such a map is impossible because an infinite regress cannot occur in a physical object. Leibniz dismissed Locke's theory of the mind because there was just such a regress within it. However, a computational procedure for representing a map can easily be contrived to call itself recursively and thus to go on drawing the map within itself on an ever diminishing scale. The procedure could in principle run for ever: the values of the variables, though too small to be physically represented in a drawing, would go on diminishing perpetually.

There is a similar computational solution to the paradox of self-awareness. Ordinarily, when you perceive the world, vision delivers to your operating system a model that makes explicit the locations and identities of the objects in the scene (Marr 1982). The operating system, however, can call on procedures that construct a model that makes explicit that it itself is perceiving the world: the contents of its working memory now contain a model representing it perceiving the particular state of affairs represented by the model constructed by perception. In other words, the visual model is embedded within a *model* of the operating system's current operation. Should you be aware that you are aware that you are perceiving the world, then there is a further embedding: the operating system's working memory contains a model of it perceiving the state of affairs represented by the model of it perceiving the world. Since the hierarchy of embedded models exists simultaneously, the operating system can be aware that it is aware of the world. Granted the limited processing capacity of the operating system and its working memory, there is no danger of an infinite regress.

In self-awareness, there is a need for an element in the model of the current state of affairs to refer to the system itself and to be known so to refer. In the formation of intentions and metacognitions, there is a more complex requirement. To have a conscious intention, for instance, the operating system must elicit a representation of a possible state of affairs, and decide that it itself should act so as to try to bring about that state of affairs. An essential part of this process is precisely an awareness that the system itself is able to make such decisions. The system has to be able to represent the fact that the system can generate a representation of a state of affairs, and decide to work towards bringing it about. At a low level, there is a program (perhaps analogous to a program in PLANNER) that can construct a model of a state of affairs, and act so as to try to achieve it. (That is all that would be necessary for unconscious intentions). But the system can construct a *model* of itself operating at this low level of performance, and it can use this model in the process of making a decision. It can also construct a model of its own performance at this level in turn, and so on . . . to any required degree of embedding. Hence conscious intentions depend on having access to an element, not merely that refers to the system itself, but that represents the specific abilities of the system, and in particular its ability to plan and to act to achieve plans.

Metacognitive abilities similarly depend on access to such *models* of the system's capabilities, predilections, and preferences. You can reason about how you reason because you have access to a model of your reasoning performance. You can even reason about your metacognitive abilities—you think about how you tend, say, to concentrate too much on your past failures in trying to solve problems when you reflect on your reasoning performance. Thus, once again, you have the ability to make recursive embeddings of mental models within mental models. (I have argued elsewhere that this same ability underlies the phenomena of free will; see Johnson–Laird 1988). The recursive aspect of this ability is hardly problematical. The crux of the problem of consciousness resides in the other requirement: the operating system must have a partial model of itself in order to begin the process of recursion underlying intentionality.

No one knows what it means to say that an automaton or computer program has a model of itself. The question has seldom been raised and certainly has yet to be answered. The notion must not be confused with self-description (*pace* Minsky 1968). It is a relatively straightforward matter to devise an automaton that can print out its own description (e.g. see Thatcher 1963). But such an automaton merely advertises its own inner structure in a way that is useful for self-reproduction, and it no more understands that description than a molecule of DNA understands genetics. A program might be devised, as Minsky (1968) argued, to use its self-description to predict its own future behaviour. Human beings, however, certainly do not have access to complete descriptions of themselves—if they did, any psychological problem could be solved by introspection. What is therefore needed is a program that has a *model* of its own high-level capabilities. This model would be necessarily incomplete, according to the present theory, and it might also be slightly inaccurate, but it would none the less be extremely useful. People do indeed know much about their own high-level capabilities: their capacity to perceive, remember, and act; their mastery of this or that intellectual or physical skill; their imaginative and ratiocinative abilities. They obviously have access only to an incomplete model, which contains no information about the inner workings of the web of parallel processors. It is a model of the major options available to the operating system.

Conclusions

The present approach to consciousness depends on putting together the three main components that I have outlined; hierarchical parallel processing, the recursive embedding of models, and the high-level model of the system itself. Self-awareness depends on a recursive embedding of models containing tokens denoting the self so that the different embeddings are accessible in parallel to the operating system. Metacognitions and conscious intentions depend on a recursive embedding of a model of elements of the self within itself, and of course the ability to use the resulting representation in thought.

This approach assumes that human behaviour depends on the computations of the nervous system. The class of procedures that I have invoked are, with the exception of a program that has a high-level model of itself, reasonably well understood. The immediate priority is therefore to attempt to construct such a program. It is often said that the computer is merely the latest in a long line of inventions—wax tablets, clockwork, steam engines, telephone switchboards—that have been taken as metaphors for the brain. What is often overlooked is that no one has yet succeeded in refuting the thesis that any explicit description of an algorithm is computable. If that thesis is true, then all that needs to be discovered is what functions the brain computes and how it computes them. The computer is the last metaphor for the mind.

Acknowledgements

This chapter is a revised version of a paper that appeared in *Cognition and Brain Theory* (Johnson-Laird 1983*b*). I am grateful to the editors of this book for their forbearance and for their helpful suggestions about how the paper could be improved. I am also indebted to Carlo Umiltà and Alan Allport for their constructive criticisms of the earlier version of the paper.

References

Adrian, E. D. and Yealland, L. R. (1919). The treatment of some common war neuroses. *Lancet*, June, 3–24.

Anderson, J. A. and Hinton, G. E. (1981). Models of information processing in the brain. In *Parallel models of associative memory*, (ed. G. E. Hinton and J. A. Anderson), pp. 9–48. Lawrence Erlbaum Associates, Hillsdale, NJ.

Hewitt, C. (1972). *Description and theoretical analysis (using schemata) of PLANNER*. Memorandum AI TR-258. MIT Artificial Intelligence Laboratory.

Hopcroft, J. E. and Ullman, J. D. (1979). *Formal languages and their relation to automata*. Addison-Wesley, Reading, MA.

John, E. R. (1976). A model of consciousness. In *Consciousness and self-regulation: advances in research, Vol. 1*, (ed. G. E. Schwartz and D. Shapiro), pp. 21–51. John Wiley, Chichester, Sussex.

Johnson-Laird, P. N. (1983*a*). *Mental models*. Cambridge University Press, Cambridge, MA.

Johnson-Laird, P. N. (1983*b*). A computational analysis of consciousness. *Cognition and Brain Theory*, **6**, 499–508.

Johnson-Laird, P. N. (1988). Freedom and constraint in creativity. In *The Nature of Creativity*, (ed. R. J. Sternberg). Cambridge University Press.

Kozdrowicki, E. W. and Theis, D. J. (1980). Second generation of vector super-computers. *Computer*, **13**, 71–83.

Kung, H. T. (1980). The structure of parallel algorithms. In *Advances in computers, Vol. 19*, (ed. M. C. Yovits), pp. 53–73. Academic Press, New York.

Mandler, G. (1975). *Mind and emotion*. John Wiley, New York.

Marcel, A. J. (1988). Phenomenal experience and functionalism. In *Consciousness in contemporary science*, (ed. A. J. Marcel and E. Bisiach), p. 121. Oxford University Press.

Marr, D. (1982). *Vision: a computational investigation in the human representation of visual information.* Freeman, San Francisco.

Mead, G. H. (1934). *Mind, self and society—from the standpoint of a social behaviorist.* (ed. C. W. Morris),. University of Chicago Press.

Minsky, M. L. (1968). Matter, mind, and models. In *Semantic information processing*, (ed. M. L. Minsky), pp. 425–32. MIT Press, Cambridge, MA.

Oatley, K. (1978). *Perceptions and representations: the theoretical bases of brain research and psychology.* Methuen, Andover, Hants.

Oatley, K. and Johnson-Laird, P. N. (1987). Towards a cognitive theory of emotions. *Cognition and Emotion*, **1**, 3–28.

Penfield, W. (1975). *The mystery of mind.* Princeton University Press.

Posner, M. I. and Boies, S. J. (1971). Components of attention. *Psychological Review*, **78**, 391–408.

Rogers, H. (1969). *Theory of recursive functions and effective computability.* McGraw-Hill, New York.

Rumelhart, D. E. and McClelland, J. L. (ed.) (1986). *Parallel distributed processing: explorations in the microstructure of cognition. Vol. 1.: Foundations.* MIT Press, Cambridge, MA.

Shallice, T. (1972). Dual functions of consciousness. *Psychological Review*, **79**, 383–93.

Simon, H. A. (1969). *The sciences of the artificial.* MIT Press, Cambridge, MA.

Thatcher, J. W. (1963). The construction of a self-describing Turing machine. In *Mathematical theory of automata*, Microwave Research Institute Symposia, Vol. 12, (ed. J. Fox), pp. 165–71. Polytechnic Press, Polytechnic Institute of Brooklyn, New York.

Weiskrantz, L., Warrington, E. K., Sanders, M. D., and Marshall, J. (1974). Visual capacity in the hemianopic field following a restricted occipital ablation. *Brain*, **97**, 709–28.

17

On changing one's mind: a possible function of consciousness

Keith Oatley

It is perfectly true, as philosophers say, that life must be understood backwards. But they forget the other proposition, that it must be lived forwards
(Kierkegaard 1843, in Dru 1938, p. 127).

Introduction

In cognitive science an important question about any mental process is:
'What function does it serve?' The answer needs to be at a level of
explanation appropriate to the function. Johnson-Laird (1983) has
proposed that functionalism is the hallmark of cognitive psychology. In
this chapter I will tentatively suggest some steps towards answering the
question: 'What kind of function might consciousness have in mental
life?'

To illustrate: one of the best known pieces of work in biology to
embody functionalism was Harvey's proposal (see 1963 edn.) that the
heart is a pump. In cognitive science, it would no longer be regarded as
adequate to explain data from an experiment in verbal learning in terms
of a short-term memory for verbal material with a capacity of around
seven items. An important question is what function such a store might
have in mental life. From the way some kinds of explanation of memory
are offered, one might think that the answer was that we humans had
been equipped with a device to hold telephone numbers while dialling
them. A satisfactory hypothesis would be an account in terms of a
function that might underlie our ability to plan complex tasks, hold
conversations, etc.

My contribution to this volume is to discuss the function of
consciousness within a broad framework of cognitive science and
phenomenology, and to relate these two approaches to each other.

From the standpoint of cognitive science, the difficulty with the idea
of consciousness is this. With the advent of computer models of mental

processes, and the role of computation as the language of choice for theories in cognitive psychology, our understanding of the role of consciousness has not become clearer. Indeed, the more we understand mental processes in computational terms, the less need there seems for consciousness. For all of the cognitive mechanisms we understand in terms of a programmed instantiation, we understand them without having to postulate consciousness. It is not clear what function consciousness might have.

The data that have prompted a renewed interest in consciousness have tended to be, as Bisiach noted in his proposal for the conference on which this volume is based, phenomena such as blindsight, perception without awareness, unilateral neglect, and so on. These data challenge explanations that assume awareness for mental processes like perception. They provide another route to the conclusion that some processes, that might be thought to depend on consciousness, can take place within mechanisms that need not be conscious at all.

A further implication of these data is that the cognitive system has a structure that is somewhat modular. Some conscious part, or parts, do not know what other parts are doing. To put this another way, communication among mental processes is incomplete. This hypothesis of a modular cognitive system is also supported by phenomena such as hysterical dissociations, difficulties in controlling oneself in dieting or giving up smoking, and the involuntary intrusions into consciousness of upsetting thoughts and emotions despite our conscious will (see Johnson-Laird 1988, this volume; Oatley and Johnson-Laird 1987). This dissociability of mental processes is highly relevant to the question of the functions of consciousness, and will be discussed in more detail later.

I suggest that until we address questions of function, and of the content of consciousness, neuropsychological data on consciousness will remain tantalizing and difficult to understand. Functions are suggested by some of the computational approaches to understanding mind, and by other, more phenomenological approaches.

I will propose a tentative taxononomy, partly of the typical phenomenology and content of consciousness and partly of the types of process that may underly these, with a view to taking some steps towards answering the question of function.

There are at least four important aspects of consciousness that may be distinguished. I will call them Helmholtz's, Woolf's, Vygotsky's and Mead's concepts of consciousness.

Four aspects of consciousness

Helmholtzian consciousness

Helmholtz (see Southall 1925) argued that conscious experiences are the conclusions of inference processes. His paradigmatic example was of seeing light when we press with a finger on the corner of a closed eyelid (p. 27).

Helmholtz referred to seeing as the psychic act of drawing conclusions from analogy (p. 4). The act itself takes place unconsciously. Even though we know that the receptors are not being stimulated by light when we press on our eyelids, when retinal receptors are stimulated we continue to experience light, not pressure on them.

In general, the underlying processes of vision, of thinking, of memory, of speech, and so on are not conscious. We are conscious only of conclusions delivered by the processes.

Helmholtz's idea is extended by Craik's (1943) proposal that to think is to draw conclusions from a model of some aspect of the world. Marcel (1983) has argued that certain perceptual processes which can affect action and judgement remain unconscious. Only conclusions based on constructive schemata which produce models of the Craikian kind, and match them to data, become conscious [e.g. see Oatley (1978) for an account of computational theories of such processes]. This type of hypothesis explains a range of phenomena of consciousness and unconsciousness (Oatley 1981). I will refer to Craikian conclusions when these are drawn from a constructive process, and Helmholtzian consciousness when we are consciously aware of such conclusions. These are not the same: many cognitive conclusions are drawn that we do not experience consciously.

Helmholtz's theory is silent on the subject of why conscious experience has the phenomenal quality it does. Merely he declares that certain percepts and other mental conclusions may be delivered to experience, and that experiencing is a particular psychological state differentiable from those other states and processes that are forever unconscious.

What his idea suggests is that there may be reasons for certain mental phenomena to be under a unified directorate, rather than being under distributed, parallel control. This kind of hypothesis makes some sense of why certain mental processes take place without awareness, while others can be conscious. Conscious awareness may occur, as Marcel (1983) has argued, when a constructive act occurs, when a schema constructs a mental model of some aspect of the world and functions to match stored and sensory data, synthesising information from different domains.

I propose the term 'Helmholtzian' for consciousness of conclusions from a constructive process driven primarily by sensory data. Sometimes there is some voluntary control over perceptual processes, as when a verbal prompt allows a particular percept to form. In general, though, Helmholtzian awareness is a perceptual awareness of something sensory, with a minimum of voluntary control.

Woolfian consciousness

A second aspect of consciousness is often now referred to in cognitive psychology as imagery. It is not clear who should deserve the eponym for this: James (1892) referred to the 'stream of consciousness', relating it to attention and coining a metaphor that has become famous. Freud (1901; see 1975 edn., p. 46) directed a friend to say 'candidly and uncritically whatever comes into . . . mind', and free association of a series of images and ideas coming into the mind, in an involuntary fashion, has become a basic procedure of psychoanalytic therapy. Some novelists including Proust and Joyce have replaced the telling of stories by streams of consciousness.

I will call this aspect Woolfian, partly because the novelist Virginia Woolf succeeded, in my view, better than other writers in portraying it, and partly because, as I shall argue, it is not to be characterized just as process, in a way that might be imagined from James's (1892) account, but as content. Here are a few lines from Woolf's *Mrs Dalloway* (1925). The protagonist, a not especially likeable society woman, is walking up London's Bond Street, looking in shop windows, noticing how all they displayed was just one roll of tweed, a few pearls, just one salmon on an ice-block . . .

'That is all' she repeated, pausing for a moment at the window of a glove shop where, before the War, you could buy almost perfect gloves. And her old Uncle William used to say a lady is known by her shoes and gloves. He had turned on his bed one morning in the middle of the War. He had said 'I have had enough'. Gloves and shoes; she had a passion for gloves, but her own daughter, her Elizabeth, cared not a straw for either of them (p. 19).

And Mrs Dalloway continues up Bond Street to a flower shop, with her thoughts turning towards her daughter's passions for her dog and for the governess whom Mrs Dalloway cordially dislikes.

Woolf portrays the intermingling of the perceptual (objects seen in shop windows), the deliberate (walking along and pausing on the way to buy flowers), the remembered (the gloves to be bought in the shop before the War), the associative (what her uncle had thought about gloves, the way her passion for gloves and being judged to be a lady contrasted with her daughter's indifference to them, as well as subsidiary associative threads such as her saying to herself 'That is all', then her uncle saying 'I have had enough'), and the pre-occupying (her dislike of her daughter's governess). This is the stream of consciousness, a succession of images and ideas each rising semi-involuntarily to the surface of the mental current.

Imagery is part of this stream, as is the contemplation involved in appreciating or creating literature and other art, and in doing science, philosophy, or other kinds of scholarship. Croce (e.g. 1917) put this idea well when he identified history as a form of art guided by evidence, although the events themselves were no longer open to direct perception, and when he proposed that to do history is to relive historical events in one's own mind and to make them one's own.

A characteristic of Woolfian consciousness is that schemata generate conscious verbal, visual, spatial, and other conclusions, but these conclusions are detached from immediate sensory input. Imagery floats free of immediate sensory control.

As phenomenologists point out, conscious experience, whether guided by perceptual processes or prompted from memory, is always of something: the object being perceived, the image, the remembered event, the problem to be solved, etc. What is typically left out of this account is any strongly implied presence of the person doing the thinking.

Vygotskyan consciousness

An aspect of consciousness in which the thinker's self becomes explicitly present is described by Vygotsky (1978) in discussing observations of children carrying out technical tasks. What Vygotsky calls the 'practical intellect' is the self directing the self in solving problems. He argued that speech substantially assists problem-solving. With the addition of self-talk, problem-solving by humans becomes different from that of other animals.

In early infancy, adult caregivers act as problem solvers for the child, and as actors on the child's behalf. It is by non-verbal gestures and then with a growing repertoire of speech acts that the child addresses these adult actors. But later, speech is directed to the self as problem solver in the same kind of way that it had previously been directed to adults.

Vygotsky gives an account, made by his colleague Levina, of a four-and-a-half-year-old girl trying to get some candy with a stick and two stools. The experimenter's descriptions are in parentheses, and the child's speech in quotation marks.

(Stands on stool, quietly looking, feeling along a shelf with stick.) 'On the stool.' (Glances at experimenter. Puts stick in other hand.) 'Is that really the candy?' (Hesitates.) 'I can get it from that other stool, stand and get it.' (Gets second stool.) 'No, that doesn't get it. I could use the stick.' (Takes stick, knocks at the candy.) 'It will move now.' (Knocks candy.) 'It moved, I couldn't get it with the stool, but the stick worked' (1978, p. 25).

Vygotsky concluded that 'children solve practical tasks with the help of their speech as well as their eyes and hands. This unity of speech and perception ... ultimately produces internalization of the visual field' (1978, p. 26). According to Vygotsky, the internalization of the visual field starts from a unification of faculties during problem-solving. We may suppose that the child imagines the unseen candy on the seen shelf and tells herself to use the second stool and then the stick.

Vygotsky's main point is not about imagery as such, but that human problem-solving abilities depart from those of other primates when the essentially social activity of speaking is added—when we talk to ourselves as we talk to others. He argues that speech 'converges' with the developing practical and technical abilities and that this convergence is the 'most significant moment in the course of intellectual development' (1978, p. 24). It involves being able consciously to re-arrange imagined objects in a simulated spatial mind-world.

It is an attractive idea that speech acts provide a paradigm of directing ourselves, and that talking to oneself as we once talked to others is important, because it gives a clue to a function of consciousness in planning. The flexibility and pragmatic qualities of language are applied to symbolic manipulation, and the idea fits neatly with Craik's (1943) idea that thinking involves the manipulation of mental models. By 'conscious' manipulation of models here, I imply 'voluntary' manipulation, in which voluntariness involves directing certain mental actions by talking to ourselves.

This manipulation is quite different from the way in which the patch of light appears involuntarily, without the possibility of mental manipulation, in Helmholtz's eye-pressing demonstration. It is different too from the semi-involuntary streaming of thoughts and images in the Woolfian mode. There are borderline cases, between the automatically guided and the voluntarily guided, but I want to draw attention to Vygotsky's account because it suggests a distinctively human function of consciousness in voluntarily manipulating schemata.

I do not wish to suggest that inner speech is necessary for all human problem-solving, as Vygotsky did suggest. There is, for instance, the evidence of Furth (1971) on the abilities in problem-solving of deaf children who were very deficient in language skills [see also the discussion of Vygotsky's work by Butterworth (1984)]. What I do mean is that Vygotsky pointed to a phenomenon, the cognitive system communicating with itself. The need for specific communication processes is implied by the modularity of the cognitive system, which in turn is implied by the evidence that the nervous system is somewhat modular. We have become aware of this aspect because children often talk to themselves. In adulthood we are conscious of more tacit direction. In this aspect of consciousness the mind is being used actively by an agent that we can identify as the self.

In normal perception constructive processes occur so that we can experience Helmholtzian conclusions. In Vygotskyan consciousness we can use this mental contructive ability to direct the process to some extent. We experience it then, not as direct perception (driven primarily by sensory input), not just as things coming to mind (perhaps triggered by an associative network), but as a simulation space in which we can plan and try things out.

In an adult (and non-verbal) kind of problem-solving, we can use this ability to infer what something might look like from the other side. The experiments of Shepard and Metzler (1971), and of their followers, seem to imply a function of being able to turn things over in one's mind. This is a kind of active simulation of certain physical properties of, and operations on, a physical object that we know a lot about. The importance of consciousness here might be that, although the inference processes themselves are not conscious, the goals of these processes of mental manipulation can be chosen. We can choose the goal of imagining a house brick, and the goal of rotating it about a particular axis. We can then become aware of the conclusions of running this mental simulation.

Meadean consciousness

The fourth aspect of consciousness I will call Mead's (e.g. 1964). It is distinguished by an explicit consciousness of self, both as director and as an object in consciousness. Like the Vygotskyan variety, this aspect of consciousness can be changed by certain kinds of thinking about it. If Vygotsky's idea of the mind was technological, a kind of direction of the means of production, Mead's was social, the experience of inner debate. Sometimes this debate takes place in anticipatory planning of some important interview; at other times it may occur in retrospective review of an upsetting one. Among the contents of consciousness is represented some version of self or the self's actions, often together with some image of another person. Moreover, Mead postulated that tacit, continuous monitoring goes on whenever we speak: as we talk to others we understand what we are saying because we are also talking to ourselves.

Mead postulated that consciousness is an internalized symbolic representation of something that started off as an external interpersonal relationship. It is a representation of our self in relation with others, where the representation is partly defined in terms of a role specification and social rules.

The psychoanalytic notion of consciousness is of the Meadean kind, i.e. essentially social. This idea has been explored in psychoanalysis by object-relations theorists, e.g. Fairbairn (1952). Moreover, Lacan (1966) supposed that the ego is an illusion of wholeness glimpsed for the first time when a child sees itself in a mirror, and, dimly recognizing that this is itself as others may see him, the child allows the image to knit together the otherwise disjointed bombardment of sensory impressions.

Mead (1964, pp. 146–7) describes how a child may acquire a sense of self:

The child can think about his conduct as good or bad only as he reacts to his own acts in the remembered words of his parents. Until this process has been developed into the abstract process of thought, self-consciousness remains dramatic, and the self which is a fusion of remembered actor and this accompanying chorus is very loosely organized and very clearly social. Later the inner stage changes into the forum and workshop of thought. The features and intonations of the dramatis personae fade out and the emphasis falls upon the meaning of the inner speech, the imagery becomes merely the barely necessary cues. But the mechanism remains social, and at any moment the process may become personal.

Here Mead has laid out the hypothesis that mind does not just emerge as a higher order property of brain function, in the same way that the properties of liquidity emerge from aggregations of molecules of water. Mind, consciousness, and the self are related notions, and are part of the cognitive structure of human social life. They arise in relation to our being born in close human interaction with at least one other person and continuing our lives interacting with others. If the sense of self is primarily social, and if fully human consciousness requires comparisons with this self, there is a serious question as to whether someone (improbably) growing up without human interaction would be conscious in the sense that consciousness is assumed in this book. We do not just spring into existence Athena-like and fully armed as individualized conscious selves who may bump into similarly constituted others in our course through life.

Natural science and social science

The distinctions between these four aspects of consciousness prompt us to reconsider the relationship of natural science to social science. If we are to admit that consciousness includes the rising of images in a stream of loosely associated thoughts, the instruction of oneself in planning, and the dialogue with oneself in the theatre of the mind, we need hypotheses, not only about brain mechanisms, but also about the content of mind.

When we talk of imagery, and particularly of self-consciousness, content becomes fundamental. A principal component of Vygotskyan and Meadean consciousness is the self, its doings, its intentions, its presence as an object of thought. This needs to be understood interpretively rather than as mechanism. The implication is that, in understanding consciousness, a rapprochement is needed between natural science with its concern for mechanism, and human or social science, as described by Dilthey (e.g. 1961), with its concern for interpretation and *Verstehen*. In the rest of this chapter I will attempt a rapproachement of this kind.

The usual assumption about consciousness is that it is an emergent property of the individual nervous system. But starting from this position leads to paradoxical questions about what causal status consciousness might have, and the unsatisfactory conclusion that it might be an epiphenomenon. My proposal is that by examining the contents of consciousness, along with the processes, we might find a better starting position, and hence avoid slipping into the rut of epiphenomenalism.

The core of my proposal is that the communicative and social functions of Vygotskyan and Meadean consciousness may also illuminate the nature and functions of the other aspects of consciousness, and consciousness in general. If the cognitive system is modular, then we may need ways of communicating to a process or structure that we take to be the self. More generally, it seems that in a post-Cartesian world psychology has taken an excessively individualistic stance on matters which more properly are social (Farr 1981). Mead's argument is that consciousness is one such issue.

The word 'consciousness' is itself quite modern. Heaton (1985) has pointed out that it is related to the term 'conscience' and etymologically connotes knowing something with others. Consciousness, then, does not mean simply knowing something—otherwise we would just say, when we saw an X, 'There is an X'. We can also say 'I see an X'. The implication of the prefix '*con*' (with) may imply a correct folk theory embodied in the English language, that the term is relational, relating 'I' to what I see.

Consciousness requires a model of self

John Donne (1628; see Simpson and Potter 1956, p. 225) put the issue nicely when he said that 'the beast does but know, but the man knows that he knows'. To be conscious involves both knowing, and knowing that we know; i.e. it implies a relation to some sense of self.

It seems likely that Vygotskyan and Meadean abilities have evolved from abilities to construct Craikian models (for this evolutionary argument, see Oatley 1985). It may be, however, that although the *mechanisms* have evolved in this way, what we experience as consciousness, the phenomenology of explicit knowing, and knowing that we know, derives from the *socially* derived experience of the sense of self as director and as part of the comparison processes of consciousness. If so, the phenomenology would be affected by, and derivative from, this sense. [Humphrey (1983) and Marcel (1988, this volume) make a similar case for consciousness being derived socially and involving a sense of self.]

Mead's conception of consciousness as involving a model of oneself involves also the postulate of Helmholtzian conclusions. Knowing that one knows must itself be a Helmholtzian conclusion, but now made within a model of oneself rather than within a model of some aspect of the physical world.

Johnson-Laird (1983, and 1988 this volume) has proposed that, in computational terms, consciousness requires that a cognitive system has an operating system with a recursive model of itself. But although the idea of having a model of oneself has entered cognitive science (e.g. Minsky 1968), Johnson-Laird (1988, this volume) points out that the question of what it means for a system to have a model of itself is almost entirely unexplored in cognitive theory. The question is made more difficult by the problem of how a system could possibly derive a model of itself.

Mead's proposal supplies a way of understanding how the human cognitive system might be able to create a model of itself in a way that neither seems paradoxical nor implies infinite regression through a series of more and more shadowy homunculi.

The model of self, according to Mead's proposal, is not a homunculus. It is not derived directly. Nor does the system just have, somehow, the ability to define itself recursively. Nor is only the bodily self glimpsed in a physical mirror as Lacan (1966) implies. Self is, as Cooley (1902) says, a 'looking-glass self'; but the looking glass is the mirror of others' actions and reactions to us. We acquire the data on which we could construct a model of self at least partly from the social world.

The model of self is not an abstract representation of certain properties of the cognitive system, but a model made up from our experience with the physical and social world. According to this idea, self is a relatively concrete distillation and internalization of our perceptions of physical effects of our actions and of the social impressions of us that people have verbally and non-verbally conveyed back to us.

If consciousness of one's self in relation to others is primary, then only secondarily do we start to conceive Helmholtzian perception and Woolfian thoughts as 'conscious'. When we do, it is by analogy with those reflections upon ourselves in relation to others. The sense of self can occur in Helmholtzian and Woolfian consciousness too, but implicitly rather than explicitly.

A consciousness involving all four of the aspects described above is the activity of simulating aspects of the world in images and metaphors which we experience in a manner comparable to our experience of percepts. It includes the possibility of manipulating some of these images in a planned Vygotskyan way, and of talking both to ourselves and to imagined others in a mind-world that is entirely analogical and primarily social. It means too that in the types of consciousness in which the self does not seem to be an explicitly active participant, in Helmholtzian perception, or Woolfian imagery, the self is none the less present implicitly, and we can be conscious by virtue of this model of self. Without it we would just see or remember. When we are conscious of seeing, we both see and can notice that we are seeing—perhaps also directing our attention to this aspect or that. When we are conscious of images or memories, the images come to mind and we notice and attend to them, allowing ourselves to be carried along on the stream of consciousness.

We are probably self-conscious for less of the time than we think. Jaynes (1976), who has proposed a set of ideas that are similar to some of those discussed here, though from a different standpoint, has suggested a metaphor. If to be conscious means that we are conscious that we are being conscious, then this would be like asking a torch (in North American, a flashlight) in a dark room if everything is illuminated. From its point of view, when it is switched on, everything is illuminated. It would not be able to know of anything that was not.

Within cognitive science, one reason why the self is necessary for a coherent conception of consciousness is that, without such a postulate, there is no counter to the argument that mechanisms of schematic construction work perfectly well without any form of consciousness whatever. Thus, for instance, model-based programs for visual scene analysis (Roberts 1965; Oatley *et al.* 1988) have properties of schematic construction and matching, but work without benefit of consciousness. The implication is that although we can agree that drawing of Craikian, schema-based conclusions is often accompanied by consciousness, this conception of schematic construction does not imply that consciousness is necessary for these operations.

The suggestion I am making [one that is similar to that proposed also by Johnson-Laird (1988) and by Marcel (1988) (both in this volume)] is that implicitly or explicitly some model of self is necessary for consciousness, in association with the constructive process.

A function for consciousness?

I have suggested ways in which consciousness occurs in mental life, with its implicitly or explicitly accompanying sense of self. But what might its function be? I want to suggest that a function might be in creating new pieces of cognitive structure.

Human intended action takes place in relation to multiple goals, with imperfect knowledge of the environment and often in co-operation with other cognitive agents—other people (e.g. see Oatley 1988). In contrast, even the most sophisticated computer planning-programs have a single main goal, work with perfect, or near perfect, knowledge of a small aspect of their environment, and operate as single agents. The human experience is that plans do not always work out and that, because of our multiple goals, conflicts occur.

What is needed is a cognitive process for rewriting plans, or for rearranging goal priorities, when the unexpected occurs. In short, we need to be able to create not just schematic models, but new pieces of cognitive structure in the form of plans that can be practised and improved, and of goals that will direct such plans.

If communication among mental processes is incomplete, I suggest that successful rewriting needs a unified and centralized direction, in which new pieces of the program can be compared with what is known of the self's abilities and goals. It is the voluntary ability to use the mind as a simulation space to try out possible actions before committing a new piece of plan to action that is hinted at by the Vygotskyan and Meadean types of consciousness. This ability to reconstruct parts of our cognitive system seems to require the kind of phenomenology I have discussed, of an implicit and sometimes explicit comparison with a model of self, and of a direction by a unified system, like an operating system, that has access to this model.

Mead (1964, p. 147) hinted at this function when he referred to those occasions when we become most acutely self-conscious:

As a mere organization of habit the self is not self-conscious. It is this self which we refer to as character. When, however, an essential problem appears, there is some disintegration in this organisation, and different tendencies appear in reflective thought as different voices in conflict with each other. In a sense the old self has disintegrated, and out of the moral process a new self arises.

Inner debate tends to occur when we find ourselves with a problem that is interpersonal; e.g. when someone has not fulfilled an expectation or a desire, and we feel angry, disappointed, sad, etc. We then find ourselves not just aware of this debate, but intruded upon as if the volume of self-conscious dialogue is turned up loud, and the phenomenology of a specifically emotional consciousness occurs.

An 'essential problem' provokes a crisis, where 'crisis' has its older meaning as a point of decision or judgement. The hypothesis is that a primary function of consciousness is to allow some reprogramming on such occasions. Mead means by 'moral process' one that involves essentially social rules (i.e. the social content of the model of self).

According to Mead, the inner debate is a process in which the self can be transformed. This can be like planned technical activity, reminiscent of comparable types of mental work on physical problems. It may occur when we replay an unsatisfactory interchange and imagine how we might have accomplished it better. Essentially this is the process of elaborating a new plan. More profound transformations occur if our model of self needs some change. This involves the re-evaluation of our actions in the eyes of others, the understanding and rearrangement of some of our goal structures.

Sussman's HACKER as a metaphor for planning

Roberts's (1965) program provides an important computational metaphor for our understanding of the process of schematic construction that may underlie Craikian construction (e.g. see Oatley 1978).

Is there a computational metaphor for the rewriting of cognitive structure that I am suggesting is an important function of consciousness? A computational metaphor of the rewriting of plans is provided by Sussman's (1975) program, HACKER. The program rewrites parts of itself as it goes along, and has a special mode in which it tries out the new pieces of program.

HACKER is a program that learns skills of building structures from toy building blocks, and encounters problems of the ordering of its actions. For instance, it can not put one block on another unless the top of that block has first been cleared. Having encountered a mistake in a sequence of actions, HACKER constructs new pieces of program (i.e. plans) prompted by comparisons between intention and outcome. The program embodies a determination to generalize from specific examples of mistakes. It draws on a knowledge of the kinds of bugs (mistakes) that can occur, and the ways in which they may be corrected by patches (new pieces of program to repair the bugs).

An important difference between HACKER and most other programs in artifical intelligence is that, in order to correct bugs and write new pieces of itself, the overall program needs to contain two types of code. One is instructions for hierarchically and temporally ordered actions. These are plans (Miller *et al.* 1960) and they correspond to our usual understanding of programs as directed sequences of actions. But in HACKER there is a second, parallel account, consisting of the goals of these plans. In more ordinary programs this second account only exists as hints in mnemonic names, in the documentation of the program, and in the minds of programmers.

In most computer programs we humans are misled by the programmer choosing names to designate procedures and variables to indicate intentions to us human users. For example, a program operating in the blocks world may contain a procedure called 'cleartop'. The name conveys to humans the intention of the programmer for the procedure. No doubt we understand it by analogy with human intention, and we can even give ourselves the impression of understanding the program in the same way. But the semantics of the name has no meaning for the program.

For an adequate theory of consciousness and selfhood, some equivalent of Sussman's second set of code (i.e. the goals and subgoals) must take on a more explicit meaning, and become part of the program's working structure. These are the goals of the program. Only with these, then, could a system make sense of mismatches between an effect of planned action and the goal which it was supposed to fulfil; between outcome and intention. Only then would it be in a position to understand backwards the effects of its actions. Only then would it be able to write new pieces of program in which patches are introduced, lower level routines integrated more intelligently. Only then could it plan how to move forwards more wisely. In fact HACKER tries out new pieces of program by running them first in so-called 'careful' mode, in which plans are compared line by line with the range of possibly interacting goals and subgoals—a procedure which one suspects Sussman introduced by analogy with conscious reflection, and because he found it necessary for any system that would learn from its mistakes.

According to the arguments of this chapter, even this is not enough: to approach a human consciousness, the program would need also to have a second-level model of its own goals and abilities, as well as an ability to reason about these, and to alter them. Such a program would provide a closer metaphor for the functions of consciousness than programs that can make schematic constructions, important though these are. We need not even begin to discuss whether a cognitive system without a model of self is conscious. Consciousness requires such a model.

Functions of the model of self

We can now ask what function this sense of self might serve in consciousness, and whether consciousness is necessary for writing new pieces of human cognitive structure.

The kind of mental function that would necessitate explicit comparisons with models of the self is that of creating new plans and goals when these have to be compatible with some existing set of goals. The basis for this comparison is our model of self, which would need to have an account of our goals and their relative priorities; let us call this a goal hierarchy. Reprogramming would then mean that new pieces of cognitive structure would need to be compatible with existing goals and priorities, and such that we could identify with any new plans and goals that were introduced. All such modifications would need, of course, to be reflected upon in some kind of 'careful' mode, since they might cause all sorts of unwanted side effects. Any changes would also involve changes in the model of self, which would represent these goals.

During development, a model of self is constructed only gradually. Partly we come to know our own abilities and the effectiveness of certain kinds of plan. These aspects of the model are derived from more and less successful applications of our plans to the world, and our attempts to rewrite them, perhaps in the way that Sussman suggests. Partly the model is derived, as Cooley and Mead suggest, from interactions with others.

It is as if plans are constructed piecemeal, starting early in life, perhaps, in the case of imitations and identifications, without any explicit goals, or perhaps with some goals in disconnection from others and based on mainly local considerations. The generalizability, the interconnections and the degrees of compatibility, among acquired pieces of cognitive structure have to be established later as new contexts occur. We also have no direct access to some aspects of the overall structure of the goal hierarchy.

But as Kierkegaard in his journal of 1843 implied (see Dru 1938), we can begin to understand our actions, and our life, by reflecting backwards. In such reflection we can infer some of our intentions, improve the compatibility of our goals, and become more realistic in our model of self. Thus, by implication, the global structure, and the interrelations among the parts of the cognitive system, become consciously clearer. Perhaps this is a bit like writing a paper, which, in my case at least, requires a good deal of writing, deleting, and re-arranging before it starts to take shape. And then it needs a lot of discussion with others, and further rewriting.

The growth of consciousness described in psychotherapy can best be characterized as the growth of a consciousness of goals and how they interact and of their relation to action. Psychotherapy might be described as a dialogue concerning the nature and repercussions of goals that are held mutually with others; concerning higher level, self-definition goals; concerning dilemmas, traps, and snags (Ryle 1982) that arise as incompatibilities in the hierarchy of goals are revealed by unanticipated features of plans.

The notion of 'meaning' that is central to human science and hermeneutics is a meaning that involves being able to give an account of human action in terms of intention. An intention is a goal that could be reached by a plausible plan to achieve it, and that is consciously compatible with a model of self. Interpretive psychotherapy is a dialogue about intention, and the consciousness that is aspired to is a consciousness in terms of intentions. In psychotherapy, becoming conscious involves adopting, as intentions, goals of which we were not conscious, but which would make sense of our actions, especiaally habitual actions. Such intentions can now become represented in our consciously available model of self, and as such become available to monitor and offer critiques of new plans and actions.

Psychotherapy starts from the proposition that, reflecting backwards on our life, many of us encounter a tale of sound and fury, signifying nothing, perhaps a muddle of actions, some deliberate, some involuntary, some with fortunate outcomes, some that seem in retrospect to be mistaken. The object of therapy is to create a sense of meaning, by taking part in a dialogue that allows actions to be connected with intentions that are conscious.

Marcel (1988, this volume) describes people with neurological damage being able to perform actions for which they can supply a personal intention, though they cannot perform the same action if the goal is supplied from the outside. The phenomena of psychotherapy are the converse of this. People may describe habits and actions for which they have not been able to recognize consciously any intentions. But by recognizing intentions, and consciously adopting them, habits and goals can be brought into communication with each other and represented in a model of self.

Problems prompting cognitive changes

When an 'essential problem' arises, giving rise to inner debate, this debate may compel conscious attention. If it is prolonged, the debate and its emotional accompaniments can reach the level of psychiatric symptoms. When it is an unexpected threatening event that has prompted it, for instance a disruption of a central role relationship (a bereavement, the breaking up of a love affair, becoming unemployed), the symptoms are predominantly of depression (Oatley and Bolton 1985). When the problem is that of an incompatibility of goals, symptoms tend predominantly to be of anxiety or attempted defences against it.

The argument I am proposing is that reflection is indeed appropriate in response to such 'essential problems' that concern the self. Because of the global quality of comparisons with a model of self, implications of such consciously debated problems would almost necessarily be pre-occupying. Their result is to prompt a series of conscious judgements and decisions for new plans to fulfil previous goals, or more radically to prompt new goals and hence acquire a revised model of self.

Conscious reflection seems to involve a mental simulation in careful mode, in which implications of possible new goals, plans, and models can be compared with past and predicted outcomes. Goals which may be local to specific roles, to specific contexts, or to specific mental modules can be examined for their generalizability. But we can never be defini-tive about this, since life is unpredictable. So the process of generaliza-tion takes place gradually throughout adult life. It is given impetus in crises which are described as breakdowns if they are severe, and it is encouraged in various cultural activities such as literature and psychotherapy.

The type of consciousness that attends an 'essential problem' is characteristically turbulent, and is different from the calmer stream of consciousness which I have called Woolfian. Nevertheless, Woolf shows throughout her later novels and indeed in the paragraph follow-ing the passage quoted from *Mrs Dalloway* earlier, that calmer mental currents very often give way to more turbulent ones. Thus, although Mrs Dalloway may be thinking rather complacently of gloves, she is borne by the stream of her thoughts towards the rapids of her repeated pre-occupation with the more problematic interpersonal issue of her dislike of the governess who she feels is subverting her daughter.

A further distinction needs therefore to be made. In Mrs Dalloway's thoughts early in the novel there is no sense that she will in any way change her conception of herself. Instead she elaborates her model of Miss Kilman, the governess. No doubt such elaboration is also part of the function of consciousness, and this function of introspection for understanding others has been described by Humphrey (1983). But the main issue raised by Mead is the possibility of elaborating or changing the model of oneself, a much more serious business. This business can, of course, be defended against in various ways—including ways that prevent anything much coming consciously to mind at all.

The more turbulent stream may surface when there is new planning to be done, either of a technical or an interpersonal kind. It need not be dysphoric. In technical and intellectural activities we often experience it as exciting and creative. Elaborating our models of others, as we do in gossip, or in some kinds of inner debate, can also be enjoyable. It is perhaps only challenge to one's model of self which is experienced as stressful. It occurs when presuppositional goals of the model of self are judged to be untenable.

Connecting action with intention

Understanding life backwards implies that we can remember our actions. But although we often know the lower level intentions, more global goals may, as I have argued, be obscure, and some actions may even appear as involuntary, e.g. as mistakes.

Although behaviour without intention, or with only partial intention, is no doubt caused by some physiological mechanism, only intended actions are experienced as freely willed (Searle 1984) and as essentially human. Though mechanism is of interest for natural scientific study of the brain, action explained by intention becomes the centre of any understanding of mind in terms of human science. Thus for any individual to understand his or her own action means understanding its intentions. More generally, for a researcher in human or social sciences (e.g. in history, sociology, psychoanalysis, etc.) to understand (*Verstehen*) an action, he or she must understand its intentions.

By consciously reflecting on intentions for our actions, we can elaborate our model of self, and make it perhaps more fitting to our social community or more coherent with itself. Although the way this might be achieved is far from clear, it would seem to involve creating new pieces of cognitive structure on the basis of our reflections.

Most radically, in a way that is not currently illuminated by artificial intelligence programs, we humans become highly self-conscious when not just the structure of plans, but some of the presuppositional (and largely social) goals themselves, are challenged. This would be as if Sussman's HACKER, having exhausted its possibilities of correcting bugs and perhaps having assembled a heterogeneous collection of patches, might realize that a more coherent program with different goals might be better.

It may be that such processes are simply uses to which a pre-existing consciousness are put. Alternatively, as I have suggested, the function of consciousness may be to draw mental conclusions from a socially constructed model of self. If so, consciousness may arise only when the self begins to take a part, as attender to an aspect of a scene, as an experiencer of images floating into awareness, in the turning over of technical plans in the mind, or in the backwards reflection on one's life in order to understand how action might be explained by intention and hence to improve the model of self.

Acknowledgements

I have much benefitted from discussions on these matters with Philip Johnson-Laird, Tony Ryle, and Steve Draper. I would like also to thank Jenny Jenkins, Edoardo Bisiach, and Anthony Marcel for reading a draft of this chapter and making helpful suggestions.

References

Butterworth, G. (1984). The relation between thought and language in young children. In *Psychology survey 5*, (ed. J. Nicholson and H. Beloff), pp. 156–74. British Psychological Society, Leicester.

Cooley, C. (1902), *Human nature and the social order*. Scribner, New York.

Craik, K. J. W. (1943). *The nature of explanation*. Cambridge University Press.

Croce, B. (1917). *Theory and history of historiography*, (trans D. Ainslie). Harrap, London.

Dilthey, W. (1961). *Meaning in history: W. Dilthey's thoughts on history and society*, (ed. and trans. H. P. Rickman). George Allen and Unwin, Hemel Hempstead, Herts.

Dru, A. (ed. and trans.) (1938). *The journals of Søren Kierkegaard*. Oxford University Press.

Fairbairn, W. R. D. (1952). *Psychoanalytic studies of the personality*. Routledge and Kegan Paul, Andover, Hants.

Farr, R. (1981). Social origins of the human mind. In *Social cognition: perspectives on everyday understanding*, (ed. J. P. Forgas), pp. 247–58. Academic Press, London.

Freud, S. (1975). *The psychopathology of everyday life*. Penguin Books, Harmondsworth, Middx.

Furth, H. (1971). Linguistic deficiency and thinking: research with deaf subjects, 1964–1969. *Psychological Bulletin*, **76**, 53–72.

Harvey, W. (1963). *The Circulation of the blood and other writings*, (ed. K. J. Kranklin). Dent, Everyman, London.

Heaton, J. (1985). Knowledge and consciousness. *Bulletin of the British Psychological Society*, **38**, A36.

Humphrey, N. (1983). *Consciousness regained*. Oxford University Press.

James, W. (1892). *Psychology: the briefer course*. Harper and Row, New York.

Jaynes, J. (1976). *The origin of consciousness in the breakdown of the bicameral mind*. Allen Lane, London.

Johnson-Laird, P. N. (1983). *Mental models: towards a cognitive science of language, inference and consciousness*. Cambridge University Press.

Johnson-Laird, P. N. (1988). A computational analyses of consciousness. In *Consciousness in contemporary science* (ed. A. Marcel and E. Bisiach), p. 357. Oxford University Press.

Lacan, J. (1966). The mirror stage as formative of the I as revealed in psychoanalytic experience. In *Ecrits*, (ed. and trans. A. Sheridan), pp. 1–7. Tavistock Publications, Andover, Hants.

Marcel, A. J. (1983). Conscious and unconscious perception: an approach to the relations between phenomenal experience and perceptual processes. *Cognitive Psychology*, **15**, 238–300.

Marcel, A. J. (1988). Phenomenal experience and functionalism. In *Consciousness in contemporary science* (ed. A. Marcel and E. Bisiach), p. 121. Oxford University Press.

Mead, G. H. (1964). The social self. In *Selected writings of George Herbert Mead*, (ed. A. J. Reck), pp. 142–49. Bobbs-Merrill, Indianapolis, IN.

Miller, G. A. Galanter, E., and Pribram, K. (1960). *Plans and the structure of behavior*. Holt, Rinehart and Winston, New York.

Minsky, M. (1968). Matter, mind and models. In M. Minsky (Ed.) *Semantic information processing* (ed. M. Minsky), pp. 425–32. MIT Press, Cambridge, MA.

Oatley, K. (1978). *Perceptions and representations: the theoretical bases of brain research and psychology*. Methuen, Andover, Hants.

Oatley, K. (1981). Representing ourselves: mental schemata, computational metaphors, and the nature of consciousness. In *Aspects of consciousness*, Vol. 2 (ed. G. Underwood and R. Stevens), pp. 85–117. Academic Press, London.

Oatley, K. (1985). Representations of the physical and social world. In D. A. Oakley (Ed.). *Brain and mind*, pp. 32–58. London, Methuen.

Oatley, K. (1988). *Best laid schemes: a cognitive psychology of emotions*. Harvard University Press.

Oatley, K. and Bolton, W. (1985). A social–cognitive theory of depression in reaction to life events. *Psychological Review*, **92**, 372–88.

Oatley, K. and Johnson-Laird, P. N. (1987). Towards a cognitive theory of emotions. *Cognition and Emotion*, **1**, 29–50.

Oatley, K., Sullivan, G. D., and Hogg, D. (1988). Drawing visual conclusions from analogy: preprocessing, cues and schemata in the perception of three-dimensional objects. *Journal of Intelligent Systems*, **1**, 97–133.

Roberts, L. G. (1965). Machine perception of three-dimensional solids. In *Optical and electro-optical information processing*, (ed. J. T. Tippett *et al.*), pp. 159–97. MIT Press, Cambridge, MA.

Ryle, A. (1982). *Psychotherapy: a cognitive integration of theory and practice.* Academic Press, London.

Searle, J. R. (1984). The freedom of the will. 1984 Reith Lecture No. 6, *The Listener*, 13 December, 10–12.

Shepard, R. N. and Metzler, J. (1971). Mental rotation of three-dimensional objects. *Science*, **171**, 701–3.

Simpson, E. M. and Potter, G. R. (ed.) (1956). *The Sermons of John Donne*, Vol. 8, pp. 219–36. University of California Press, Berkeley CA.

Southall, J. P. C. (ed.) (1925). *Helmholtz's treatise on physiological optics. Vol. 3.* Dover, New York.

Sussman, G. J. (1975). *A computer model of skill aquisition.* American Elsevier, New York.

Vygotsky, L. S. (1978). Tool and symbol in children's development. In *Mind and society*, (ed. M. Cole, V. John-Steiner, S. Scribner, and E. Souberman), pp. 19–30. Harvard University Press.

Woolf, V. (1925). *Mrs Dalloway.* Hogarth Press, London.

AUTHOR INDEX

SUBJECT INDEX